Recycled Polymers: Eco-Design, Structure/Property Relationships and Compatibility

Recycled Polymers: Eco-Design, Structure/Property Relationships and Compatibility

Editor

Didier Perrin

Basel • Beijing • Wuhan • Barcelona • Belgrade • Novi Sad • Cluj • Manchester

Editor
Didier Perrin
Polymers Composites
Hybrids (PCH) Team
IMT Mines Ales
Ales
France

Editorial Office
MDPI
St. Alban-Anlage 66
4052 Basel, Switzerland

This is a reprint of articles from the Special Issue published online in the open access journal *Polymers* (ISSN 2073-4360) (available at: www.mdpi.com/journal/polymers/special_issues/recycled_polymers_ecodesign_structure_property_compatibility).

For citation purposes, cite each article independently as indicated on the article page online and as indicated below:

Lastname, A.A.; Lastname, B.B. Article Title. *Journal Name* **Year**, *Volume Number*, Page Range.

ISBN 978-3-7258-0002-5 (Hbk)
ISBN 978-3-7258-0001-8 (PDF)
doi.org/10.3390/books978-3-7258-0001-8

© 2024 by the authors. Articles in this book are Open Access and distributed under the Creative Commons Attribution (CC BY) license. The book as a whole is distributed by MDPI under the terms and conditions of the Creative Commons Attribution-NonCommercial-NoDerivs (CC BY-NC-ND) license.

Contents

Charles Signoret, Pierre Girard, Agathe Le Guen, Anne-Sophie Caro-Bretelle,
José-Marie Lopez-Cuesta, Patrick Ienny, et al.
Degradation of Styrenic Plastics during Recycling: Accommodation of PP within ABS after
WEEE Plastics Imperfect Sorting
Reprinted from: *Polymers* **2021**, *13*, 1439, doi:10.3390/polym13091439 1

Paul J. Freudenthaler, Joerg Fischer, Yi Liu and Reinhold W. Lang
Polypropylene Post-Consumer Recyclate Compounds for Thermoforming Packaging
Applications
Reprinted from: *Polymers* **2023**, *15*, 345, doi:10.3390/polym15020345 29

Youssef El Bitouri and Didier Perrin
Compressive and Flexural Strengths of Mortars Containing ABS and WEEE Based Plastic
Aggregates
Reprinted from: *Polymers* **2022**, *14*, 3914, doi:10.3390/polym14183914 44

Victor S. Cecon, Greg W. Curtzwiler and Keith L. Vorst
Influence of Biofillers on the Properties of Regrind Crystalline Poly(ethylene terephthalate)
(CPET)
Reprinted from: *Polymers* **2022**, *14*, 3210, doi:10.3390/polym14153210 58

Damayanti Damayanti, Latasya Adelia Wulandari, Adhanto Bagaskoro, Aditya Rianjanu and
Ho-Shing Wu
Possibility Routes for Textile Recycling Technology
Reprinted from: *Polymers* **2021**, *13*, 3834, doi:10.3390/polym13213834 73

Dorota Czarnecka-Komorowska, Jagoda Nowak-Grzebyta, Katarzyna Gawdzińska,
Olga Mysiukiewicz and Małgorzata Tomasik
Polyethylene/Polyamide Blends Made of Waste with Compatibilizer: Processing, Morphology,
Rheological and Thermo-Mechanical Behavior
Reprinted from: *Polymers* **2021**, *13*, 2385, doi:10.3390/polym13142385 104

Inès Meyer zu Reckendorf, Amel Sahki, Didier Perrin, Clément Lacoste, Anne Bergeret,
Avigaël Ohayon, et al.
Chemical Recycling of Vacuum-Infused Thermoplastic Acrylate-Based Composites Reinforced
by Basalt Fabrics
Reprinted from: *Polymers* **2022**, *14*, 1083, doi:10.3390/polym14061083 121

Ahmad Mamoun Khamis, Zulkifly Abbas, Raba'ah Syahidah Azis, Ebenezer Ekow Mensah
and Ibrahim Abubakar Alhaji
Effects of Recycled Fe_2O_3 Nanofiller on the Structural, Thermal, Mechanical, Dielectric, and
Magnetic Properties of PTFE Matrix
Reprinted from: *Polymers* **2021**, *13*, 2332, doi:10.3390/polym13142332 143

Junghee Joo, Heeyoung Choi, Kun-Yi Andrew Lin and Jechan Lee
Pyrolysis of Denim Jeans Waste: Pyrolytic Product Modification by the Addition of Sodium
Carbonate
Reprinted from: *Polymers* **2022**, *14*, 5035, doi:10.3390/polym14225035 162

Anoma Thitithammawong, Sitisaiyidah Saiwari, Subhan Salaeh and Nabil Hayeemasae
Potent Application of Scrap from the Modified Natural Rubber Production as Oil Absorbent
Reprinted from: *Polymers* **2022**, *14*, 5066, doi:10.3390/polym14235066 171

Jacopo De Tommaso and Jean-Luc Dubois
Risk Analysis on PMMA Recycling Economics
Reprinted from: *Polymers* **2021**, *13*, 2724, doi:10.3390/polym13162724 . . . . **188**

Paul J. Freudenthaler, Joerg Fischer and Reinhold W. Lang
Assessment of Commercially Available Polyethylene Recyclates for Blow Molding Applications by a Novel Environmental Stress Cracking Method
Reprinted from: *Polymers* **2023**, *15*, 46, doi:10.3390/polym15010046 . . . . **221**

Article

Degradation of Styrenic Plastics during Recycling: Accommodation of PP within ABS after WEEE Plastics Imperfect Sorting

Charles Signoret [1], Pierre Girard [1], Agathe Le Guen [1], Anne-Sophie Caro-Bretelle [2], José-Marie Lopez-Cuesta [1], Patrick Ienny [2] and Didier Perrin [1,*]

[1] Polymers Composites and Hybrids (PCH), IMT Mines Ales, 30100 Ales, France; charles.van.signoret@orange.fr (C.S.); pierre.girard@mines-ales.org (P.G.); agathe.le-guen@mines-ales.org (A.L.G.); jose-marie.lopez@mines-ales.fr (J.-M.L.-C.)

[2] LMGC, IMT Mines Ales, Université Montpellier, CNRS, 30100 Ales, France; anne-sophie.caro-bretelle@mines-ales.fr or anne-sophie.caro@mines-ales.fr (A.-S.C.-B.); patrick.ienny@mines-ales.fr (P.I.)

* Correspondence: didier.perrin@mines-ales.fr; Tel.: +33-4-66-78-53-69

Citation: Signoret, C.; Girard, P.; Guen, A.L.; Caro-Bretelle, A.-S.; Lopez-Cuesta, J.-M.; Ienny, P.; Perrin, D. Degradation of Styrenic Plastics during Recycling: Accommodation of PP within ABS after WEEE Plastics Imperfect Sorting. *Polymers* **2021**, *13*, 1439. https://doi.org/10.3390/polym13091439

Academic Editor: Dimitrios Bikiaris

Received: 28 March 2021
Accepted: 26 April 2021
Published: 29 April 2021

Publisher's Note: MDPI stays neutral with regard to jurisdictional claims in published maps and institutional affiliations.

Copyright: © 2021 by the authors. Licensee MDPI, Basel, Switzerland. This article is an open access article distributed under the terms and conditions of the Creative Commons Attribution (CC BY) license (https://creativecommons.org/licenses/by/4.0/).

Abstract: With the development of dark polymers for industrial sorting technologies, economically profitable recycling of plastics from Waste Electrical and Electronical Equipment (WEEE) can be envisaged even in the presence of residual impurities. In ABS extracted from WEEE, PP is expected to be the more detrimental because of its important lack of compatibility. Hence, PP was incorporated to ABS at different rates (2 to 8 wt%) with a twin-screw extruder. PP was shown to exhibit a nodular morphology with an average diameter around 1–2 µm. Tensile properties were importantly diminished beyond 4 wt% but impact resistance was decreased even at 2 wt%. Both properties were strongly reduced as function of the contamination rate. Various potential compatibilizers for the ABS + 4 wt% PP system were evaluated: PPH-g-MA, PPC-g-MA, ABS-g-MA, TPE-g-MA, SEBS and PP-g-SAN. SEBS was found the most promising, leading to diminution of nodule sizes and also acting as an impact modifier. Finally, a Design Of Experiments using the Response Surface Methodology (DOE-RSM) was applied to visualize the impacts and interactions of extrusion temperature and screw speed on impact resistance of compatibilized and uncompatibilized ABS + 4 wt% PP systems. Resilience improvements were obtained for the uncompatibilized system and interactions between extrusion parameters and compatibilizers were noticed.

Keywords: mechanical recycling; impact resistance; WEEE; polymer compatibilization; design of experiments; surface response methodology

1. Introduction

The consumer society led to the disposal of 4.9 billion metric tons of polymeric waste in the environment from 1950 to 2015 according to Geyer et al. [1]. Authors anticipated, with optimistic measures taken in favor of recycling and incineration, that the corresponding cumulative waste stock should stabilize around 12 billion in 2050. Furthermore, Robinson [2] reported that E-waste (electronic waste) is estimated at around 20–25 million tons of worldwide production per year. Waste Electrical and Electronic Equipment (WEEE) is a more generic term and also includes traditionally strictly electrical equipment such as refrigerators or kettles. Plastics are reported to represent from 10 to 35 wt% of WEEE [3–8] even if it varies much between studies and WEEE types (small/large appliance for instance).

Recycling is viewed as one of the main routes to prevent plastic pollution, and, in the same way, to produce new materials [9,10]. However, WEEE plastics (WEEP) are more complex, diverse and contain generally more additives than other plastics [5,11–13]. Stenvall et al. [5] found that HIPS, ABS and PP were the major constituents of their studied WEEP samples, respectively, at 42, 38 and 20 wt%. Maris et al. [14] found 26, 29 and 22 wt% for these three categories of plastics. Additionally, an important fraction of WEEP is dark-colored, preventing an efficient sorting thanks to current technologies such as NIR-HSI [15,16]. Yet, sorting is mandatory to achieve interesting properties since most polymers are said incompatible, leading to immiscibility and poor mechanical properties [11,16,17]. However, several technologies are currently in development, such as MIR-HSI [18–20], Terahertz spectroscopy [21,22] or LIBS [23–25], which could then enable profitable recycling of WEEP.

Nevertheless, sorting alone can be insufficient as perfect purities are impossible to achieve, especially in an economic way [15]. However, scientific literature, about recycled materials containing a small fraction of impurities close to what an automatized sorting could produce, is scarce. A consistent review from Vilaplana and Karlsson [6] shows that most studies concern systems where the second component is present above 10 wt% or, more frequently, above 20 wt%. Thus, Perrin et al. [26] studied the influence of different possible contaminants, PS, ABS and PP, on the mechanical properties of HIPS. Impact properties were the most reduced. They concluded that ABS was tolerable upon 4 wt% and PP upon 1 wt% only.

Since ABS/HIPS blends were thoroughly studied [27–29] as well as contaminated HIPS [26], another interesting case would be ABS contaminated with PP. Especially, ABS presents several interesting properties compared to HIPS (higher chemical resistance, Young's Modulus, resilience and stress at peak [27–30] and thus could present better added-value upon recycling.

ABS is basically a copolymer of styrene and acrylonitrile (SAN) where polybutadiene (PB) nodules are present to dramatically improve impact properties through the slowdown of crazes propagation, preventing fast crack formation [31–34]. Thus, differences in PB concentrations lead to very different impact properties, but also viscosities, between several commercial grades. PPC stands for polypropylene copolymer. It is differentiated from PPH (Homopolymer) as ethylene is present within its chemical formula. However, it is not always clear if it is by copolymerization [35,36] through EPR or EPDM introduction [36,37]. In either case, it results in improved impact properties and lower modulus. Sadly, the difference between PPH and PPC is rarely highlighted during WEEP characterization even though it is easily possible through FTIR [30]. In WEEE, PPC seems more adequate, however.

The addition of adequate compatibilizers, often copolymers or grafted polymers, leads to morphology reduction thanks to interfacial tension decrease (and thus capillary number increase) and/or coalescence prevention through steric stabilization as compatibilizer "shells" can form around the dispersed phase droplets [17,38,39]. In terms of ABS/PP systems, most authors worked on PP-rich systems, often around 70/30 or 80/20 [40–47]. Kum et al. [43] tried PP-g-SAN on ABS/PP systems by 10 wt% steps and found reduced nodular morphology and enhanced mechanical properties on most of the range even if the overall is still far below pure ABS. Patel et al. [46] added home-made PP-g-2-HEMA (2-hydroxyethyl methacrylate) and PP-g-AA (acrylic acid) to ABS/PP at different ratios and highlighted morphological refinement. Tensile and flexural properties, but also impact resistance, were improved in most of the range, with the notable exception of ABS with 10 wt% PP on the last property. Deng et al. [41] compared the effects of PP-g-MA and home-made PP-graft-cardanol on a PP70/ABS30 system and found that the first one was more efficient to improve mechanical properties, mainly tensile and flexural strengths, thanks to morphological refinement. Bonda et al. [40] showed that ABS impact and tensile properties were strongly diminished by 10 wt% PP and that addition of PP-g-MA, SEBS-g-MA and EAO (ethylene α-olefin) could improve impact properties of a PP80/ABS20 blend through morphological refinement to the detriment of tensile properties. They proposed a dipole–dipole interaction between a C=O group of maleic anhydride sites and an ABS nitrile group to explain the improvement in properties by the decrease in interfacial tension. Ibrahim et al. [42] later proposed a real reactivity between the two moieties where the cycle of maleic anhydride could open up to form an eight-membered cycle with acrylonitrile, thus creating in situ copolymers. Lee et al. [44] evaluated SEBS and SEBS-g-MA on a PP70/ABS30 system and found improved impact properties but decreased tensile strength. The second compatibilizer was found to be more efficient, especially after thermal ageing where it preserved properties contrary to the first one. Finally, Tostar et al. [47] worked on unsorted WEEP as described by Stenvall et al. [5] of the same research group (42 wt% HIPS, 38 wt% ABS, 10 wt% PP, the rest being PE, PUR, PC, etc.). They found SEBS much more efficient than SEBS-g-MA, EPDM-g-SAN, PP-g-MA or gamma irradiation to improve mechanical properties, especially strain at break. It also greatly improved the impact properties.

For other polymer blends, several authors highlighted the importance of mixing parameters on phase dispersion. Plochocki et al. [48] studied an LDPE/PS (2/1) blend realized with different industrial processes and found that PS domain size versus mixing energy gave a hyperbolic curve. They reported that dispersion followed a first order rule with shear rate, but coalescence a second order one as shearing also increase droplets collision probability and thus coalescence. Sundaraj and Macosko [39] studied different polymer blends realized in a batch mixer and also found a hyperbolic curve for dispersed phase diameters versus shear rate in the case of uncompatibilized systems. They explained this phenomenon rather by different rheological responses of both phases toward shear rate. They also highlighted that copolymers, especially formed in situ, enables interfacial tension decrease and coalescence prevention. Therefore, mixing parameters can have an important impact on blends morphologies, and thus properties, especially toward impact [49].

It appears from the literature that, in a WEEP recycling context, ABS/HIPS and contaminated HIPS have been already studied. Additionally, many authors studied PP-rich blends to find that most properties of pure ABS were lost by blending, especially impact properties. Compatibilization effectiveness was often explained through morphological refinement but was mainly operant in PP-rich systems. Thus, in a first part, the present study aims to evaluate the impact of PPC within ABS with incorporation rates from 2 to 8 wt% through twin-screw extrusion. The second part relates to compatibilization trials on an ABS/PP (96/4) system, with the purpose to retrieve properties close to virgin ABS. Morphological assessment, through SEM, and impact properties, through instrumented Charpy impact, are at the core of considerations. Tensile properties were also evaluated. In the literature, PP-g-MA (here differentiated between PPH-g-MA and PPC-g-MA), SEBS and PP-g-SAN were found to have positive effects and are applied here to an ABS-rich system. TPE-g-MA and ABS-g-MA were also evaluated. Finally, the third part is about a Design Of Experiments Response Surface Methodology (DOE-RSM) to model impact resistance of ABS/PP (96/4) as a function of speed and temperature extrusion, with or without selected compatibilizers. Ultimately, this was used to localize maximal resistances.

2. Materials and Methods

2.1. Materials

ABS used for this work was Terluran GP22 provided by Styrolution and PPC was PHC27 from Sabic. Five compatibilizers were kindly provided by Velox: PPH-g-MA (Bondyram 1001), PPC-g-MA (Bondyram 1001 CN), ABS-g-MA (Bondyram 6000) and TPE-g-MA (Bondyram 7108) and SEBS (Tuftec P2000). The last one was PP-g-SAN (Modiper A3400) kindly supplied by Modiper. Figure 1 shows their chemical structures.

Maleic anhydride (MA) grafting rates were about 1 wt% for every grafted polymer used here. As reported in the introduction, MA-grafted polymers were chosen since this moiety could promote compatibility towards acrylonitrile. More precisely, PP-g-MA grades were chosen because of results found in the literature and as it is a very common compatibilizer for many systems (Maris et al., 2018). PPH and PPC were differentiated in case. PPC, which is less common, could show better compatibility with the minority phase. The TPE used here was an olefinic thermoplastic elastomer with a strong ethylene base. It could show compatibility or even miscibility with polyolefins. It could also act as an impact modifier. Finally, a soft interphase between PP nodules and the matrix could be created. SEBS, here with a weight S/EB ratio of 67/33, can also fulfil these roles, even to a larger extent. This specific ratio was chosen as ABS is the major component. Available SEBS-g-MA grades were all EB rich, so these compatibilizers were not evaluated. Finally,

PP-g-SAN appears theoretically to be the optimal polymer for such a system as it could be placed at the interface and bring adhesion.

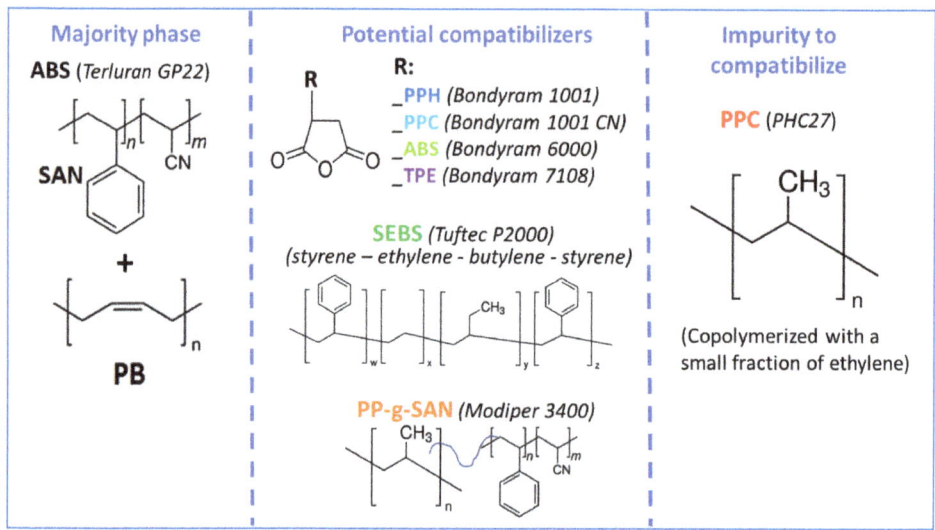

Figure 1. Materials chemistry.

2.2. Polymer Processing

ABS was dried at 80 °C for at least 16 h before processing. Pellets were mixed by hand and were then processed with a 1200 mm Clextral twin-screw extruder of type BC21 (L/d = 48) at 250 rpm and 220 °C along the screw, unless otherwise specified. The nozzle was 5 mm and feed speed was at 6 kg/h thanks to a K-Tron KGx-2 weighing dozer from Coperion. Extrudate was then pelletized for further transformations.

Pellets were then injected into dogbone specimens corresponding to ISO 3167 thanks to a Krauss-Maffei 180/CX 50 molding injection press at 230 °C for plasticization and mold maintained at 40 °C. Pressure was maintained at 450 bars for 12 s following injection, then 400 bars for 12 s again before ejection. Specimens for Charpy impact, $80 \times 10 \times 4$ mm^3, were cut off from dogbones.

Amounts of PP are given in wt% as this simulates what recyclers receive and should deal with. Amounts of compatibilizers are given in phr as this is what they should add to the material to give it interesting properties.

As processing itself can affect the material, each study included at least one uncontaminated ABS batch that went through the same transformations as the others. This batch is identified as "Vext" or "0 wt%" to differentiate it from a really virgin batch, "V" which consists of directly injected pellets.

2.3. Mechanical Characterization

Results from both Charpy impact and tensile tests were processed thanks to Matlab 2018a from MathWorks and then plotted with OriginPro 9.1 from OriginLab Corp, Northampton, MA, USA. Boxplots plotted in these works represent quartiles for box limits, medians as a horizontal line, means as crosses and minima and maxima as whiskers. Values are also plotted as dots at the left of each corresponding box.

2.4. Charpy Impact

Instrumented impact tests were carried out thanks to an impact drop tester Ceast 9340 from Instrom. Both notched (1eA) and unnotched (1eU) Charpy impact tests were performed according to ISO 179. The two tests give different but complementary information about the material. Whereas unnotched tests cumulate crack initiation and propagation, notched tests focus on the second part. Additionally, whereas unnotched tests are very sensitive to skin effects, notched tests really evaluate the core of the material. *Figure S1—Supporting information* shows force–displacement curves of notched specimens at two different velocities which are chosen by changing drop height. 2.9 m/s, the velocity recommended by ISO 179, generates relatively important vibrations compared to 0.8 m/s because impact energy is too important. The break energy is calculated by integration. However, force does not really return to 0 N. Therefore, 15% of maximal force was chosen in the present work as a cut-off parameter. However, vibrations which can be irregular between batches can falsely shorten integration at 2.9 m/s; 0.8 m/s was thus chosen for all notched Charpy impact tests in this study.

2.5. Tensile Tests

Tensile tests were performed according to the ISO 527 standard and thanks to a Z010 Zwick press equipped with a 2.5 kN load cell and a Clip-on extensometer. All tensile properties were measured at a 10 mm/min crosshead speed. Young's moduli were determined between 0.05% and 0.25% of deformation. Ultimate strains were measured according to the crosshead displacement.

2.6. Characterization of Morphologies

Observations of blend morphologies were performed thanks to a scanning electron microscope (Quanta FEG 200, Thermo-Fisher Scientific, Waltham, MA, USA) on cryo-fractured dogbones samples (orthogonally or 45°) and on post-mortem Charpy impact fragments after carbon metallization at magnifications of ×5000 and ×25,000. Only secondary electrons were considered as chemical differences are too weak between considered polymeric materials.

2.7. Design of Experiments

Design of Experiments (DOE) builds and visualizations were carried out using the Nemrodw® 2015 software. Two responses (Y) were considered: mean break energies from notched Charpy impact, and from unnotched Charpy impact. The DOE was applied to an uncompatibilized ABS + 4 wt%PP system, or with SEBS (1, 2 or 3 phr), PP-g-MA (0.5, 1.0 or 1.5 phr) or PP-g-SAN (0.5, 1.0 or 1.5 phr). Twenty experiments were performed, each resulting in 15 Charpy's impact tests performed on notched and unnotched specimens. A Response Surface Methodology (RSM) was applied to evaluate the influence of several process parameters, extrusion temperature, screw speed and compatibilizer addition rate, on the two responses. The applied interaction mathematical linear model is given in Equation (1):

$$Y = b_0 + b_1 \cdot X_1 + b_2 \cdot X_2 + b_3 \cdot X_3 + b_{1-2} \cdot (X_1 \cdot X_2) + b_{1-3} \cdot (X_1 \cdot X_3) + b_{2-3} \cdot (X_2 \cdot X_3) \quad (1)$$

where X_1 is the extrusion temperature (200–240 °C), X_2 the screw speed (200–300 rpm) and X_3 the additive loading rate (phr, depends on considered additive) and different b_n constants (Equation (1): Interaction mathematical linear model equation).

This model represents influence of different parameters at first order and 1st order interactions.

3. Results and Discussion

3.1. Impact of PP on ABS for Various Incorporation Rates

To evaluate the tolerance of PPC impurities from ABS, PPC was added to ABS through extrusion at increasing rates: 2, 4, 6 and 8 wt%.

3.1.1. Impact Properties

For both notched and unnotched impact tests, a first slight diminution is due to processing alone (from black to blue on Figure 2). Then, impact resistance progressively decreases with PPC incorporation rate, from about 16.5 to below 10.0 kJ/m^2 for notched impact (-40%), and about 80 ± 20 kJ/m^2 to about 15 kJ/m^2 for unnotched impact (-80%). Interestingly enough, 2 wt% does not affect unnotched impact resistance much as it really begins to decrease at 4 wt%, which also corresponds to a more important drop in notched impact resistance.

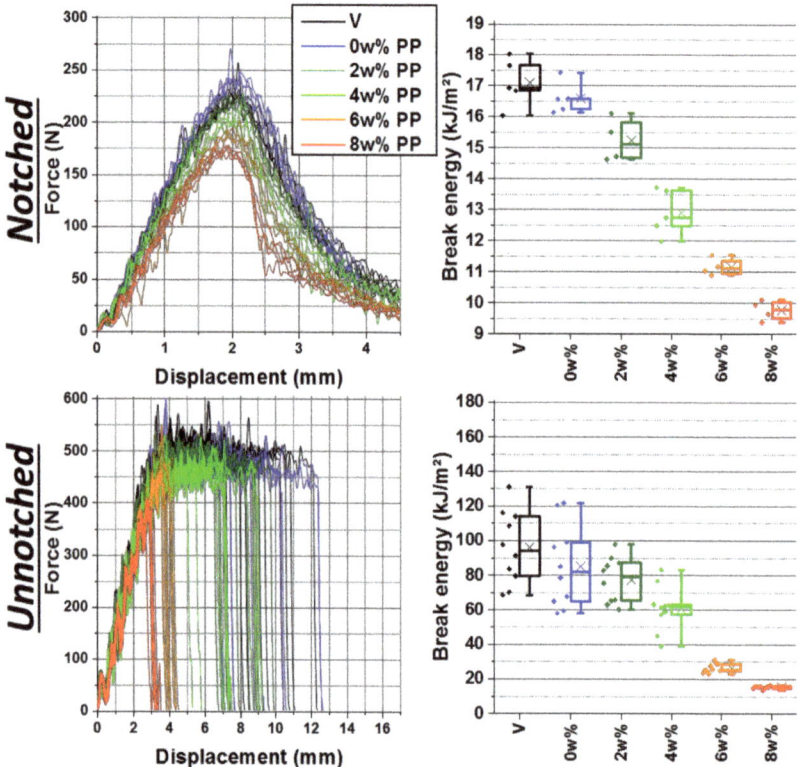

Figure 2. Force–displacement curves (**left**) and break energies boxplots (**right**) of notched (**top**) and unnotched (**bottom**) Charpy impact tests on ABS contaminated at increasing rates of PP —"V" for directly injected ABS, "0 wt%" went through extrusion.

On notched force–displacement curves of Figure 2, the whole bell-shaped curve gradually collapses, with the maximal force decreasing from about 230 N to about 175 N. Especially on the 8 wt% PP batch, the bell shape gets more asymmetrical, with a more vertical force drop after the maximum. It hints at a more brittle behavior. On unnotched impact curves, the behavior is quite different. The force first increases to reach a plateau in load. On virgin batches, the drop of force is dispersed in displacement values because of a more statistical occurrence of fracture. The whole plateau in load is linked to irreversible behavior, the material undergoing damaged plastic behavior and developing a multitude of crazes, resulting in visible whitening shown further below. As the break energy is calculated by the integration of the shown curves, this break displacement dispersion results in break energy dispersions as seen in the boxplots. Progressively with PP contamination, this plateau shortens and so failure occurs in a narrowed displacement range, especially from 6 wt%. Thus, break energies become smaller and closer. Finally, at 8 wt% of contamination, force does not even reach the plateau level and curves are perfectly superimposed, adopting a brittle behavior similar to what can be obtained with SAN (*Figure S2 of supporting information*) or PMMA [50]. Especially, 6 wt% contaminated ABS is even less resistant than SAN Luran 368 R. Finally, for both notched and unnotched impact, the initial force slope gets slightly weaker with contamination. As the Charpy impact is basically a high-speed flexion test, a flexion modulus can be extracted from these curves as performed by other authors [30] on photodegraded and recycled HIPS. However, in the present case, comparisons to Young's moduli measured from tensile test were not conclusive even if general trends are more or less in agreement (*Figure S3 of supporting information*).

Figure 3 shows broken specimens after unnotched Charpy impact for the same batches. Hardly visible on the picture, the general hue of specimen gets slightly whiter with PP contamination.

Figure 3. Picture of unnotched Charpy impact broken specimens of ABS gradually contaminated with PP—arrow shows direction of impact; rectangle points out dephasing at macro-scale.

As seen on bottom specimens, the whitening behavior during impact develops at the opposite of the impactor, in the tensile part of the bending test specimen (from the underside to the neutral line), displaying a rather triangular whitened shape. The fracture is rather planar and directly orthogonal to the direction of uniaxial stresses. With PP, the whitening progressively fades, almost invisible at 4 wt%, and is totally nonexistent beyond this, as corroborated by the absence of plateau a (Figure 2). The fracture surface is also getting more and more uneven and more often oblique. Additionally, on several fragments, little spikes protruding from the center of break surfaces can be seen as framed in Figure 3. Beginning at 6wt%, this heterogeneity is even more important at 8 wt%, represented by rods up to 2 mm thick as shown and discussed with SEM characterization in 0.

Whitening is commonly explained in impact modified styrenics by light diffusion due to important craze netting originating from the boundary region of PB to the matrix (Yokouchi et al., 1983). The absence of whitening proves that energy dissipation from PB is inhibited. The rougher aspect of break surface underlines the heterogeneity of the sample discussed above.

3.1.2. Tensile Properties

Tensile properties are presented in Figure 4. The magnification on stress peak shows its progressive collapse beginning clearly from 4 wt%. As with unnotched Charpy impact, strain at break is very dispersed because of plastic damage behavior.

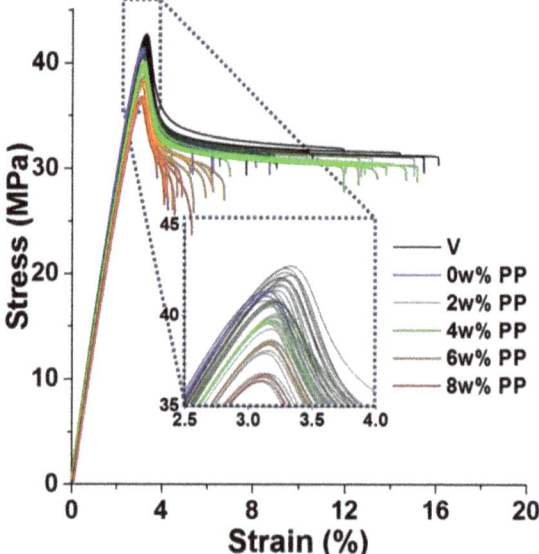

Figure 4. Stress–strain curves of tensile tests on ABS contaminated at increasing rates of PP—magnification on stress peak "V" for directly injected ABS, "0 wt%" went through extrusion.

Figure 5 gives values extracted from these curves. Overall, the Young modulus is slightly impacted but its decrease is more pronounced at 8 wt%. Maximal stress diminished almost linearly with PP concentration, except at 0 wt%. Whereas strain at break is rather dispersed and does not display clear evolution, stress at break steadily decreases, especially above 2 wt% and becomes more and more dispersed. At 6 and 8 wt% however, the strain at the break is significantly smaller, as is clearly visible on the full curves as well (Figure 4). This can be linked to the heterogeneities seen in Figure 3 beginning at these concentrations.

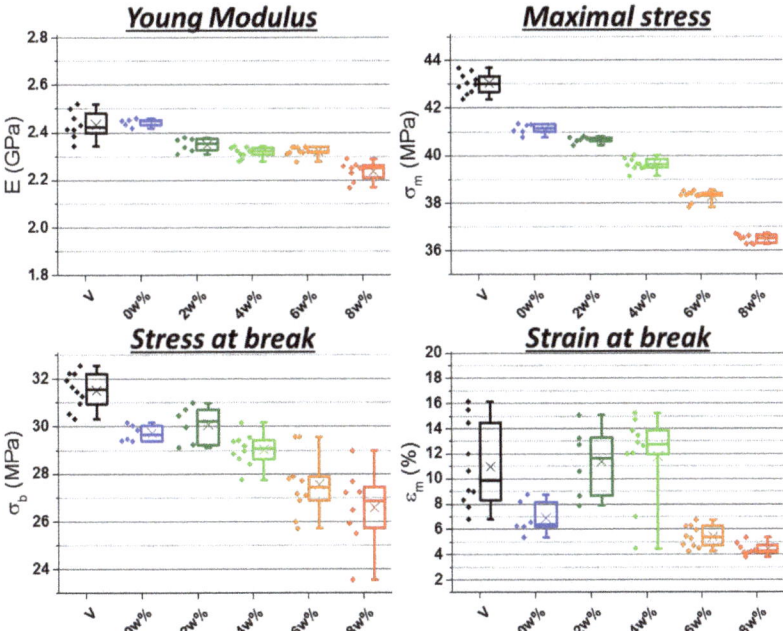

Figure 5. Young Modulus, maximal stress, stress and strain at break from tensile tests on ABS contaminated at increasing rates of PP—"V" for directly injected ABS, "0 wt%" went through extrusion.

3.1.3. Morphology

At the studied concentrations, PP adopts a nodular morphology (Figure 6). Cryofractures at 45° (*Figure S4 of supporting information*) confirms the spherical shape as pictures are very similar from both break angles. At 2 wt%, nodules are very scarce and their diameters are comprised between 0.5 and 1.5 µm. At 4 wt%, nodules are more notably more numerous and general diameters go up to 0.5–2.0 µm and some especially big nodules are measured between 3.0 and 3.5 µm. At 6 wt%, the nodules concentration continues to increase but measured diameters are in the same ranges.

As seen in Figure 3, macroscopic heterogeneity is visible to the naked eye beginning at 6 wt% PP. The phenomenon is even more important at 8 wt% as shown in Figure 7. A central part (pink arrow) was protruding from the fracture surfaces. On some specimens, it is a simple spike as in Figure 3 and on others, it is an ellipsoidal cylinder.

Figure 8 shows SEM pictures on cryofractured samples. The right bottom corner SEM picture represents the sample at a magnification low enough for its full thickness (4 mm) to be visible. Fractures can be seen near the surfaces, magnified on the right-bottom corner, separating "core" from the "skin". The "core" displays different fracture behaviors and the part indicated in pink seems to result from a more fragile break than the areas around it. Stronger magnifications indicate that nodules could be coarser and more numerous in the central part (top-left picture) than in the rest of the sample (top-right picture). Due to the rough fracture surface due to cryofracture, image analysis could be not performed to corroborate this.

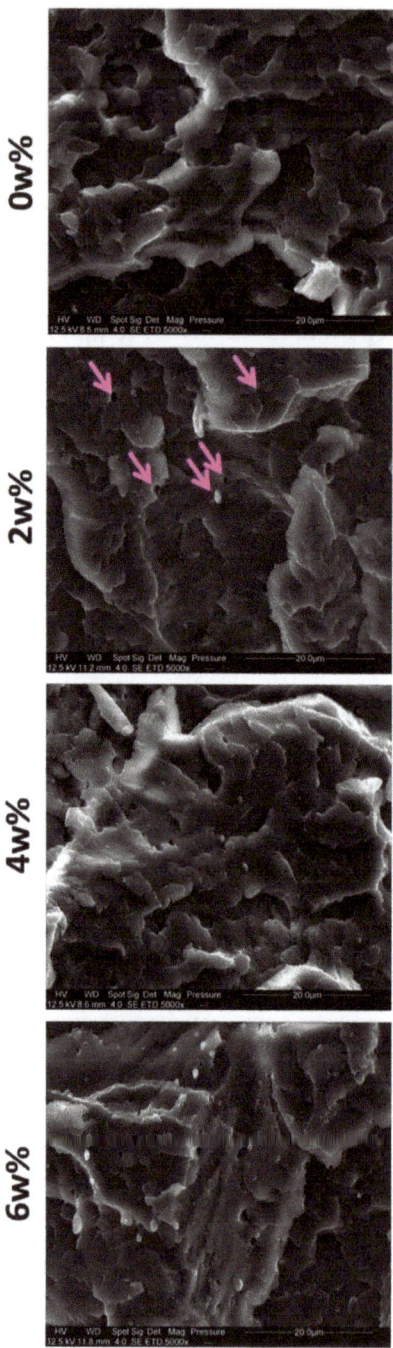

Figure 6. SEM pictures of cryofractured ABS dogbones contaminated at 0, 2, 4 and 6 wt% of PP—5000× magnification.

Figure 7. Picture of a Charpy post-mortem dogbone of ABS + 8 wt% PP—arrows indicate different areas.

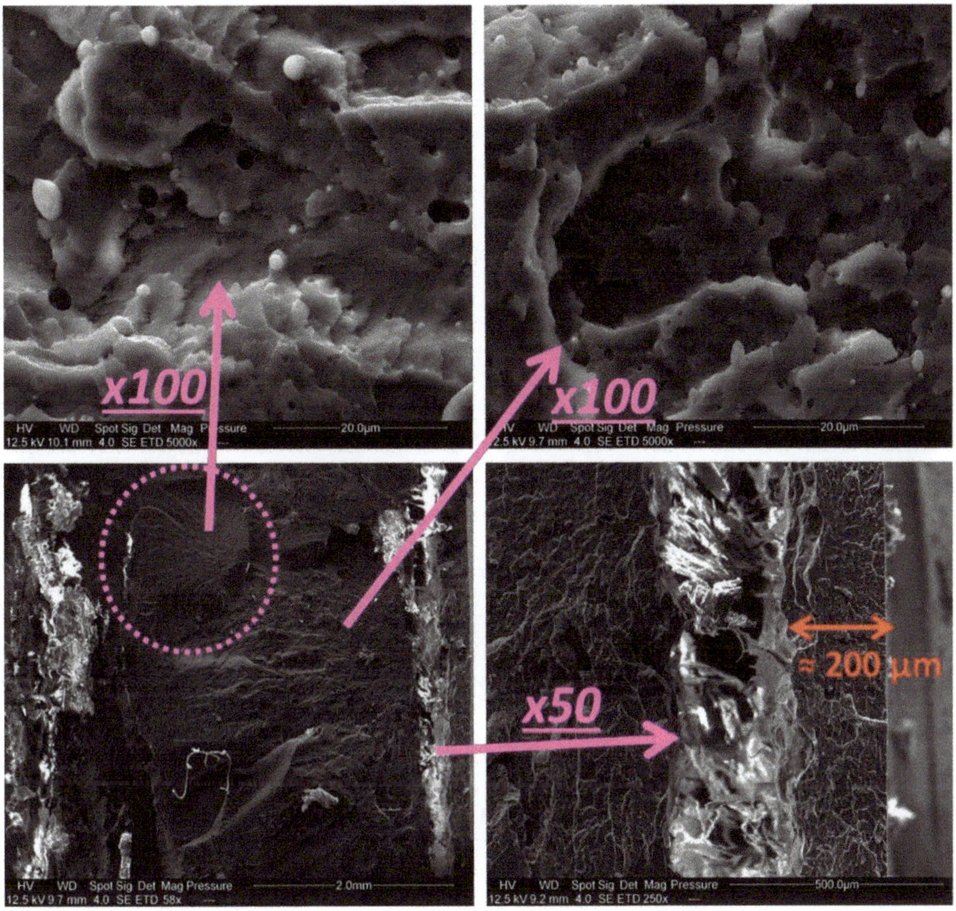

Figure 8. SEM pictures of cryofractured 8 wt% PP contaminated ABS dogbone—magnified views of different parts.

Fellahi et al. [51] described a complex system in HDPE/PA6 blends, as skin–subskin–core. However, the "subskin" of their system was of the same order of magnitude as the "skin". Other authors [52–57] also described "skin-core" as rather progressive, with intermediate layers, issued from important differences in cooling and shearing during injection-molding.

It was found through higher magnifications that the skin is devoid of any nodules. On their impact-modified PP, Karger-Kocsis and Csikai [52] found a "Surface Matrix Layer" but differentiated it from the skin which they defined asfIGURE6 displaying fibrillar morphology. The "skin" can be measured in Figure 8 at roughly 200 μm thick on the SEM picture, coherently with micrometer measurements performed on the few broken specimens from both tensile and impact tests where the "skin" was protruding from the fragments. FT-IR analysis (ATR, 4 cm^{-1} resolution, 16 scans, Vertex 70, Bruker) confirmed that skin was ABS (spectra presented in *Figure S5 of supporting information* and Appendix A). Other contaminated specimens also display "skins" of pure ABS, roughly from 100 to 200 μm thick, from both SEM and micrometer measures. Especially, they keep fragments attached after notched Charpy impact at low speed in the form of a flexible hinge (noted "H" in ISO 179 standard).

3.1.4. Conclusions on PP Contamination of ABS

It is confirmed through these first works that ABS and PP are incompatible as PP forms a nodular dispersion, which can result in heterogeneities at the millimeter scale for highest contamination rates. Tensile properties are mostly preserved below 4 wt% contamination, with just a small maximal stress decrease. Beyond this, elongation at break is also strongly diminished. Impact resistance is reduced in a stronger measure, by about 40% for notched impact, and 80% for unnotched impact. This drop also begins at lower concentrations, at 2 wt% for notched and at 4 wt% for unnotched. From 6 wt%, the force–displacement curves of unnotched impact demonstrate a brittle behavior. Industrial sorting nominal purity is recommended at 5 wt% max as reported by Beigbeder et al. [15], which is strongly justified by present results. However, 4 wt% contaminated samples correspond already to a drop in impact properties and the beginning of morphology coarsening. Thus, this concentration was chosen for the following compatibilization trials.

3.2. Compatibilizers Preselection

Six potential compatibilizers, PPH-g-MA, PPC-g-MA, ABS-g-MA, TPE-g-MA, SEBS and PP-g-SAN, were tested on ABS contaminated with 4 wt% of PP by additions at 1, 2 and 3 phr to check the effect of additive concentration.

3.2.1. Impact Properties

Figure 9 represents the notched Charpy impact on these samples. The first two boxplots correspond to references, virgin but extruded ABS and ABS with a contamination of 4 wt% PP. All compatibilizers led roughly to intermediate properties between these two references at most concentrations, except for PP-g-SAN. This last compatibilizer seems to entail a degradative behavior of PP at 2 and 3 phr. All maleic anhydride compatibilizers present effects similar to each other, around 14.5 kJ/m^2. A negative effect of concentration is perceived for all of the g-MA compatibilizers, especially PP-g-MA, homopolymer or copolymer. Even worse, some samples are far below the others in terms of energy. Force–displacement curves (*Figures S6 and S7 of supporting information*) inspection proves that it is due to a premature break, in a more brittle manner. All these concentration-related negative effects can be explained as the compatibilizer is added in excess and creates another non-dissipating phase, playing the role of a contaminant.

ABS-g-MA has thus a limited negative effect, as it should not be incompatible or even immiscible with the matrix. However, it is surprising that even TPE-g-MA has this negative effect as it is an elastomer and could simultaneously play an impact modifier role. It could be that its impact properties and/or its compatibility toward the matrix are insufficient, leading to the formation of a separated phase. Finally, SEBS displays the best results, just slightly above the others. Impact reinforcement increases with SEBS concentration, improbably because of its compatibilization effect due to the presence of styrenic moieties, but also because of its elastomeric nature. Results are also less dispersed.

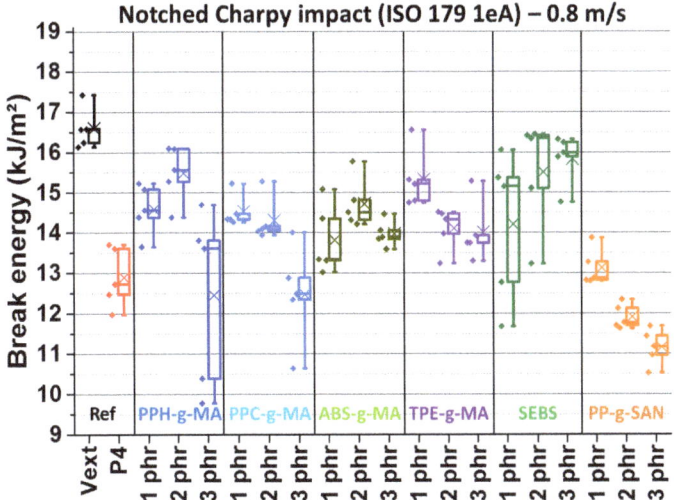

Figure 9. Notched Charpy impact break energies—virgin extruded ABS batch "Vext", 4 wt% PP contaminated batch "P4"and separately added batches with 1, 2 and 3 phr of PPH-g-MA, PPC-g-MA, ABS-g-MA, TPE-g-MA, SEBS and PP-g-SAN.

Figure 10 represents unnotched counterparts of a previous figure. Here, all candidates, except for PP-g-SAN again, improve ABS-PP properties back to virgin reprocessed ABS level and even above. Again, 1 phr PP-g-SAN is rather equivalent of the non-compatibilized batch and then PP-g-SAN extend properties degradation, going as low as 25 kJ/m^2, the equivalent of 6 wt% PP contaminated ABS (see Figure 3).

On other batches, the concentration effect seems nonexistent, maybe except for ABS-g-MA. Here, TPE-g-MA seems to be slightly above the reference, indicating an impact reinforcement effect. The effect is more evident and more pronounced with SEBS, especially as almost half of the samples from all batches (circled in pink) led to partial break (noted "P" in ISO 179 standard). A partial break of the specimen means that a solid and stiff ligament, here at least 1 mm thick, links the two fragments whereas a hinge is flexible, easily broken and very thin, here about 100–200 μm, "skin" thickness. As start velocity was at 2.9 m/s and total mass of the impactor at 3.14 kg, impact energy is theoretically 13.2 J. However, maximal measured energies here are about 7 J, which is significantly inferior. This means that specimens were ejected before the impactor could communicate the energy needed to break it.

Force–displacement curves of unnotched Charpy impact on references and SEBS and PP-g-SAN additive batches are presented in Figure 11. The equivalent for all batches and both impact test types is given in *supporting information* (*Figures S6 and S7*). SEBS lengthens the plateau further than the virgin reference, explaining the higher break energies of Figure 10. A slight force decrease in the plateau is even observed. About half of presented curves are linked to partial break but do not present a different appearance. These curves are just generally longer. Separate curves for each batch and a picture of a partially broken specimen are available in *supporting information* (*Figure S8*). Interestingly, some totally broken specimens have curves very close to some partially broken ones. Thus, their "break" energies are close as well. Other compatibilizers (*Figures S6 and S7 of supporting information*) led to curves reminiscent of virgin ABS, except for ABS-g-MA, slightly less performant.

Finally, the PP-g-SAN (Figure 11) effect on unnotched impact curves is very similar to the effect of PP, explaining obtained energies.

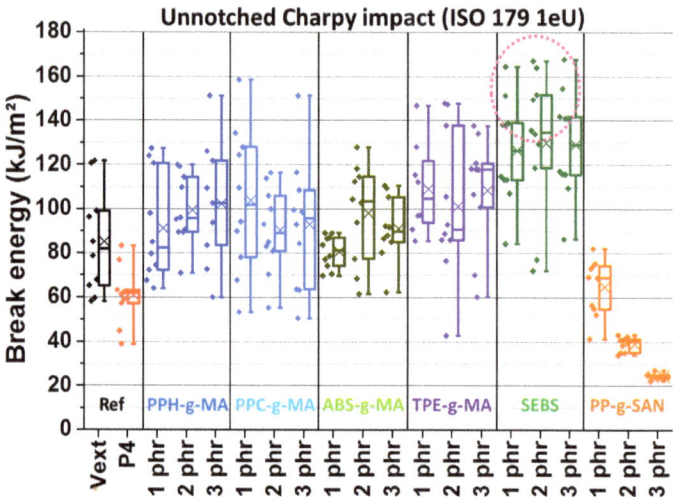

Figure 10. Unnotched Charpy impact break energies—same batches and denominations as previous figure—circle for partial breaks, respectively, 4, 5 and 4 specimens for 1, 2 and 3 phr of SEBS.

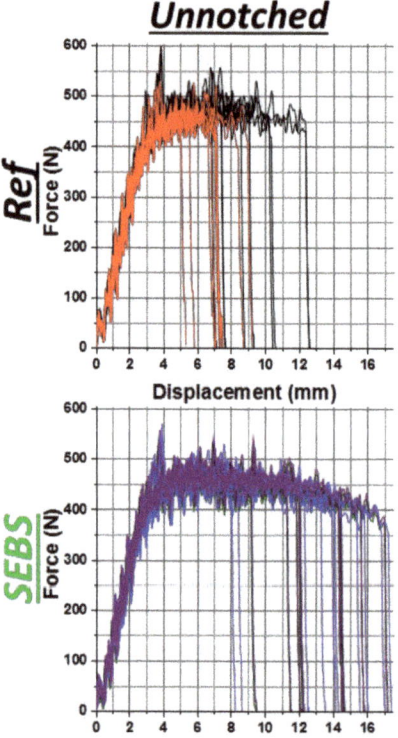

Figure 11. *Conts.*

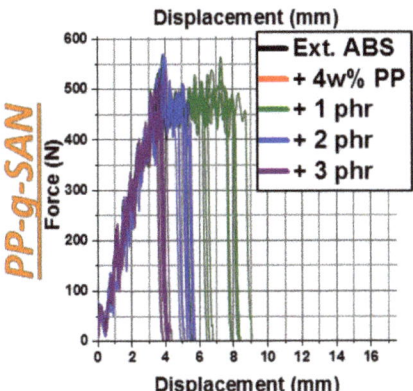

Figure 11. Force–displacement curves of unnotched Charpy impact of ABS—virgin and 4 wt% PP contaminated (1st line), added with 1, 2 and 3 phr of SEBS (2nd line) or PP-g-SAN (3rd line)—color associated with additive rate.

3.2.2. Morphology

SEM pictures of cryofractured dogbones of references and 3 phr additive systems are displayed in Figure 12 (5000× magnifications here, 25,000× available in *Figure S9 of supporting information*). As explained before, image analysis is challenged by the rough fracture surface. However, it seems that PP nodules are smaller with ABS-g-MA and TPE-g-MA than on uncompatibilized ABS/PP. Simple measurements gave diameters around 1–2 µm on ABS/PP. With ABS-g-MA and TPE-g-MA, sizes were found around 0.5–1.5 µm. With PP-g-MA and SEBS, nodules were harder to see as they are smaller. They were measured at, respectively, 0.5–1.0 µm and 0.2–0.6 µm. Size diminution could be imputable to interfacial tension decrease as these additives play a tensioactive role as their chemical nature is dual. Considering chemical affinities, it is rather logical that PP-g-MA and SEBS display the best results. Finally, with PP-g-SAN, PP nodules seem surprisingly unaffected.

Figure 13 displays a closer magnification of 1 and 3 phr PP-g-SAN additive system. The SEM pictures were taken within 150 µm from the surface, thus within the skin. Elongated phases were found exclusively in all analyzed samples containing PP-g-SAN. This phase is thus probably PP-g-SAN which was segregated from ABS. These elongated phases were found parallel to the surface and close to it. They were typically found less than 150 µm from it. Interestingly enough on this picture, this new phase seems to have contained what could have been nodules. Since it is mainly found near the surface of specimens, it is more detrimental to unnotched tests since in standard notched tests the notch is 2 mm deep. In this way, one could assume that it could not play a weakening role.

It is surprising to see the PP-g-SAN demixing in this way. Since PP nodules do not appear to be affected by the presence of PP-g-SAN, it could be presumed that most of this additive is found in these segregated phases. However, PP-g-SAN should be compatible to ABS considering its chemistry. Thus, the mixing failure is probably rooted in two different viscosities because of very poor solubility of the additive system in the ABS phase. Modulations of process parameters are performed in the next part to try improving this polymer incorporation.

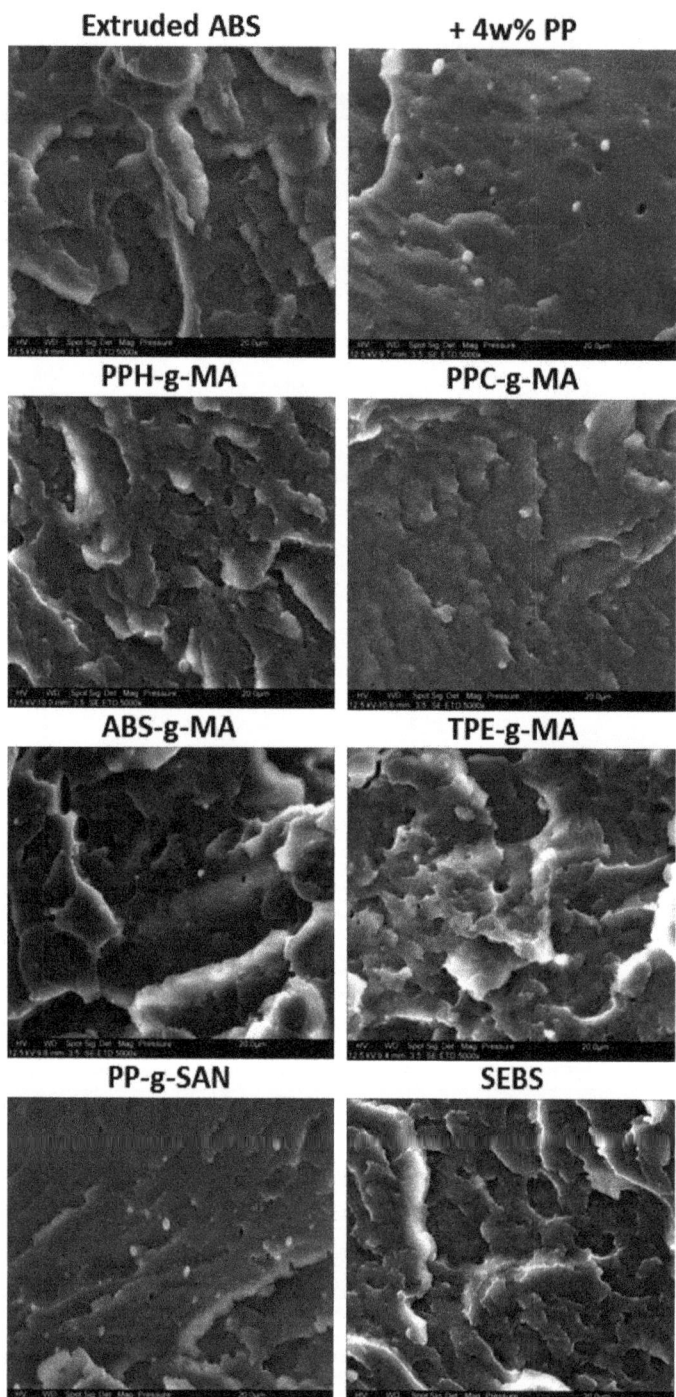

Figure 12. SEM pictures of cryofractuted dogbones—effect of compatibilizers at 3 phr on a 4 wt% PP contamination—5000 × magnification.

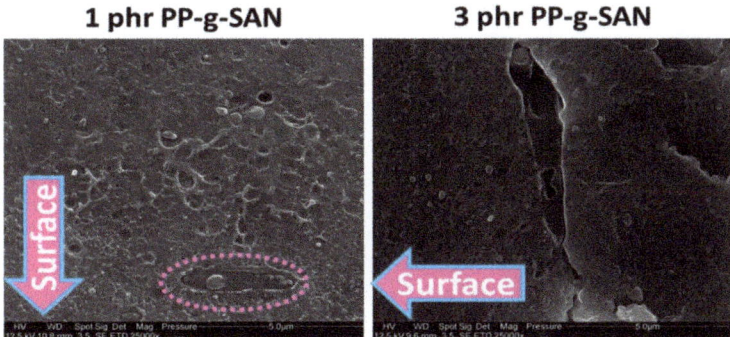

Figure 13. SEM pictures of cryofractured dogbones of ABS contaminated with 4 wt% PP and added with PP-g-SAN at 1 and 3 phr—25,000× magnifications—pictures taken within 100 µm from the surface—arrows indicates surface direction—supposedly PP-g-SAN phase.

3.2.3. Tensile Properties

From impact-resistance values and morphology refinement, it was found that PP-g-MA (homopolymer and copolymer undifferentiated) and SEBS are the most promising materials. The second one is more performant but the first one is more common and cheaper. However, tensile properties could be negatively modified, especially by SEBS. They were then assessed for these two additives at different applied concentrations (results in Figure 14 and force–displacement curves in *Figure S10 of supporting information*). Most batches led to results very similar to references which were already close to each other. Notably, properties do not visibly change with the concentration.

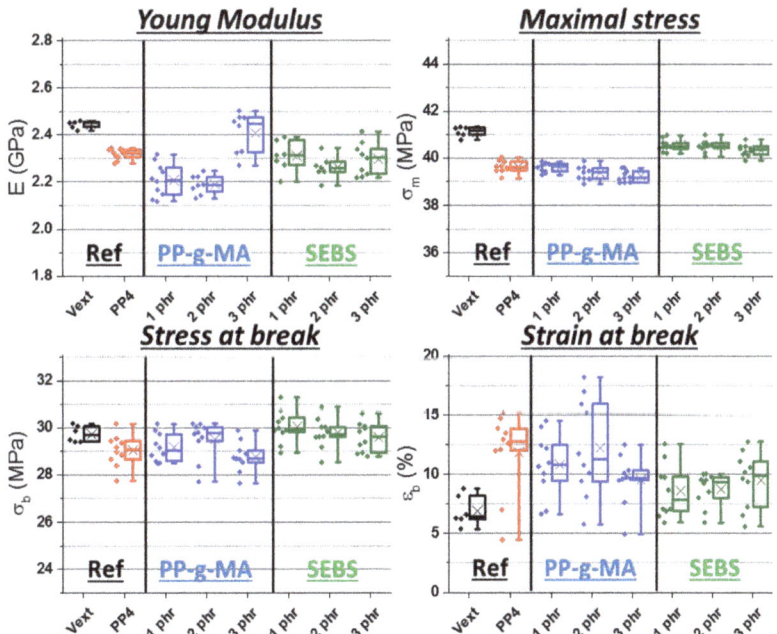

Figure 14. Young's modulus, maximal stress, stress and strain at break from tensile tests on ABS contaminated with 4 wt% of PP—"Vext" for virgin and extruded ABS, other batches with 4 wt% PP—phr rates correspond to PP-g-MA and SEBS.

3.2.4. Conclusions on Compatibilization Trials

Except for PP-g-SAN, all compatibilizers proved to have an interesting effect, partially restoring impact resistance through morphological refinement, probably due to interfacial tension decrease and coalescence prevention. PP-g-SAN was found to be segregated to the skin, devoid of PP. Additionally, PP nodules seemed unaffected. Finally, it rendered the material even more brittle than with PP alone. All polymers grafted with maleic anhydride had similar effects. In particular, PPH-g-MA and PPC-g-MA behaved very similarly. SEBS had the most important effect of morphological refinement and its quality of impact modifier even led to a partial break during unnotched Charpy impact tests. For the sake of simplicity, PPH-g-MA and SEBS were kept as interesting candidates because of their widespread industrial use, compared to ABS-g-MA for instance. Tensile properties were also shown to be almost unmodified by the compatibilizer presence.

3.3. DOE-RSM: Influence of Process Parameters and Interactions with Compatibilizers toward Impact Properties

The following results relate a Design of Experiment (DOE) with Response Surface Methodology (RSM) applied to the previous system to evaluate possible influences and interactions with selected compatibilizers from process parameters on impact properties as they are the most sensible properties of this system. As SEBS and PP-g-MA gave the most promising results and PP-g-SAN was found to be poorly incorporated into the matrix, they were selected for this DOE. The purposes were to observe if properties could be further improved for PP-g-MA and SEBS, and if PP-g-SAN could be better incorporated. As easily tunable parameters, extruding temperature and screws rotation speed were chosen as variables. For the sake of simplicity, extrusion temperature was kept homogenous along the screw, except for the feeding zone, kept at 80 °C. Extrusion temperatures and screw speed limits were, respectively, 200 to 240 °C and 200 to 300 rpm. Following previous results, which showed that PP-g-SAN and PP-g-MA were surely added in excess and that the SEBS effect was only marginally dependent on its loading rate, 0.5 to 1.5 phr were chosen for the first ones and 1.0 to 3.0 phr for the last one as DOE limits. Experimental results were used to build the models as averages and standard deviations. Obtained response surfaces are presented below in 2D projections of isoresponse lines. In *Figures S16 and S17 of supporting information*, response surfaces are given in 3D projections, one figure by impact test type (notched or unnotched), enabling better visualization but forbidding objective interpretation.

The Nemrodw® analysis of the RSM study showed that the polynomial model with interaction terms is significant (with a significant p-value of 1.69 for the six responses). The coefficient of multiregression equation was calculated at 0.966 (close to 1) with an average standard deviation between 0.07 and 0.22 kJ/m^2 for notched impact and between 1.5 and 4.9 kJ/m^2 for unnotched impact. Averages were, respectively, between 11.47 and 13.69 kJ/m^2 for notched, and between 49.5 and 100.3 kJ/m^2 for unnotched. The data were then fitted to a first order polynomial equation with interaction between the terms X_i and $X_i.X_j$, which validated the model overall response surface of resilience 1eA and 1eU. Finally, this allows us to intrapolate (within the model domain, whereas interpolate is strictly) between experimental points) a real simulation of the parameters of extrusion temperature (X1) and screw speed (X2) on the whole of the DOE for the three ratios of PP-g-MA, SEBS and PP-g-SAN compatibilizers. The most important interactions were globally observed between temperature and screw speed and in a lesser measure with compatibilizer rate, even insignificant in some cases. Therefore, it was chosen to represent response surfaces at fixed concentrations, with temperature and screw speed as variables.

3.3.1. Uncompatibilized System

Parameter modulation already leads to impact resistances modifications in uncompatibilized systems (Figure 15). For notched impact, break energies are at 12.4–12.6 kJ/m^2 for lowest screw speed on all ranges, and highest temperature for all ranges as well. Then, energy increases progressively by following the diagonal line, with decreasing temperature and increasing screw speed, going up to 13.4 kJ/m² in the top left-handed corner. High screw speed can lead to better dispersion as shear rate increase leads to capillary number increase. Lower temperature can prevent material degradation. A lower temperature can also affect viscosity ratios of both phases but a complete rheological assessment is necessary to support this statement. Morphological characterization via SEM (*Figure S11 of supporting information*) was however not conclusive on a potential refinement.

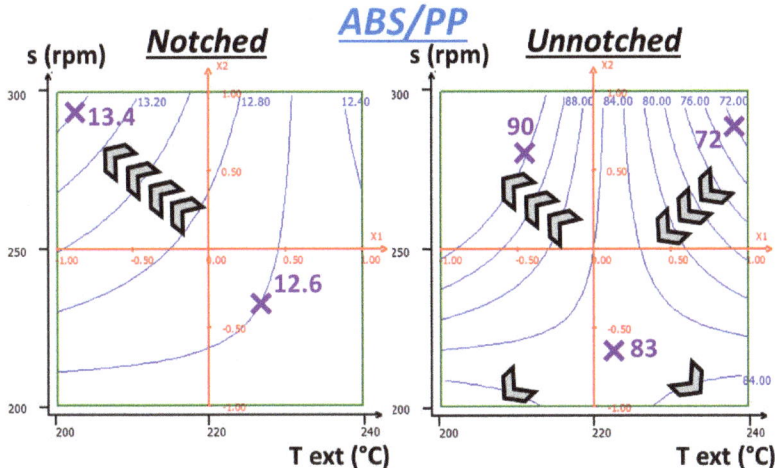

Figure 15. RSM applied to notched and unnotched Charpy break energies (kJ/m²) of ABS + 4 wt% PP—extruder temperature (200–240 °C) and screw speed (200–300 rpm) as variables—chevrons indicate increasing energy direction.

For unnotched Charpy impact, the overall behavior is quite different as evolutions are not homogeneous on the whole ranges, going from 72 kJ/m² from the top right corner to 83 near the center to 90 kJ/m² on a large area at the top left corner. On the right-handed half, corresponding to 220–240 °C, energies increase by diminishing screw speed and temperature simultaneously whereas on the other half, 200–220 °C, energies are increased by diminishing temperature while increasing screw speed. At the extreme bottom of the range (very low screw speed), increases change directions and go toward the corners. These values are close to the domain limits and the validity of the model is thus questionable whereas the top left corner displays a large area with homogenous response.

Unnotched Charpy impact adds skin effects and crack initiation to what is observed in notched impact. These supplementary behaviors are surely due to different skin effects. Indeed, a skin devoid of PP was systematically observed. Its thickness is probably a consequence of the blend dispersion conditions, especially as ABS-PP system viscosity is probably the result of PP domain sizes. Since stress is concentrated at the bottom of the specimen during the impact, as in a flexion test, a thicker skin could enable better properties. To the contrary, poor cohesion between the skin and the core, as seen on Figure 8, could have a negative influence on properties.

Nonetheless, optimization of both properties is achieved by going to the top left corner, with the highest screw speed and lowest temperature. Thus, 3D surfaces (*Figures S16 and S17 of supporting information*) display very similar overall shapes for uncompatibilized system for both types of impact.

3.3.2. PP-g-SAN

Figure 16 shows the same response surfaces as those previously seen on ABS/PP/PP-g-SAN systems at 0.5, 1.0 and 1.5 phr. All energies are globally still below the uncompatibilized system, ranging from 8.5 to 12.5 kJ/m² for notched impact instead of 12.6–13.4. This shows a global failure at incorporation attempts. A common break energy maximum about 12.1–12.5 kJ/m^2 for all rates can be seen for maximal temperature (240 °C) and minimal screw speed (200 rpm). At 0.5 phr, energy is almost independent from temperature and overall is rather constant as only 0.5 kJ/m^2 evolutions are seen. With higher loading rates, lower temperatures and higher screw speeds lead to stronger decreases down to 8.5 kJ/m^2 for the highest speed and concentration and lowest temperature. Additionally, the more PP-g-SAN is added to the system, the worse the issue is, probably linked to more segregation occurring in the system.

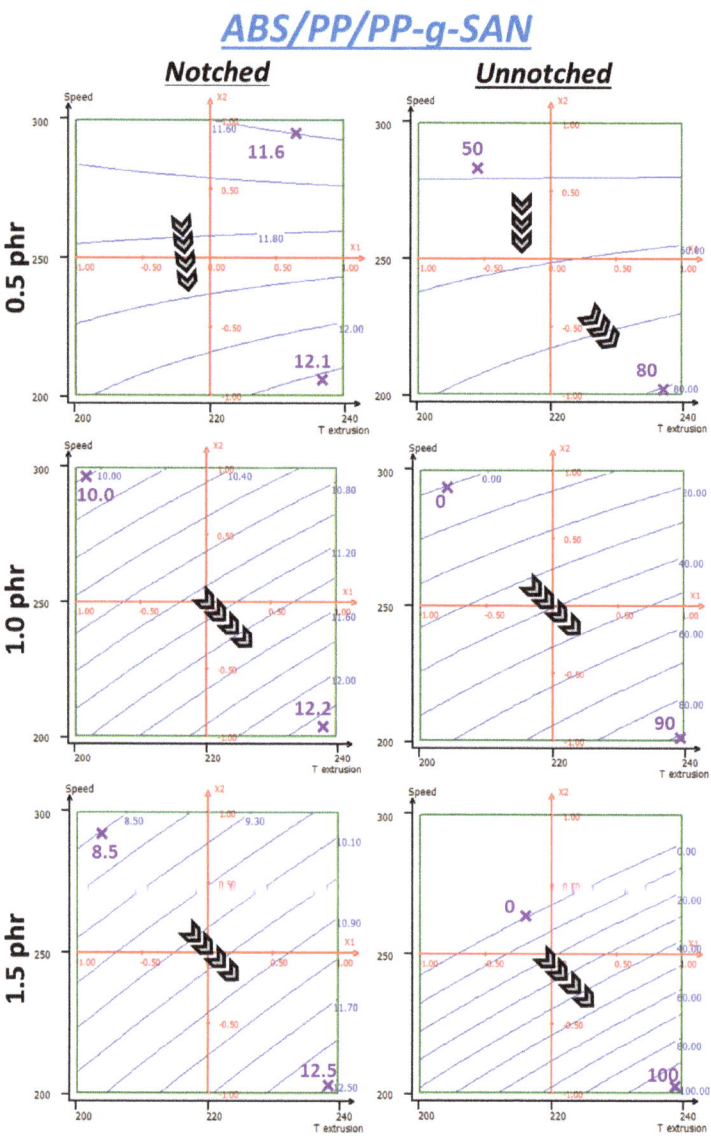

Figure 16. RSM applied to unnotched Charpy break energies (kJ/m^2) of ABS + 4wt% PP + PP-g-SAN—same variables as those used previously.

Similar behavior is seen on the unnotched test, with stronger energy variations on the studied ranges, from 50 to 80 for 0.5 phr. At 1 phr, the top left corner is at 0 kJ/m^2 whereas the maximum rises from 80 to 90 at the bottom right corner. Null and negative values have no physical signification, and this proves that such a model cannot be applied to such a complex system with such material heterogeneity and so few experimental data, especially when approaching model boundaries. Additionally, even at 0.5 phr, 240 °C and 200 rpm, which is supposed to be optimal here, elongated phases were found near specimens' surface (*Figures S12 and S13 of supporting information*). At 1.5 phr, invalid values occupy roughly a third of the range. A maximum is found at 100 kJ/m^2 at the lowest speed and highest temperature.

Eventually, even experimental results (40–70 kJ/m^2) are not satisfactory and do not suggest significant ameliorations above the uncompatibilized system through process alone. Especially, temperature should be increased, which could lead to the material degradation. PP-g-SAN itself should be modified, in molecular weight and/or PP/SAN ratio, to tune its rheological behavior. Otherwise, work on screw profiles could enable better dispersion.

3.3.3. PP-g-MA

PP-g-MA additive systems display rather different behavior (Figure 17) and trends greatly change with incorporation rate, especially beyond 0.5 phr, maybe indicating it is then in excess. Interestingly, notched impact energy ranges are the same for all three rates, roughly 12.2–13.0 kJ/m^2 despite behavior differences. At 0.5 phr, energy mainly increases following lower temperatures. Similar to what was found with unnotched impact on uncompatibilized ABS/PP, screw speed lowers at the 220–240 °C range but increases at the 200–220 °C range, with a maximum in the top left corner. At 1.0 phr, parameter influences are permuted and screw speed becomes more important and should be lowered. By doing so, the maximum is relocated at the bottom right corner. It can be seen that PP-g-MA notched impact properties' surface responses are globally lower than those of the uncompatibilized system, whereas experimental data (given in *Figures S14 and S15 of supporting information*) seem to indicate the opposite. These contradictory trends are to be linked to the relatively important dispersions of PP-g-MA systems. The Nemrodw® software takes the standard deviation as a value modulation range to reach the best correlation. Thus, here, it was to the disadvantage of PP-g-MA.

For unnotched impact, behaviors at different rates are more homogenous and the general trend is toward low speed and low temperature with rather equal importance of both parameters. However, for 0.5 phr, isoresponse curves go along the axis, respectively, for lowest temperatures and lowest screw speed. As for the uncompatibilized system, this corresponds to the model limits and its reliability is thus questionable. Here, increasing compatibilizer rates improves resistance, whereas results from the previous part shows that 3 phr represent an excess. Optimal rate of PPH-g-MA for 4 wt% PP is therefore probably between 1.5 and 2.0 phr. An optimal compromise between notched and unnotched properties is located around 210 °C and 200 rpm where both values are at the maximum at 1.5 phr, reaching, respectively, 13 kJ/m^2 and 116 kJ/m^2. Temperature and screw speed are to be kept at minimum here probably to prevent ABS degradation. PP dispersion could be good enough because of compatibilization and higher shearing would not affect it further as interfacial tension is significantly low enough. Otherwise, Plochocki et al. [48] suggested that shear rate could have an important effect on collision probability, and thus coalescence, in uncompatibilized systems. However, it could be expected from selected compatibilizers that they prevent coalescence through steric stabilization by being situated at the interface (Sundararaj and Macosko, 1995).

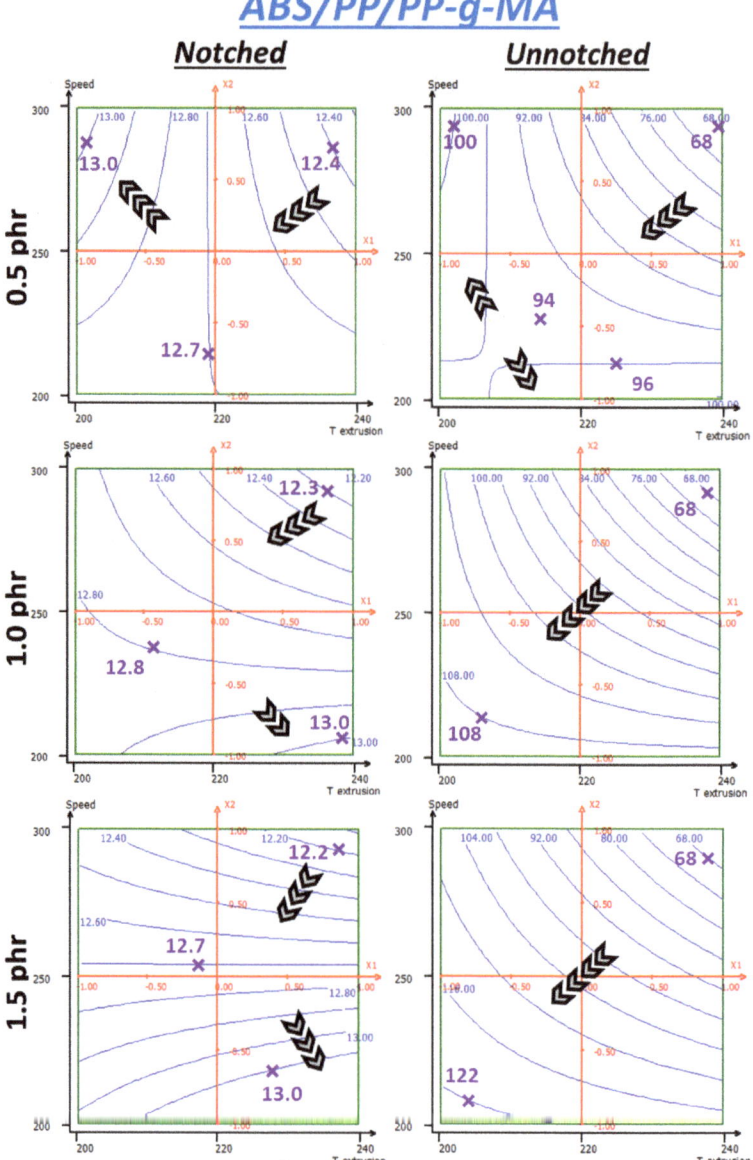

Figure 17. RSM applied to notched and unnotched Charpy break energies (kJ/m2) of ABS + 4wt% PP + PP-g-MA—same variables as those used previously.

3.3.4. SEBS

SEBS results (Figure 18) show behaviors closer to what the uncompatibilized system had, especially for notched impact. Top-left corner is optimal for notched properties and this is amplified by the loading rate. Additionally, values on the whole range are improved due to the high compatibilizer content. Consequently, values go from 13.0 to 14.2 kJ/m^2 for 1 phr, but from 14.0 to 16.0 kJ/m^2 at 3 phr. Unnotched impact also displays a strong sensibility to temperature, perhaps to be related SEBS thermal sensitivity. Response toward shear is quite different. At 1 phr, the behavior is reminiscent of what was seen on uncompatibilized ABS-PP and 0.5 PP-g-MA. Since uncompatibilized experimental data were used to build these models as well, it could be its repercussion which is seen at lower rates. For higher loading rates, the behavior is very similar to PP-g-MA systems, with higher values on the whole range. Thus, similar behavior explanations could be applied to this system. Values increments are very similar for both compatibilizers, rate by rate. Additionally, a notable effect of compatibilizer loading rate is seen, stronger than with PP-g-MA, with a maximum at 122 kJ/m^2. This is also very visible on 3D projections in *supporting information (Figures S16 and S17)*.

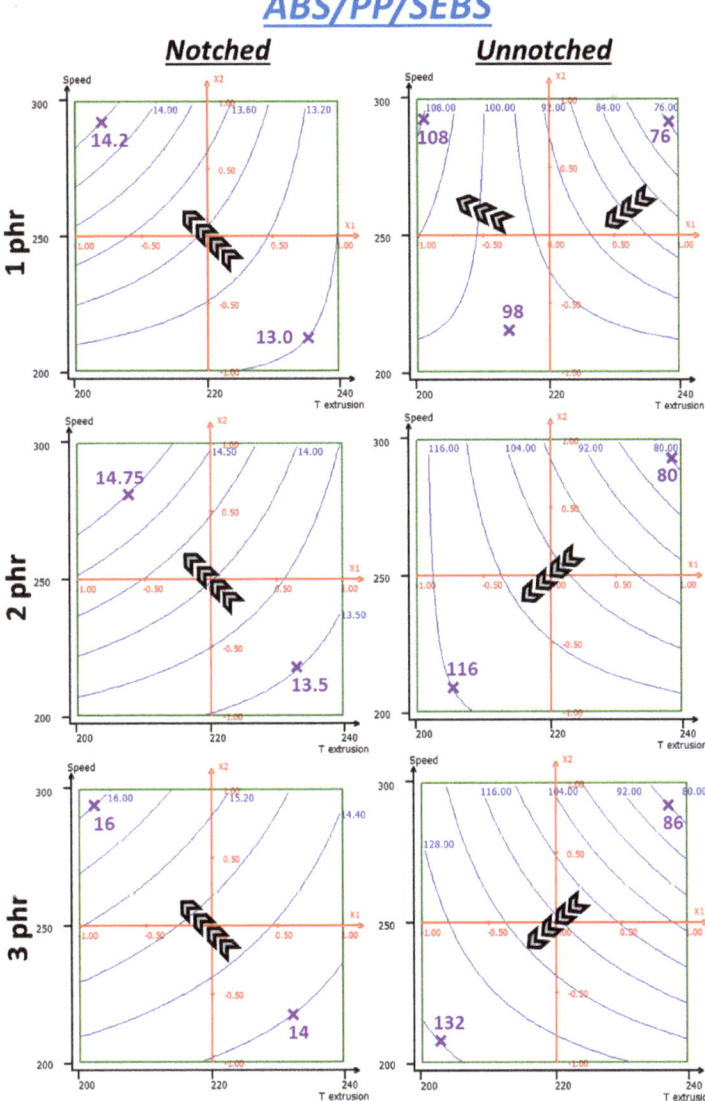

Figure 18. RSM applied to notched and unnotched Charpy break energies (kJ/m^2) of ABS + 4 wt% PP + SEBS—same variables as those used previously.

Compromise between notched and unnotched is more delicate than with PP-g-MA since notched impact has here a higher sensibility to temperature, probably due to SEBS decomposition which is more sensible to temperature than PP-g-MA. Indeed, isoresponse curves are almost diagonal instead being vertical. The temperature of 200 °C is indubitably optimal but screw speed should be chosen in a between 225 and 275 rpm, depending on which property is preferred.

3.3.5. Conclusions on RSM Results

These experiments showed that impact resistance could be modified by the modulation of extrusion parameters, temperature and screw speed, and compatibilizer type and loading rates. An uncompatibilized ABS + 4 wt% PP system could be optimized with high screw speed and lower temperature. The first one promotes finer dispersion of the minor phase through shear rate and capillary number increases. Temperature minimization can avoid material degradation. PP-g-SAN was not noticeably better dispersed and its properties remained lower than the uncompatibilized system. Rheological assessment could enable us to understand this failure. PP-g-MA and SEBS showed different results regarding notched and unnotched impact resistance. For notched, high screw speed and low temperature are to be preferred whereas low screw speed and low temperature are better for unnotched. This highlights difference in crack initiation and potentially skin effect. However, lower shear and temperature can prevent ABS degradation. Overall SEBS is confirmed as the most performant candidate, imputable to its dispersing role, but also its impact modifier role. In-depth rheological assessment of this system could be interesting to better highlight the occurring phenomena. For the sake of the economy of the experiment, all systems were incorporated into the same DOE. In the future, quadratic models (individual factors taken at 2nd degree) could be applied to these systems for better representativity of physico-chemical reality, especially for PP-g-SAN because of its lack of significance.

4. Conclusions

PP was incorporated in ABS through twin-screw extrusion at different rates: 2, 4, 6 and 8 wt%, in order to evaluate and manage the effect of residual impurities in WEEE plastics. It induced important losses of impact resistance, from both notched and unnotched Charpy impact. Tensile properties were unmodified at 2 wt% but they decrease above, with a noticeable weakening above 6 wt%. Morphology assessment showed a nodular structure. Additionally, important heterogeneities were observed at the millimetric scale above 6 wt% of PP.

Several potential compatibilizers were then tested on ABS contaminated with 4 wt% PP since properties began to be severely affected from this fraction. PP-g-MA and SEBS were found to be the most interesting, the first one which is relatively usual led to interesting properties, the second one enabled obtaining the best results of the study. Both entailed an important morphological refinement effect as proven by SEM observations. Tensile properties were unaffected by both compatibilizers. PPC-g-MA, ABS-g-MA and TPE-g-MA had similar effect than PPH-g-MA but were less interesting as they are less usual and commercially available. PP-g-SAN was found to be segregated. It did not modify the morphology and even led to a weakening of the material.

Design Of Experiment (DOE) were applied on ABS + 4wt% PP system with compatibilizer (none, PP-g-MA, SEBS, PP-g-SAN), compatibilizer rate, extrusion temperature (200–240 °C) and screw speed (200–300 rpm) as variables and impact properties (both notched and unnotched) as responses. The experimental data were fitted to a first-order and second-order interactions polynomial equation, which validated the model overall response surface of notched and unnotched impact resistances. Response Surface Methodology (RSM) revealed that properties of an uncompatibilized system could be enhanced, but still far from virgin ABS, with stronger screw speed and lower temperature. PP-g-MA remained segregated, resulting in poor results. Blends with PP-g-MA and SEBS were found to be optimized, especially for unnotched impact. Lower temperature and milder screw speed are generally to be advised probably to prevent ABS degradation. Since PP could be already adequately dispersed, stronger shearing could be unnecessary.

This study was based on virgin samples from the same grade. In a real recycling context, different grades with various formulations and different degradation degrees are mixed together, leading to specific viscosities and mechanical properties. As highlighted with ABS contaminated with 8 wt% PP, multi-scale segregation can occur, probably inducing mechanical weakness. Several research and development pathways can follow these works. Additionally, different types of SEBS exist, with different S/EB ratios, rheological behavior; they also impact modification ability. Other process parameters could be considered in DOE as feeding rates and models could be refined through more experimental data. Finally, it could be interesting to better assess created morphologies with other sample preparation methods and then evaluate their durability through annealing tests.

Supplementary Materials: The following are available online at https://www.mdpi.com/article/10.3390/polym13091439/s1, Figure S1: Comparison of force-displacement curves of instrumented notched Charpy impact at 0.8 m/s (left) and 2.9 m/s (right)—uncontaminated and contaminated with 4w% PP ABS—5 specimens / batch; Figure S2: Force-displacement curves of unnotched Charpy impact on virgin ABS, SAN (both directly injected from pellets) and ABS contaminated with 6w% PP; Figure S3: G-Moduli extracted from unnotched Charpy test versus E-Moduli from tensile tests of ABS progressively contaminated with PP—dots represent means on 6 to 10 specimens, errors bars represent standard deviations; Figure S4: SEM pictures ABS + 4 w% PP dogbones cryofractured at 90° (top) & 45° (bottom)—5000 × and 25,000 × magnifications; Figure S5: FT-IR ATR spectra of ABS (black), PP (blue) and ABS + 8w% PP (red); Figure S6: Force-displacement curves of notched Charpy impact—virgin extruded ABS batch "Vext", 4w% PP contaminated batch "P4" and separately additived batches with 1, 2 & 3 phr of PPH-g-MA, PPC-g-MA, ABS-g-MA, TPE-g-MA, SEBS and PP-g-SAN; Figure S7: Force-displacement curves of unnotched Charpy impact—same batches and denomination as previous figure; Figure S8: Force-displacement curves of unnotched Charpy impact of ABS contaminated with 4w% PP and additived with 1, 2 & 3 phr SEBS (a plot for each rate) and picture of 2 phr SEBS specimens after impact—green curves and frame for partially broken specimens; Figure S9: SEM pictures of cryofractured dogbones—effect of compatibilizers at 3 phr on a 4w% PP contamination—25000 × magnification; Figure S10: Tensile tests of compatibilization trials with PP-g-MA and SEBS of ABS contaminated with 4w% of PP—magnification on stress peak-"Vext" for virgin and extruded ABS, other batches with 4w% PP—phr rates correspond to PP-g-MA and SEBS; Figure S11: SEM pictures of cryofractured dogbones of ABS contaminated with 4 w% PP—different extrusion temperatures and screw speeds; Figure S12: SEM pictures of cryofractured dogbones of ABS + 4 w% PP—0.5 phr PP-g-SAN, 200 °C and 200 rpm—near the surface, elongated phases framed in pink; Figure S13: SEM pictures of cryofractured dogbones of ABS + 4 w% PP—0.5 phr PP-g-SAN, 240 °C and 300 rpm—near the surface, elongated phases framed in pink; Figure S14: Notched Charpy impact break energies—D.O.E.-R.S.M. experimental values—ABS + 4w% PP without compatibilizer or with SEBS, PP-g-SAN or PP-g-MA—abscissa axis indicates extrusion temperature (°C), screw speed (rpm) and compatibilizer loading rate (phr); Figure S15: Unnotched Charpy impact break energies—D.O.E.-R.S.M. experimental values—samples and designation identical to previous figure; Figure S16: Response Surface Methodology applied to notched Charpy break energy of ABS + 4w% PP system—uncompatibilized, additived with SEBS, PP-g-MA or PP-g-SAN—additive concentration (1–3 or 0.5–1.5 phr), extruder temperature (200–240 °C) and screw speed (200–300 rpm) as variables—lower color scale for PP-g-SAN (values in orange); Figure S17: Response Surface Methodology applied to unnotched Charpy break energy of ABS + 4w% PP system—uncompatibilized, additived with SEBS, PP-g-MA or PP-g-SAN—additive concentration (1–3 or 0.5–1.5 phr), extruder temperature (200–240 °C) and screw speed (200–300 rpm) as variables—lower color scale for PP-g-SAN (values in orange).

Author Contributions: Conceptualization, C.S., P.G. and A.L.G.; methodology, C.S.; software, D.P. and C.S.; validation, D.P., P.I., J.-M.L.-C. and A.-S.C.-B.; formal analysis, C.S.; investigation, C.S., P.G. and A.L.G.; resources, C.S., P.G. and A.L.G.; data curation, D.P., P.I., J.-M.L.-C. and A.-S.C.-B.; writing—original draft preparation, C.S., P.G. and A.L.G.; writing—review and editing, C.S. and D.P.; visualization, C.S.; supervision, D.P.; project administration, D.P. and P.I.; funding acquisition, D.P. and P.I. All authors have read and agreed to the published version of the manuscript.

Funding: This work was supported by BPI France via the FUI 20 (Fonds Unique interministériel) grant.

Institutional Review Board Statement: Not applicable.

Informed Consent Statement: Not applicable.

Data Availability Statement: Not applicable.

Acknowledgments: The authors would like to thank Benjamin Gallard, Alexandre Cheron, and Jean-Claude Roux for technical support, respectively, for polymer processing, mechanical testing and SEM experiments. Pellenc ST and Suez are gratefully acknowledged for partnership in this work.

Conflicts of Interest: The authors declare no conflict of interest.

Abbreviations

ABS:	Acrylonitrile Butadiene Styrene
DOE:	Design Of Experiments
ELV:	End-of-Life Vehicles
-g-MA:	Grafted with Maleic anhydride
RSM:	Response Surface Methodology
phr:	Parts per Hundred of Resin
PP:	Polypropylene
PPC:	Polypropylene copolymer
PPH:	Polypropylene homopolymer
SEBS:	Styrene-Ethylene-Butylene-Styrene
SEM:	Scanning Electron Microscope
TPE:	Thermoplastic elastomer
WEEE:	Waste Electrical and Electronic Equipment
WEEP:	WEEE plastics

Appendix A. Highlights

- Impact properties of ABS decreases from 2 wt% PP for notched, 4 wt% for unnotched;
- Tensile properties are diminished beyond 4 wt% of PP;
- SEBS is the most promising compatibilizer improving the dissipative behavior and refining morphologies;
- DOE-RSM showed an influence of process parameters on impact properties;
- Response is different with or without compatibilizer.

References

1. Geyer, R.; Jambeck, J.R.; Law, K.L. Production, use, and fate of all plastics ever made. *Sci. Adv.* **2017**, *3*, e1700782. [CrossRef] [PubMed]
2. Robinson, B.H. E-waste: An assessment of global production and environmental impacts. *Sci. Total Environ.* **2009**, *408*, 183–191. [CrossRef]
3. Bovea, M.D.; Pérez-Belis, V.; Ibáñez-Forés, V.; Quemades-Beltrán, P. Disassembly properties and material characterisation of household small waste electric and electronic equipment. *Waste Manag.* **2016**, *53*, 225–236. [CrossRef]
4. Menad, N.; Björkman, B.; Allain, E.G. Combustion of plastics contained in electric and electronic scrap. *Resour. Conserv. Recycl.* **1998**, *24*, 65–85. [CrossRef]
5. Dimitrakakis, E.; Tustai, J.; Boldizar, A.; Foreman, M.R.S.; Moller, K. An analysis of the composition and metal contamination of plastics from waste electrical and electronic equipment (WEEE). *Waste Manag.* **2013**, *33*, 915–922. [CrossRef]
6. Vilaplana, F.; Karlsson, S. Quality concepts for the improved use of recycled polymeric materials: A review. *Macromol. Mater. Eng.* **2008**, *293*, 274–297. [CrossRef]
7. Wang, R.; Xu, Z. Recycling of non-metallic fractions from waste electrical and electronic equipment (WEEE): A review. *Waste Manag.* **2014**, *34*, 1455–1469. [CrossRef]
8. Martinho, G.; Pires, A.; Saraiva, L.; Ribeiro, R. Composition of plastics from waste electrical and electronic equipment (WEEE) by direct sampling. *Waste Manag.* **2012**, *32*, 1213–1217. [CrossRef]
9. Wäger, P.A.; Hischier, R. Life cycle assessment of post-consumer plastics production from waste electrical and electronic equipment (WEEE) treatment residues in a Central European plastics recycling plant. *Sci. Total Environ.* **2015**, *529*, 158–167. [CrossRef] [PubMed]
10. WRAP. *LCA of Management Options for Mixed Waste Plastics, Waste Resource Action Programme (WRAP)*; WRAP: Banbury, UK, 2008.
11. Maris, J.; Bourdon, S.; Brossard, J.-M.; Cauret, L.; Fontaine, L.; Montembault, V. Mechanical recycling: Compatibilization of mixed thermoplastic wastes. *Polym. Degrad. Stab.* **2018**, *147*, 245–266. [CrossRef]

12. Peeters, J.R.; Vanegas, P.; Kellens, K.; Wang, F.; Huisman, J.; Dewulf, W.; Duflou, J.R. Forecasting waste compositions: A case study on plastic waste of electronic display housings. *Waste Manag.* **2015**, *46*, 28–39. [CrossRef] [PubMed]
13. Schlummer, M.; Gruber, L.; Mäurer, A.; Wolz, G.; van Eldik, R. Characterisation of polymer fractions from waste electrical and electronic equipment (WEEE) and implications for waste management. *Chemosphere* **2007**, *67*, 1866–1876. [CrossRef]
14. Maris, E.; Botané, P.; Wavrer, P.; Froelich, D. Characterizing plastics originating from WEEE: A case study in France. *Miner. Eng.* **2015**, *76*, 28–37. [CrossRef]
15. Beigbeder, J.; Perrin, D.; Mascaro, J.-F.; Lopez-Cuesta, J.-M. Study of the physico-chemical properties of recycled polymers from waste electrical and electronic equipment (WEEE) sorted by high resolution near infrared devices. *Resour. Conserv. Recycl.* **2013**, *78*, 105–114. [CrossRef]
16. Ragaert, K.; Delva, L.; Van Geem, K. Mechanical and chemical recycling of solid plastic waste. *Waste Manag.* **2017**, *69*, 24–58. [CrossRef] [PubMed]
17. Utracki, L.A. Compatibilization of Polymer Blends. *Can. J. Chem. Eng.* **2002**, *80*, 1008–1016. [CrossRef]
18. Kassouf, A.; Maalouly, J.; Rutledge, D.N.; Chebib, H.; Ducruet, V. Rapid discrimination of plastic packaging materials using MIR spectroscopy coupled with independent components analysis (ICA). *Waste Manag.* **2014**, *34*, 2131–2138. [CrossRef] [PubMed]
19. Rozenstein, O.; Puckrin, E.; Adamowski, J. Development of a new approach based on midwave infrared spectroscopy for post-consumer black plastic waste sorting in the recycling industry. *Waste Manag.* **2017**, *68*, 38–44. [CrossRef]
20. Signoret, C.; Caro-Bretelle, A.-S.; Lopez-Cuesta, J.-M.; Ienny, P.; Perrin, D. MIR spectral characterization of plastic to enable discrimination in an industrial recycling context: II. Specific case of polyolefins. *Waste Manag.* **2019**, *98*, 160–172. [CrossRef]
21. Fraunhofer-Gesellschaft, 2016. Sorting Black Plastics according to Type [WWW Document]. Available online: https://www.fraunhofer.de/en/press/research-news/2016/June/sorting-black-plastics-according-to-type.html (accessed on 28 September 2017).
22. Küter, A.; Reible, S.; Geibig, T.; Nüßler, D.; Pohl, N. THz imaging for recycling of black plastics. *Tech. Mess.* **2018**, *85*, 191–201. [CrossRef]
23. Barbier, S.; Perrier, S.; Freyermuth, P.; Perrin, D.; Gallard, B.; Gilon, N. Plastic identification based on molecular and elemental information from laser induced breakdown spectra: A comparison of plasma conditions in view of efficient sorting. *Spectrochim. Acta Part B At. Spectrosc.* **2013**, *88*, 167–173. [CrossRef]
24. Roh, S.-B.; Park, S.-B.; Oh, S.-K.; Park, E.-K.; Choi, W.Z. Development of intelligent sorting system realized with the aid of laser-induced breakdown spectroscopy and hybrid preprocessing algorithm-based radial basis function neural networks for recycling black plastic wastes. *J. Mater. Cycles Waste Manag.* **2018**, 1–16. [CrossRef]
25. Wagner, F.; Peeters, J.R.; De Keyzer, J.; Janssens, K.; Duflou, J.R.; Dewulf, W. Towards a more circular economy for WEEE plastics—Part B: Assessment of the technical feasibility of recycling strategies. *Waste Manag.* **2019**, *96*, 206–214. [CrossRef] [PubMed]
26. Perrin, D.; Mantaux, O.; Ienny, P.; Léger, R.; Dumon, M.; Lopez-Cuesta, J.M. Influence of impurities on the performances of HIPS recycled from Waste Electric and Electronic Equipment (WEEE). *Waste Manag.* **2016**, *56*, 438–445. [CrossRef] [PubMed]
27. Brennan, L.B.; Isaac, D.H.; Arnold, J.C. Recycling of acrylonitrile-butadiene-styrene and high-impact polystyrene from waste computer equipment. *J. Appl. Polym. Sci.* **2002**, *86*, 572–578. [CrossRef]
28. De Souza, A.M.C.; Cucchiara, M.G.; Ereio, A.V. ABS/HIPS blends obtained from WEEE: Influence of processing conditions and composition. *J. Appl. Polym. Sci.* **2016**, *133*, 1–7. [CrossRef]
29. Vazquez, Y.V.; Barbosa, S.E. Compatibilization Strategies for Recycling Applications of High Impact Polystyrene/Acrylonitrile Butadiene Blends. *J. Polym. Environ.* **2017**, *25*, 903–912. [CrossRef]
30. Signoret, C.; Edo, M.; Lafon, D.; Caro-Bretelle, A.-S.; Lopez-Cuesta, J.-M.; Ienny, P.; Perrin, D. Degradation of styrenic plastics during recycling: Impact of reprocessing photodegraded material on aspect and mechanical properties. *J. Polym. Environ.* **2020**, *28*, 2055–2077. [CrossRef]
31. Bucknall, C.B.; Cote, F.F.P.; Partridge, I.K. Rubber toughening of plastics—Part 9 Effects of rubber particle volume fraction on deformation and fracture in HIPS. *J. Mater. Sci.* **1986**, *21*, 301–306. [CrossRef]
32. Hall, R.A. Computer modelling of rubber-toughened plastics: Random placement of monosized core-shell particles in a polymer matrix and interparticle distance calculations. *J. Mater. Sci.* **1991**, *26*, 5631–5636. [CrossRef]
33. Han, Y.; Lach, R.; Grellmann, W. The Charpy impact fracture behaviour in ABS materials. *Die Angew. Makromol. Chemie* **1999**, *270*, 13–21. [CrossRef]
34. Yokouchi, M.; Seto, S.; Kobayashi, Y. Comparison of polystyrene, poly(styrene/acrylonitrile), high-impact polystyrene, and poly(acrylonitrile/butadiene/styrene) with respect to tensile and impact properties. *J. Appl. Polym. Sci.* **1983**, *28*, 2209–2216. [CrossRef]
35. Ramírez-Vargas, E.; Margarita Huerta-Martínez, B.; Javier Medellín-Rodríguez, F.; Sánchez-Valdes, S. Effect of heterophasic or random PP copolymer on the compatibility mechanism between EVA and PP copolymers. *J. Appl. Polym. Sci.* **2009**, *112*, 2290–2297. [CrossRef]
36. Roumeli, E.; Markoulis, A.; Chrissafis, K.; Avgeropoulos, A.; Bikiaris, D. Substantial enhancement of PP random copolymer's thermal stability due to the addition of MWCNTs and nanodiamonds: Decomposition kinetics and mechanism study. *J. Anal. Appl. Pyrolysis* **2014**, *106*, 71–80. [CrossRef]
37. Bouvard, J.L.; Denton, B.; Freire, L.; Horstemeyer, M.F. Modeling the mechanical behavior and impact properties of polypropylene and copolymer polypropylene. *J. Polym. Res.* **2016**, *23*, 70. [CrossRef]
38. Koning, C.; Duin, M.; van Pagnouille, C.; Jerome, R. Strategies for compatibilization of polymer blends. *Prog. Polym. Sci.* **1998**, *23*, 707–757. [CrossRef]

39. Sundararaj, U.; Macosko, C.W. Drop Breakup and Coalescence in Polymer Blends: The Effects of Concentration and Compatibilization. *Macromolecules* **1995**, *28*, 2647–2657. [CrossRef]
40. Bonda, S.; Mohanty, S.; Nayak, S.K. Influence of compatibilizer on mechanical, morphological and rheological properties of PP/ABS blends. *Iran. Polym. J.* **2014**, *23*, 415–425. [CrossRef]
41. Deng, Y.; Mao, X.; Lin, J.; Chen, Q. Compatibilization of polypropylene/Poly(acrylonitrile-butadiene-styrene) blends by polypropylene-graft-cardanol. *J. Appl. Polym. Sci.* **2015**, *132*. [CrossRef]
42. Ibrahim, M.A.H.; Hassan, A.; Wahit, M.U.; Hasan, M.; Mokhtar, M. Mechanical properties and morphology of polypropylene/poly(acrylonitrile–butadiene–styrene) nanocomposites. *J. Elastomers Plast.* **2017**, *49*, 209–225. [CrossRef]
43. Kum, C.K.; Sung, Y.-T.; Kim, Y.S.; Lee, H.G.; Kim, W.N.; Lee, H.S.; Yoon, H.G. Effects of compatibilizer on mechanical, morphological, and rheological properties of polypropylene/poly(acrylonitrile-butadiene-styrene) blends. *Macromol. Res.* **2007**, *15*, 308–314. [CrossRef]
44. Lee, Y.K.; Lee, J.B.; Park, D.H.; Kim, W.N. Effects of accelerated aging and compatibilizers on the mechanical and morphological properties of polypropylene and poly(acrylonitrile-butadiene-styrene) blends. *J. Appl. Polym. Sci.* **2013**, *127*, 1032–1037. [CrossRef]
45. Luo, Z.; Lu, Q.; Ma, F.; Jiang, Y. The effect of graft copolymers of maleic anhydride and epoxy resin on the mechanical properties and morphology of PP/ABS blends. *J. Appl. Polym. Sci.* **2014**, *131*. [CrossRef]
46. Patel, A.C.; Brahmbhatt, R.B.; Sarawade, B.D.; Devi, S. Morphological and mechanical properties of PP/ABS blends compatibilized with PP-g-acrylic acid. *J. Appl. Polym. Sci.* **2001**, *81*, 1731–1741. [CrossRef]
47. Tostar, S.; Stenvall, E.; Foreman, M.; Boldizar, A. The Influence of Compatibilizer Addition and Gamma Irradiation on Mechanical and Rheological Properties of a Recycled WEEE Plastics Blend. *Recycling* **2016**, *1*, 101. [CrossRef]
48. Plochocki, A.P.; Dagli, S.S.; Andrews, R.D. The interface in binary mixtures of polymers containing a corresponding block copolymer: Effects of industrial mixing processes and of coalescence. *Polym. Eng. Sci.* **1990**, *30*, 741–752. [CrossRef]
49. Favis, B.D.; Chalifoux, J.P. The effect of viscosity ratio on the morphology of polypropylene/polycarbonate blends during processing. *Polym. Eng. Sci.* **1987**, *27*, 1591–1600. [CrossRef]
50. Landrein, P.; Lorriot, T.; Guillaumat, L. Influence of some test parameters on specimen loading determination methods in instrumented Charpy impact tests. *Eng. Fract. Mech.* **2001**, *68*, 1631–1645. [CrossRef]
51. Fellahi, S.; Favis, B.D.; Fisa, B. Morphological stability in injection-moulded high-density polyethylene/polyamide-6 blends. *Polymer* **1996**, *37*, 2615–2626. [CrossRef]
52. Karger-Kocsis, J.; Csikai, I. Skin-Core morphology and failure of injection-molded specimens of impact-modified polypropylene blends. *Polym. Eng. Sci.* **1987**, *27*, 241–253. [CrossRef]
53. Li, Z.-M.; Yang, W.; Yang, S.; Huang, R.; Yang, M.-B. Morphology-tensile behavior relationship in injection molded poly(ethylene terephthalate)/polyethylene and polycarbonate/polyethylene blends (I) Part I Skin-core Structure. *J. Mater. Sci.* **2004**, *39*, 413–431. [CrossRef]
54. Li, Z.-M.; Xie, B.-H.; Yang, S.; Huang, R.; Yang, M.-B. Morphology-tensile behavior relationship in injection molded poly(ethylene terephthalate)/polyethylene and polycarbonate/polyethylene blends (II) Part II Tensile behavior. *J. Mater. Sci.* **2004**, *39*, 433–443. [CrossRef]
55. Li, Z.M.; Qian, Z.Q.; Yang, M.B.; Yang, W.; Xie, B.H.; Huang, R. Anisotropic microstructure-impact fracture behavior relationship of polycarbonate/polyethylene blends injection-molded at different temperatures. *Polymer* **2005**, *46*, 10466–10477. [CrossRef]
56. Charoen, N.; Leong, Y.W.; Hamada, H. Determination of different morphological structures in PC/ABS open spiral injection moldings. *Polym. Eng. Sci.* **2008**, *48*, 786–794. [CrossRef]
57. Zhong, G.-J.; Li, Z.-M. Injection molding-induced morphology of thermoplastic polymer blends. *Polym. Eng. Sci.* **2005**, *45*, 1655–1665. [CrossRef]

Article

Polypropylene Post-Consumer Recyclate Compounds for Thermoforming Packaging Applications

Paul J. Freudenthaler [1,*], Joerg Fischer [1], Yi Liu [2] and Reinhold W. Lang [1]

[1] Institute of Polymeric Materials and Testing, Johannes Kepler University Linz, Altenberger Straße 69, 4040 Linz, Austria
[2] Borealis Polyolefine GmbH, Innovation Headquarters, St. Peterstraße 25, 4021 Linz, Austria
* Correspondence: paul.freudenthaler@jku.at; Tel.: +43-732-2468-6620

Abstract: Polypropylene (PP) plastic packaging waste consists of a variety of different plastic packaging products with a great span in rheological and mechanical behavior. Therefore, the resulting post-consumer recyclates usually show melt mass-flow rates (MFR) in the region of injection molding grades and intermediate mechanical properties. High-quality packaging applications demand a distinct property profile that is met by tailor-made PP grades and cannot be met by recyclates with intermediate performance. One such application with high market volume is high-stiffness thermoforming trays. The aim of this research was to blend intermediate-performance recyclates with a virgin PP grade to obtain compounds that fulfill the rheological and mechanical demands of this application. Three commercially available PP post-consumer recyclates were acquired and compounded with different blending ratios with a high stiffness, low MFR virgin PP grade. As the pure recyclates show different rheological properties, the blending ratios had to be adapted for each of them to fit into the MFR range of 2–4 g/10 min which is desirable for thermoforming applications. The resulting PP recyclate compounds show a distinct correlation of recyclate content with rheological and mechanical performance. However, the resulting property profile was directly dependent on the performance of the originally used recyclate. The best-performing recyclate could be used in a blending ratio of 65 m% recyclate content while adhering to both property limits, the MFR of 2–4 g/10 min and the lower bound tensile stiffness of 1500 MPa.

Keywords: plastic recycling; compounding; thermoforming; melt mass-flow rate; mechanical properties; polypropylene; post-consumer; recyclate

1. Introduction

Polypropylene (PP) is a popular polymer used in high amounts for a multitude of applications [1]. This comes partly due to its low price [2], but also because of its ability to be tailored to specific applications by using co-polymers, additives, or fillers [3–5]. Single-use packaging products profit highly from both these properties, making PP one of the most used packaging materials [1]. Nevertheless, the different packaging applications have different demands in terms of stiffness, toughness, and rheology, hence the PP grades used vary greatly [6]. This leads to the problem that the gathered PP packaging waste stream, even if other polymers are sorted out, consists of a myriad of different PP products with their properties ranging from stiff and brittle to soft and tough. The resulting commercially available PP recyclates (rPP) from packaging waste streams usually have melt mass-flow rates (MFR) which are typical for processing via injection molding [7]. Due to the high variability of the feedstock, only intermediate mechanical properties are achieved, hence these recyclates are usually not favorable to be used in any specific use case. Furthermore, PP shows disadvantages compared to polyethylene terephthalate (PET) when it comes to decontamination, hence recycling of food contact rPP is more challenging and has yet to be developed and validated [8,9]. To usefully employ these mediocre-performing recyclates, combination with high-quality virgin materials could present a solution.

The research presented here aims to investigate the usage of rPP for thermoforming applications. Thermoforming is a high-volume, low-cost production method that is used for the production of a broad range of products [6,10,11]. There is published research regarding the use of rPP in sheet applications [12], but only the processing of multi-layer sheets was investigated and no mechanical characteristics of the material combinations were tested. Another possibility for modifying rPP could be compounding with virgin polymers. While there is research available for other applications [13–15], no preceding research could be found for creating and assessing thermoforming compounds with post-consumer recyclate content.

Virgin PP thermoforming grades are highly specialized and often make use of the advantages of co-polymerization to make them more transparent or tougher, e.g., for low-temperature food packaging [3–5,16,17]. The stiffest grades are used for containers and trays, with and without food contact, e.g. As there are many applications for containers and trays apart from food packaging, the specific goal of this research was to create thermoforming recyclate compounds using commercially available PP recyclates for high-stiffness applications. Hence, application-specific properties and limits were determined and tested together with additional analytical methods. The results gathered from these tests are used to assess the applicability of the created recyclate compounds for their use as thermoforming grades.

2. Experimental Section

2.1. Materials

Three different commercially available recyclate grades were provided by three European recycling companies. All of them are designated as "rPP" for recycled PP and are made from post-consumer PP packaging waste. All three recyclates were delivered as pellets and will henceforth be called rPP-A, rPP-B, and rPP-C. While the main feedstock for all of these recyclates was the "yellow-bag" system, which is a separate collection stream for plastic packaging products in Germany and Austria [18], the recyclates passed different sorting, washing, and recycling steps and differ in their properties, which can be seen in Table 1, as well as in color. rPP-A is anthracite-colored, rPP-B is white-colored, and rPP-C is without pigment and thus translucent.

Table 1. Datasheet values of all used materials.

Datasheet Values	vPP [1]	rPP-A [2]	rPP-B [2]	rPP-C [2]
melt mass-flow rate (230 °C/2.16 kg; g/10 min)	0.25	≥ 11	10.1–15.0	18
tensile modulus (MPa)	2000	≥ 1100	1100–1400	1200
yield stress (MPa)	36	≥ 25	23–27	28
strain at break (%)	-	≥ 180	-	-
Charpy notched impact strength (kJ/m^2)	30	≥ 6	4–8	5.5

[1] Low melt mass-flow rate (MFR) and high modulus virgin polypropylene (PP) grade. [2] Recyclates derived from PP packaging waste.

For thermoforming applications, the most requested properties are (1) a melt mass-flow rate (MFR) within the range of the currently used virgin thermoforming materials, (2) a high tensile modulus of above 1500 MPa, and (3) a Charpy notched impact strength of above 3.5 kJ/m^2. To make blends with the three high MFR rPPs, a virgin blending partner with very low MFR is crucial [19]. Furthermore, a high tensile modulus and Charpy notched impact strength are also beneficial to achieve high recyclate contents while still fulfilling the requirements. Therefore, a low MFR, high modulus virgin PP grade was acquired for compounding and will be called vPP from here on.

All compounds were produced on a Leistritz ZSE MAXX 18 twin-screw extruder (Leistritz Extrusionstechnik GmbH, Nuremberg, Germany) with an L/D ratio of 40D, co-rotating screws, a screw speed of 400 rpm, and a mass throughput of around 8–10 kg/h. Three gravimetric feeders, two Brabender DSR28 for the pellets, and one Brabender Minitwin for stabilizer powder (Brabender Technologie GmbH & Co. KG, Duisburg, Germany) were used to ensure a consistent ratio of the virgin material, the recyclate, and the stabilization mixture to ensure no further degradation was occurring during the compounding of the materials. The stabilization used for every compound was 0.15 m% Irganox® 1010 (BASF SE, Ludwigshafen am Rhein, Germany), a phenolic primary antioxidant for long-term thermal stabilization [20], and 0.15 m% Irgafos® 168 (BASF SE, Ludwigshafen am Rhein, Germany), a hydrolytically stable phosphite secondary antioxidant for processing stabilization [21]. The stabilization of the materials should not influence the mechanical properties measured within the scope of this paper, as no aging was applied to the specimens and no lengthy tests in media and/or at elevated temperatures were conducted. Nevertheless, the effect of the applied stabilization on the resistance to thermal oxidation was investigated.

Pretests were performed to discover the individual MFRs of the blending partners and conclude starting blending ratios (one per recyclate) to fit within the desired property window. After characterization of the first compound for each recyclate, additional blending ratios were defined, compounded, and the resulting compounds were characterized. This step-by-step approach was chosen because of the limited amount of recyclate available. Therefore, not all mixtures use the same recyclate content. Blends containing 50 m% and 60 m% of rPP-A will be called A50 and A60, blends containing 55 m%, 60 m%, and 65 m% of rPP-B will be called B55, B60, and B65, and lastly, the blend containing 55 m% of rPE-C will be called C55. A list of all compounds together with the blending ratios is presented in Table 2.

Table 2. List of compounds and their respective amounts of blending partners. To all compounds, additional 0.15 m% Irganox® 1010 long-term stabilizer [20], and 0.15 m% Irgafos® 168 processing stabilizer [21] were added during compounding.

	vPP [1]	rPP-A [2]	rPP-B [2]	rPP-C [2]
	m%	m%	m%	m%
A50	50	50	-	-
A60	40	60	-	-
B55	45	-	55	-
B60	40	-	60	-
B65	35	-	65	-
C55	45	-	-	55

[1] Low MFR and high modulus PP grade. [2] Recyclates derived from PP packaging waste.

2.2. Methods

2.2.1. Melt Mass-Flow Rate Measurements

The MFR measurements were conducted at 230 °C under a 2.16 kg static load on a Zwick/Roell Mflow melt flow indexer (ZwickRoell GmbH & Co. KG, Ulm, Germany) according to ISO 1133-1 [22] and ISO 19069-2 [23]. Cuts were made every 3 mm piston movement. The time between cuts was measured and each extrudate was weighed on an ABS 220-4 electronic balance (Kern & Sohn GmbH, Balingen-Frommern, Germany). The extra- and interpolation to 10 min calculated the MFR in g/10 min for each cut. For each material, one measurement was conducted. Within one measurement, 6 cuts were made and used for the calculation of average values.

2.2.2. Specimen Production

According to ISO 19069-2 [23], tensile and impact test specimens for PP should be injection-molded. Therefore, all multipurpose specimens (MPS) and bar specimens were produced via injection molding according to ISO 294-1 [24], ISO 20753 [25], and ISO 19069-2 on an Engel Victory 60 (Engel Austria GmbH, Schwertberg, Austria) with a 25 mm cylinder. The processing temperature to be used is dependent on the MFR of the material. Therefore, as vPP has an MFR of below 1.5 g/10 min, its specimens were injection-molded with 255 °C melt temperature, the three rPPs with MFRs above 7 g/10 min were injection-molded at 200 °C melt temperature, and as the compounds should all fall within the MFR range of between 1.5 g/10 min and 7 g/10 min, a melt temperature of 230 °C was used for injection molding of these specimens as prescribed by ISO 19069-2. All specimens were conditioned at 23 °C and 50% relative humidity for 3–5 days. After conditioning, the MPS were used for tensile testing and the bar specimens were notched and used for Charpy notched impact testing in accordance with ISO 179-1 [26]. The Charpy Type A notches, i.e., notches with a 0.25 mm notch radius, were produced with a Leica RM2265 microtome (Leica Biosystems Nussloch GmbH, Nussloch, Germany) according to ISO 179-1 and measured with an Olympus SZX16 stereomicroscope (Olympus, Tokyo, Japan).

2.2.3. Differential Scanning Calorimetry Measurements

Differential scanning calorimetry (DSC) measurements were carried out on a PerkinElmer differential scanning calorimeter DSC 8500 (PerkinElmer Inc., Waltham, MA, USA). Samples were cut from shoulders of injection-molded MPS and encapsulated in perforated aluminum pans. The average sample weight was around 5 mg. The procedure consisted of an initial heating phase, subsequent cooling, and a second heating phase, each in the temperature range of 0 °C to 200 °C with a constant heating/cooling rate of 10 K/min with nitrogen as purge gas with a flow rate of 20 mL/min. The DSC measurements were accomplished to determine the melting peak in the second heating phase, which is characteristic of the semi-crystallinity achieved under controlled cooling in a DSC device. To determine the melting enthalpy of PP, first, the whole area of the melting peak was integrated in the temperature range from around 90 °C to 175 °C. Since all recyclates were contaminated with polyethylene (PE), the small PE peak was integrated from around 100 °C to 130 °C, which gives the PE melting enthalpy. Subtraction of the PE melting enthalpy from the above-mentioned whole area enthalpy gives the PP melting enthalpy. For each material, five samples, each cut from an individual MPS, were used for the calculation of average values and standard deviations. Measurements were made according to ISO 11357-1 [27] and ISO 11357-3 [28].

2.2.4. Oxidation Induction Temperature (Dynamic OIT) Measurements

Thermal degradation is not a critical failure mechanism in packaging materials. Nevertheless, sufficient stabilization is still important for processing without degrading the material. For the experiments in this work a differential thermal analysis instrument from PerkinElmer, the DOC 1000, was utilized to characterize the oxidation induction temperature (dynamic OIT) according to ISO 11357-6 [29]. Samples were cut from shoulders of injection-molded MPS and encapsuled in perforated aluminum pans. The average sample weight was around 5 mg. A single heating step between 23 °C and 300 °C was performed with a heating rate of 10 K/min with synthetic air as purge gas with a flow rate of 20 mL/min. The point of intersection of the slope before oxidation and during oxidation gives the onset of oxidation or the oxidation induction temperature in °C and gives an indication of stabilization [30]. For each material, five samples, each cut from an individual MPS, were used for the calculation of average values and standard deviations.

2.2.5. Tensile Tests

The tensile properties (tensile modulus, yield stress, and strain at break) were examined with a universal testing machine Zwick/Roell AllroundLine Z020 equipped with a Zwick/Roell multi-extensometer strain measurement system with MPS. Test parameters and MPS were used according to ISO 527-1 [31], ISO 527-2 [32], and ISO 19069-2 [23] with a testing speed of 1 mm/min for tensile modulus determination until a strain of 0.25% and after that 50 mm/min until failure. Calculations of tensile modulus, yield stress, and strain at break were performed in accordance with ISO 527-1. Therefore, the tensile modulus was calculated as the slope of the stress/strain curve between 0.05% and 0.25% via regression, the yield stress was the stress at the first occurrence of strain increase without a stress increase, and the strain at break was the strain when the specimen broke. The strain was recorded via a multi-extensometer until yield. From there, the nominal strain was calculated via Method B according to ISO 527-1 with the aid of the crosshead displacement. This process is integrated and automated in the testing software TestXpert III (v1.61, Zwick-Roell GmbH & Co. KG, Ulm, Germany). For each material, five MPS were tested for the calculation of average values and standard deviations.

2.2.6. Charpy Impact Tests

Impact properties were determined according to ISO 179-1 [26] on a Zwick/Roell HIT25P pendulum impact tester. After pretests to determine the suitable pendulum size (absorbed energy between 10% and 80% of the available energy at impact), a 2 Joule pendulum, the pendulum with the highest available energy that still conforms to these requirements, was chosen for testing all materials. Test conditions were 23 °C test temperature with Type 1 specimen, edgewise blow direction, and notch Type A, i.e., a 0.25 mm notch radius, or short ISO 179-1/1eA, which is one of the preferred methods of ISO 19069-2 [23]. For each material, ten specimens were tested for the calculation of average values and standard deviations.

3. Results

3.1. Melt Mass-Flow Rates

Virgin thermoforming grades are found in the MFR range of between 1.2 g/10 min and 5 g/10 min, but most of the grades range between 2 g/10 min and 4 g/10 min [16]. Therefore, this property window was decided upon to be met with the compounds.

The recyclates rPP-A, rPP-B, and rPP-C and the virgin grade vPP were measured together with the compounds for an accurate comparison. vPP shows a value of 0.23 g/10 min and is shown at 0 m% recyclate content in Figure 1. The three rPP grades show much higher values of 15.8 g/10 min (rPP-A), 13.3 g/10 min (rPP-B), and 17.4 g/10 min (rPP-C) and are shown at 100 m% recyclate content in Figure 1. The compounds show a rising MFR with rising recyclate content with good correlations. Therefore, linear approximations within the logarithmic graph show high R^2 values of 0.996 (A compounds), 0.975 (B compounds), and 0.999 (C compound). As predicted by the MFRs of the blending partners, compounds containing the lower MFR recyclates generally show lower MFRs at the same recyclate contents. Not all the produced compounds fit the property window perfectly, as A50 (1.83 g/10 min) and B55 (1.90 g/10 min) fail to deliver MFRs of above 2 g/10 min. Linear approximations of the compounding-series A, B, and C are plotted as dashed lines and the respective R^2 values are shown in matching colors in Figure 1. The desired property window for compounds is highlighted in yellow in Figure 1.

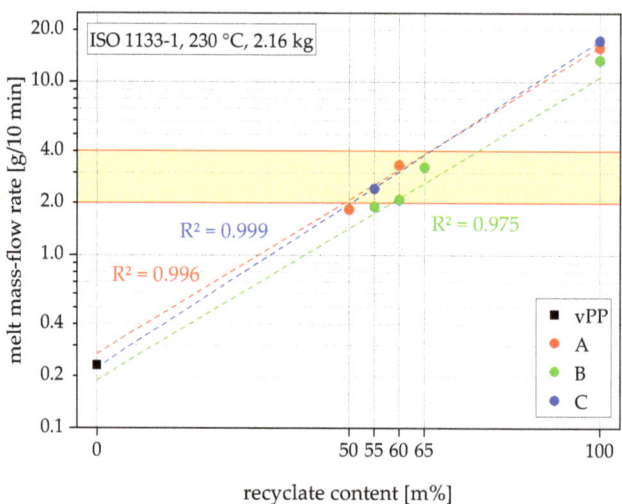

Figure 1. Graphical illustration of the MFR values of all materials.

3.2. Melting Behaviors

The DSC measurements are used to assess the melting peaks and enthalpies of the materials and detect foreign polymers. It is known that contaminations by other polymers can affect the performance of polyolefins [33–35]. All thermograms shown in Figure 2a (vPP as well as the recyclates rPP-A, rPP-B, and rPP-C) are dominated by a melting peak around 165 °C, identifying them as PP materials [36,37]. The PP melting peaks differ substantially in shape and peak temperature, which indicates different PP copolymer compositions. Especially rPP-C shows a comparatively low PP melting peak, indicating smaller crystal lamellae thicknesses [36], and would suggest a higher amount of PP random copolymer [4]. In addition to that, the three rPPs show a pronounced endothermic peak at around 125 °C, indicating contamination with PE [33]. When looking closely at the PE melting peak region (see Figure 2b), even vPP shows a small but quantifiable PE melting peak, which identifies the material as a PP block-copolymer with PE phases [36]. This result points out a high enough PE content within all materials to form PE crystals [4]. Furthermore, there is an obvious difference in melting temperature and melting enthalpy (area of peak), which will be explained in Figure 3. For better comparability, the thermograms were shifted to be shown stacked instead of on top of each other. The arrows indicate PE melting peaks in all four blending partners. Scales are added on the left side of each graph for comparison of magnitudes between Figure 3a,b.

The PP melting temperatures follow the trend of lower PP melting temperature with rising recyclate contents, as shown in Figure 3a. vPP shows the highest PP melting temperature of 167 °C. The recyclates show much lower PP melting temperatures of 161 °C (rPP-A), 162.1 °C (rPP-B), and 159.1 °C (rPP-C). Higher melting temperatures in PP indicate a higher crystal lamellae thickness [36]. The compounds provide melting temperatures in the range of 163–165 °C representing temperatures in between those of the blending partners. When comparing compounds with the same recyclate content, the compounds containing the recyclate with the higher melting temperature also show the higher melting temperature. The melting enthalpies show a different trend, as shown in Figure 3b. While vPP and the recyclates maintain high melting enthalpies, the melting enthalpies of the compounds are lower. This effect could arise from antagonistic effects between the vPP and the recyclates. Furthermore, the compounds follow the trend of lower melting enthalpy with higher recyclate content. A higher PP melting enthalpy indicates a higher crystallinity [38]. The vertical bars in Figure 3 show the sample standard deviations.

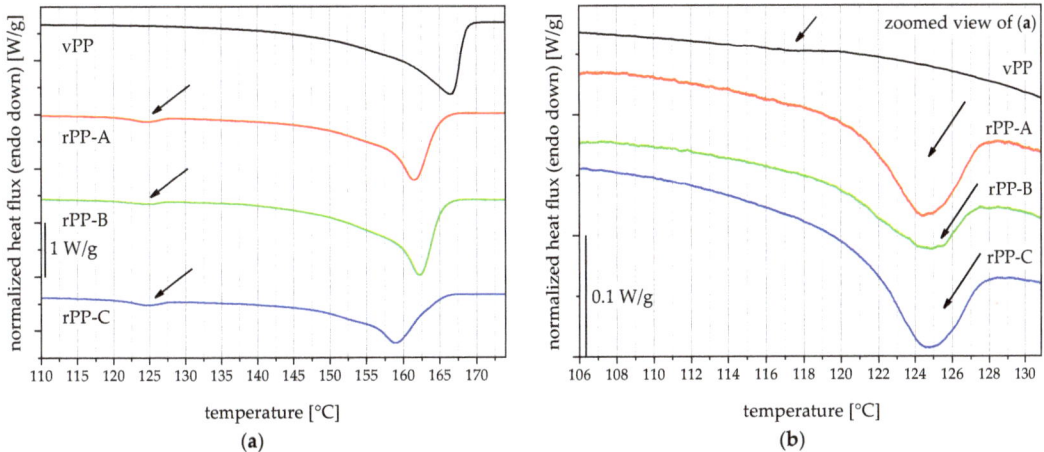

Figure 2. Thermograms of the used blending partners vPP, rPP-A, rPP-B, and rPP-C (**a**), and a detailed view of the polyethylene (PE) melting peaks in the same materials (**b**).

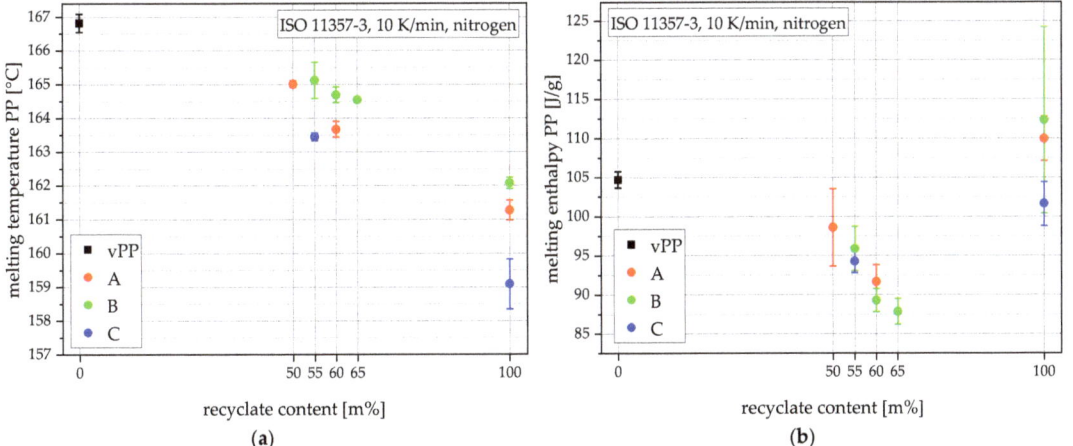

Figure 3. PP melting temperatures (**a**), and PP melting enthalpies (**b**) of all materials.

All three recyclates have similarly high PE melting temperatures of around 125 °C while vPP shows a much lower 117 °C. The recyclate compounds PE melting peaks lie at the level of the pure recyclates between 124 °C to 125 °C, as seen in Figure 4a. The PE melting enthalpies show a clear trend, as seen in Figure 4b, starting with a very low PE melting enthalpy of 0.03 J/g for vPP, rising with the recyclate content, and ending with the high melting enthalpies of the recyclates from 1.76 J/g (rPP-B), 2.19 J/g (rPP-C), up to 2.23 J/g (rPP-A). This suggests a rising level of PE contamination with rising recyclate content. Compounds containing the recyclate with lower PE melting enthalpy (rPP-B) also show lower PE melting enthalpies from 0.65 J/g (B55) to 0.71 J/g (B65), and compounds using the recyclates with higher melting enthalpies (rPP-A and rPP-C) show higher melting enthalpies like 0.90 J/g (A50), 1.02 J/g (C55), and 1.14 J/g (A60). The vertical bars in Figure 4 show the sample standard deviations. Linear approximations of the compounding-series A, B, and C are plotted as dashed lines and the respective R^2 values are shown in matching colors.

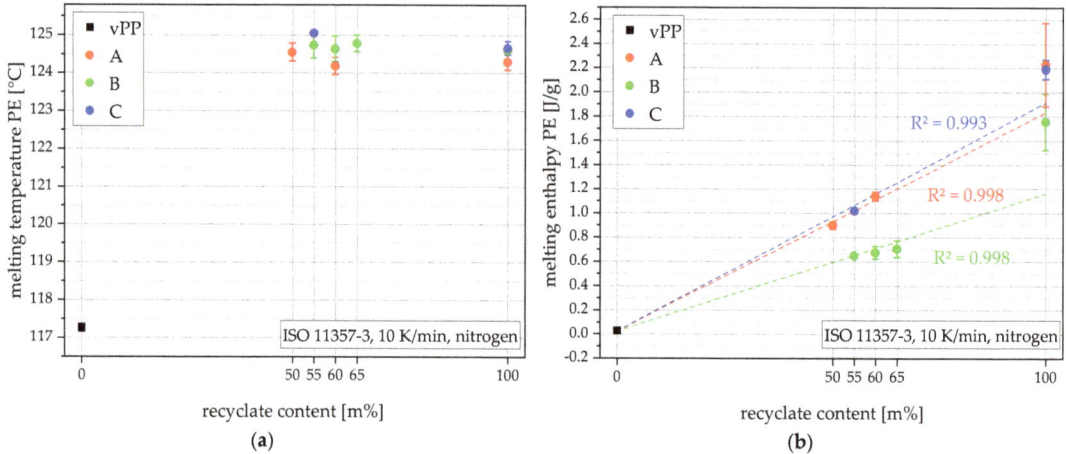

Figure 4. PE melting temperatures (**a**), and PE melting enthalpies (**b**) of all materials.

3.3. Oxidation Induction Temperatures (Dynamic OITs)

The results of the oxidation induction temperature experiments, shown in Figure 5, primarily depict the effectiveness of the stabilizers which were inherent within the blending partners and were added during compounding. rPP-A has the lowest dynamic OIT with 210 °C, followed by rPP-B with 212 °C, and rPP-C with 217 °C. These values are comparable to the literature values for rPP [39]. During compounding, added stabilizers combined with up to 50 m% vPP content led to some compounds gaining better than virgin results, hence higher dynamic OITs than that of vPP (261 °C). The more recyclate in the compounds, the lower the dynamic OIT. Therefore, the lowest dynamic OIT within the compounds is shown by B65 with 251 °C, followed by the two 60 m% recyclate content compounds A60 (258 °C) and B60 (258 °C), the two 55 m% recyclate content compounds B55 (260 °C) and C55 (264 °C), and lastly the compound with the highest dynamic OIT and the lowest recyclate content, A50 with 265 °C. These results verify the stability against oxidation of all produced compounds [30]. The vertical bars in Figure 5 show the sample standard deviations.

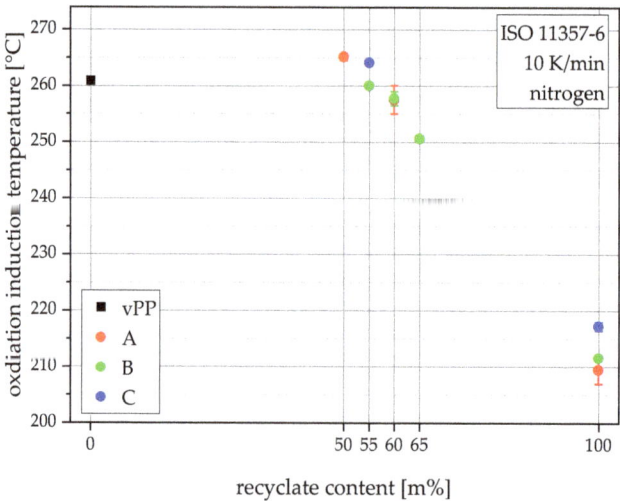

Figure 5. Graphical illustration of oxidation induction temperatures (dynamic OIT) of all materials.

3.4. Tensile Properties

Apart from the MFR for processability, the tensile properties and especially the tensile modulus are important for the function of a thermoforming product [40]. As described before, the virgin PP grade vPP was used as a blending partner because of its low MFR and because of its very high tensile modulus, which was measured with 1850 MPa as can be seen in Figure 6a. The three recyclates vary strongly in tensile modulus in the range of 1090 MPa (rPP-C) to 1320 MPa (rPP-B). rPP-A shows an intermediate value of 1170 MPa. rPP-B also showed the highest PP melting enthalpy of the recyclates, as shown in Figure 3b, hence offering the highest crystallinity [38,41] which explains the high tensile modulus [42,43]. The compounds show values as expected according to a linear mixing rule and therefore show high R^2 values of 0.998 for the B compounds, 0.995 for the A compounds, and a lower 0.982 for the C compound. This also means that the compounds containing the stiffer recyclate show higher tensile moduli at comparable recyclate contents. Therefore, the B compounds excel with tensile moduli between 1520 MPa and 1590 MPa, followed by the A compounds with values between 1420 MPa to 1540 MPa, and lastly, C55 shows a relatively low tensile modulus of 1340 MPa.

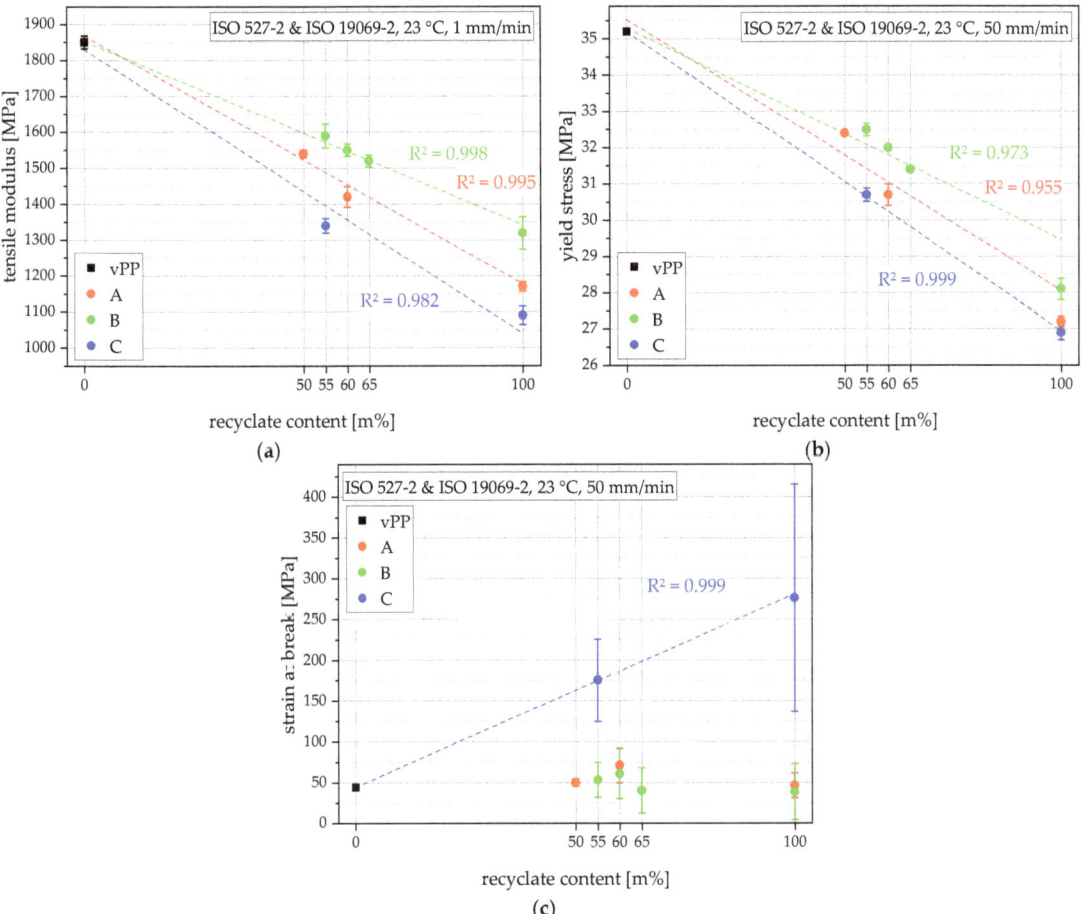

Figure 6. Graphical illustration of tensile modulus (**a**), yield stress (**b**), and strain at break values (**c**) for all materials.

The yield stress is usually not a critical design parameter for thermoforming products which is indicated by its absence on summary data sheets [16,17,44]. Nevertheless, it was investigated within this paper for comparison and is shown in Figure 6b. vPP again leads the field with 35.2 MPa yield stress while the recyclates show much lower values of 28.1 MPa (rPP-B), 27.2 MPa (rPP-A), and 26.9 MPa (rPP-C). While all compounding-series still uphold the trend of lower yield stress with higher recyclate content, the A and B compounds generally perform better as the in each case used recyclate would suggest leading to lower R^2 values for these two compounding-series (0.955 for A and 0.973 for B) while C55 fits in the expected trend and shows a high R^2 of 0.999. Furthermore, the decrease in yield stress over the recyclate content, or rather the negative slope within the compounds of each compounding-series is higher than the slope suggested by the linear approximation. The A compounds show yield stresses of 32.4 MPa (A50) and 30.7 MPa (A60), the B compounds show values of 32.5 MPa (B55), 32.0 MPa (B60), and 31.4 MPa (B65), and lastly, the C55 shows a yield stress of 30.7 MPa.

For warm drawing or thermoforming, the strain at break (which would relate more to a cold drawing) is also not a critical design parameter. Furthermore, the strain at break value highly depends on the injection molding conditions and MFR or rather orientation due to injection molding [13,14], as well as inclusions within the specimen which lead to premature fracture at lower strains and give an indication of the contamination of the compounds. vPP, rPP-A, and rPP-B, as well as all compounds from these materials, show relatively low strain at break values of between 44% and 71%, as can be seen in Figure 6c. Only rPP-C (277%) and the compound C55 (176%) achieve much higher strain at break values suggesting fewer unknown inclusions in rPP-C than in rPP-A and rPP-B.

The vertical bars in Figure 6 show the sample standard deviations. Linear approximations of the compounding-series A, B, and C are plotted as dashed lines and the respective R^2 values are shown in matching colors.

3.5. Charpy Notched Impact Strengths

The Charpy notched impact strength is a critical design parameter, as filled thermoforming containers must be able to survive a drop without critical failure and leakage of its contents. Semi-crystalline polymers naturally have high impact strengths and the design parameter used in high-performance thermoforming applications highly depends on its application. Depending on its application temperature and optical requirements (low temperature, high temperature, transparent), different PP grades are used. While PP homopolymer grades show the highest moduli, their Charpy notched impact strengths are generally low (3–6 kJ/m^2), PP random copolymer grades achieve higher 9–22 kJ/m^2 and PP block copolymers reach values up to 43 kJ/m^2 [16,17] but with significantly lower tensile moduli compared to PP homopolymers. The particular effects of different copolymerization methods on PP material properties are also explained in the literature [4,5]. As the developed recyclate compounds are intended to be used as surrogates for high stiffness homo PP, the lowest range from 3–6 kJ/m^2 should be used as a comparison.

As can be seen in Figure 7, the virgin blending partner vPP shows an exceptionally high Charpy notched impact strength of 33.2 kJ/m^2, especially in combination with its high tensile modulus. In comparison, all recyclates show low values of 6.2 kJ/m^2 (rPP-A), 6.8 kJ/m^2 (rPP-B), and 6.1 kJ/m^2 (rPP-C) which are similar to rPP values found in the literature [45]. However, these values are generally higher than those of most high-modulus homopolymer virgin thermoforming grades [16]. The compounds hardly benefit from the high performance of vPP, staying at values between 8.7 kJ/m^2 and 6.8 kJ/m^2, making them suitable for applications where virgin PP homopolymers are used. The vertical bars in Figure 7 show the sample standard deviations.

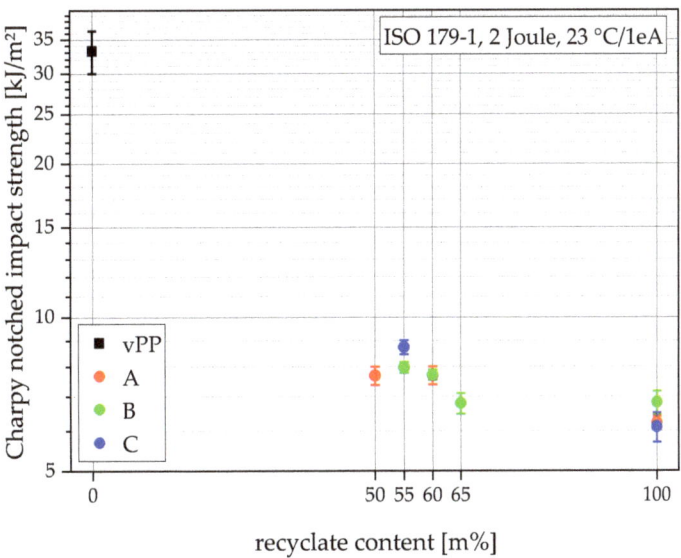

Figure 7. Graphical illustration of Charpy notched impact strengths of all materials.

4. Discussion

While many results in the former chapter were shown for comprehensiveness and presentation of the created compounds, the initial research goal was to create compounds that fit into a small property window defined by their MFR, tensile modulus, and impact toughness measured via Charpy notched impact strength. The impact toughness of virgin PP homopolymers used for thermoforming is already achieved by all pure recyclates [16,17], most probably since these recyclates come from a multitude of different PP grades (homopolymers, random and block copolymers), and also benefit from their PE contamination. The compounds containing 35 m% to 50 m% vPP achieve slightly higher impact strengths, so this mechanical property is seen as fulfilled for all compounds.

The recyclate content determines the MFR of the compounds and through that, the MFR and tensile modulus are, for these compounds, dependent on each other. Higher recyclate contents lead to higher MFRs and lower tensile moduli and both properties are furthermore dependent on the initial values of the respective recyclate (see Figures 1 and 8). To find the sweet spot of appropriate MFR with high enough tensile modulus the right recyclate content with the right recyclate must be found. The correlation of MFR with tensile modulus, seen in Figure 8, shows the desired property window (2–4 g/10 min MFR and above 1500 MPa tensile modulus) highlighted in yellow and values for all materials. The vertical bars show the sample standard deviations. Linear approximations of the compounding-series A, B, and C are plotted as dashed lines and the respective R^2 values are shown in matching colors. Only two compounds, B60, and B65 abide by these two criteria. B55 and A50 miss the property windows closely by delivering too low MFRs and A60 and C55 deliver too low tensile moduli.

Furthermore, when looking at the linear approximation of the A compounds, a small portion of the approximation, which corresponds to about 52 m% recyclate content, lies within the property window. The linear approximation of the C compounding-series does not cross the property window, meaning that no concentration compounds made from these two blending partners (vPP and rPP-C) lie within this property window. The two compounds that meet all three criteria are also summarized in Figure 9.

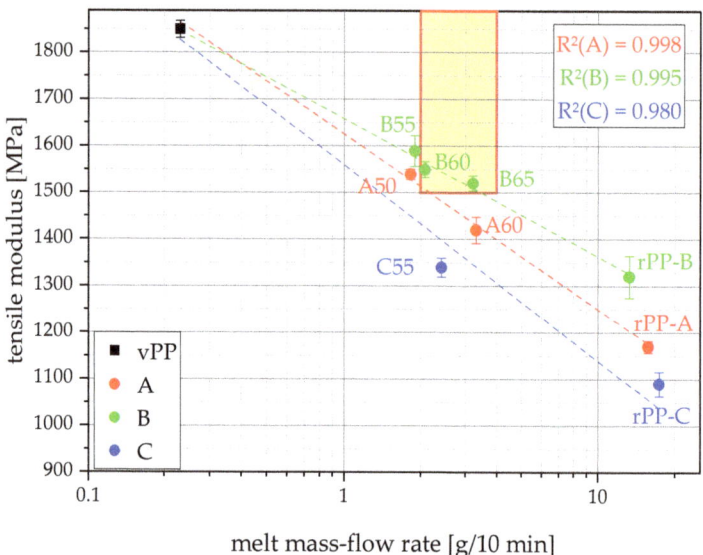

Figure 8. Correlation between MFR and tensile modulus for all materials. The desired property window (MFR: 2–4 g/10 min & tensile modulus above 1500 MPa) is highlighted in yellow.

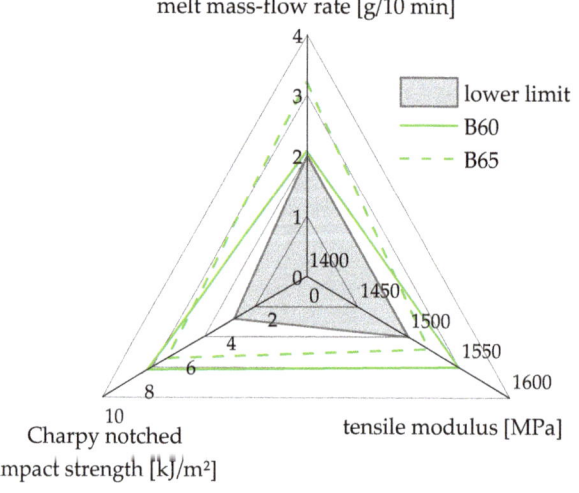

Figure 9. Summary of the two compounds that meet the chosen specifications for MFR (2–4 g/10 min), tensile modulus (>1500 MPa), and Charpy notched impact strength (>3 kJ/m^2).

5. Conclusions

This study provides an outlook of how the usage of highly-specialized PP virgin material compounded with commercially available PP recyclates can achieve reasonable mechanical properties of the resulting compounds, suitable for high-stiffness thermoforming applications. As expected, the properties of the compounds depend highly on the blending partners used. Therefore, not all recyclates are suitable for use within compounds for thermoforming applications, as e.g., a toolow recyclate stiffness will lead to the compound's underperformance in this property. Since not only stiffness but also an MFR within a certain property window is necessary, the blending options are different for each recyclate. rPP-B performed the best, as it showed the lowest MFR and the highest stiffness, therefore compounds made with rPP-B could take up the highest amount of recyclate (65 m%) while still maintaining a tensile modulus of above 1500 MPa and an MFR of lower than 4 g/10 min.

While this study shows the influences and limits of just blending different commercially available recyclates with a suitable virgin blending partner, other ways of optimizing mechanical performance like the use of additives or fillers were not investigated. The use of pure recyclate in combination with virgin materials e.g., a multi-layer composition [46] was also not investigated within this paper. Nevertheless, multi-layer film extrusion would be even more challenging in terms of processing. This brings us to the crucial point of processing these compounds. While the authors investigated the rheological properties in terms of MFR and fit the compounds within the common property window for thermoforming materials, the thermoforming process demands other investigated properties like melt stability and warm drawing capabilities not investigated within this paper also. The investigation of the rheological and mechanical properties within this paper, therefore, should be seen as the first step in the development of suitable thermoforming compound candidates which can be submitted to sheet extrusion and thermoforming trials.

Author Contributions: Conceptualization, P.J.F. and J.F.; methodology, P.J.F. and J.F.; investigation, P.J.F.; resources, P.J.F., J.F. and R.W.L.; data curation, P.J.F.; writing—original draft preparation, P.J.F.; writing—review and editing, P.J.F., Y.L. and J.F.; visualization, P.J.F.; supervision, J.F.; project administration, J.F., Y.L. and R.W.L. All authors have read and agreed to the published version of the manuscript.

Funding: This research was funded by the Austrian Research Promotion Agency (FFG), grant number 867 431. Open Access Funding by the University of Linz.

Acknowledgments: The authors would like to express their gratitude for the recyclate material donations used in this work. Support by Moritz Mager (JKU Linz, Austria) in specimen production is highly appreciated. Furthermore, the contribution and help within this research and the funding of the encompassing research project of Borealis Polyolefine GmbH, Austria is appreciated.

Conflicts of Interest: All authors (Paul J. Freudenthaler, Joerg Fischer, Reinhold W. Lang and Yi Liu) are inventors of a pending patent application (EP22172300.0) on this work. The patent applicant is BOREALIS AG. Yi Liu is an employee of Borealis Polyolefine GmbH.

References

1. PlasticsEurope. Plastics—The Facts 2021: An Analysis of European Plastics Production, Demand and Waste Data. Available online: https://plasticseurope.org/wp-content/uploads/2021/12/AF-Plastics-the-facts-2021_250122.pdf (accessed on 31 August 2022).
2. plasticker. Prices in the Plasticker Material Exchange: Overview about Pellets. Available online: https://plasticker.de/preise/preise_monat_en.php?group=gran (accessed on 31 August 2022).
3. Gahleitner, M.; Paulik, C. Polypropylene and other polyolefins. In *Brydson's Plastics Materials*; Elsevier: Amsterdam, The Netherlands, 2017; pp. 279–309.
4. Gahleitner, M.; Tranninger, C.; Doshev, P. Polypropylene Copolymers. In *Polypropylene Handbook*; Karger-Kocsis, J., Bárány, T., Eds.; Springer International Publishing: Cham, Switzerland, 2019; pp. 295–355. ISBN 978-3-030-12902-6.
5. Gahleitner, M.; Paulik, C. Polypropylene. In *Ullmann's Encyclopedia of Industrial Chemistry*; Wiley-VCH Verlag GmbH & Co. KGaA: Weinheim, Germany, 2000; pp. 1–44. ISBN 9783527306732.
6. Selke, S.E.M. *Plastics Packaging: Properties, Processing, Applications, and Regulations*, 3rd ed.; Hanser: München, Germany, 2015; ISBN 978-1-56990-443-5.
7. Akhras, M.H.; Fischer, J. Sampling Scheme Conception for Pretreated Polyolefin Waste Based on a Review of the Available Standard Procedures. *Polymers* **2022**, *14*, 3450. [CrossRef] [PubMed]
8. Cecon, V.S.; Da Silva, P.F.; Curtzwiler, G.W.; Vorst, K.L. The challenges in recycling post-consumer polyolefins for food contact applications: A review. *Resour. Conserv. Recycl.* **2021**, *167*, 105422. [CrossRef]
9. European Food Safety Authority. Plastics and Plastic Recycling. Available online: https://www.efsa.europa.eu/en/topics/topic/plastics-and-plastic-recycling (accessed on 3 November 2022).
10. Kalpakjian, S.; Schmid, S.R. *Manufacturing Processes for Engineering Materials in SI Units*, 6th ed., Global ed.; Pearson Education Limited: Harlow, UK, 2022; ISBN 9781292254388.
11. Throne, J.L. *Understanding Thermoforming*, 2nd ed.; Hanser: München, Germany, 2008; ISBN 978-3-446-40796-1.

12. Wittmann, L.-M.; Drummer, D. Two Layer Sheets for Processing Post-Consumer Materials. *Polymers* **2022**, *14*, 1507. [CrossRef] [PubMed]
13. Freudenthaler, P.J.; Fischer, J.; Liu, Y.; Lang, R.W. Polypropylene Pipe Compounds with Varying Post-Consumer Packaging Recyclate Content. *Polymers* **2022**, *14*, 5232. [CrossRef] [PubMed]
14. Freudenthaler, P.J.; Fischer, J.; Liu, Y.; Lang, R.W. Short- and Long-Term Performance of Pipe Compounds Containing Polyethylene Post-Consumer Recyclates from Packaging Waste. *Polymers* **2022**, *14*, 1581. [CrossRef] [PubMed]
15. Stanic, S.; Koch, T.; Schmid, K.; Knaus, S.; Archodoulaki, V.-M. Improving Rheological and Mechanical Properties of Various Virgin and Recycled Polypropylenes by Blending with Long-Chain Branched Polypropylene. *Polymers* **2021**, *13*, 1137. [CrossRef] [PubMed]
16. Borealis AG. Solutions for Thermoforming—Summary Data Sheet. Available online: https://www.borealisgroup.com/storage/Polyolefins/Consumer-Products/Solutions_for_Thermoforming.pdf (accessed on 10 August 2022).
17. Braskem Netherlands B.V. Polypropylene for Rigid Packaging Applications. Available online: https://www.braskem.com.br/cms/europe/Catalogo/Download?CodigoCatalogo=275 (accessed on 1 September 2022).
18. Umweltbundesamt: Präsidialbereich/Presse- und Öffentlichkeitsarbeit, Internet. Verpackungen. Available online: https://www.umweltbundesamt.de/themen/abfall-ressourcen/produktverantwortung-in-der-abfallwirtschaft/verpackungen#undefined (accessed on 15 February 2022).
19. Traxler, I.; Marschik, C.; Farthofer, M.; Laske, S.; Fischer, J. Application of Mixing Rules for Adjusting the Flowability of Virgin and Post-Consumer Polypropylene as an Approach for Design from Recycling. *Polymers* **2022**, *14*, 2699. [CrossRef] [PubMed]
20. SpecialChem S.A. Irganox® 1010 Technical Datasheet Supplied by BASF. Available online: https://polymer-additives.specialchem.com/product/a-basf-irganox-1010#related-documents (accessed on 11 May 2022).
21. SpecialChem S.A. Irgafos® 168 Technical Datasheet Supplied by BASF. Available online: https://polymer-additives.specialchem.com/product/a-basf-irgafos-168#related-documents (accessed on 11 May 2022).
22. ISO/TC 61/SC 5 Physical-Chemical Properties. ISO 1133-1:2011 Plastics—Determination of the Melt Mass-Flow Rate (MFR) and Melt Volume-Flow Rate (MVR) of Thermoplastics—Part 1: Standard Method, 1st ed. 2011. Available online: https://www.iso.org/standard/44273.html (accessed on 10 March 2022).
23. ISO/TC 61/SC 9 Thermoplastic Materials. ISO 19069-2:2016 Plastics—Polypropylene (PP) Moulding and Extrusion Materials—Part 2: Preparation of Test Specimens and Determination of Properties, 1st ed. 2016. Available online: https://www.iso.org/standard/66828.html (accessed on 2 August 2022).
24. ISO/TC 61/SC 9 Thermoplastic Materials. ISO 294-1:2017 Plastics—Injection Moulding of Test Specimens of Thermoplastic Materials—Part 1: General Principles, and Moulding of Multipurpose and Bar Test Specimens, 2nd ed. 2017. Available online: https://www.iso.org/standard/67036.html (accessed on 10 August 2022).
25. ISO/TC 61/SC 2 Mechanical Behavior. ISO 20753:2018 Plastics—Test Specimens, 2nd ed. 2018. Available online: https://www.iso.org/standard/72818.html (accessed on 10 August 2022).
26. ISO/TC 61/SC 2 Mechanical Behavior. ISO 179-1:2010 Plastics—Determination of Charpy Impact Properties—Part 1: Non-Instrumented Impact Test, 2nd ed. 2010. Available online: https://www.iso.org/standard/44852.html (accessed on 2 August 2022).
27. ISO/TC 61/SC 5 Physical-Chemical Properties. ISO 11357-1:2016 Plastics—Differential Scanning Calorimetry (DSC)—Part 1: General Principles, 3rd ed. 2016. Available online: https://www.iso.org/standard/70024.html (accessed on 2 August 2022).
28. ISO/TC 61/SC 5 Physical-Chemical Properties. ISO 11357-3:2018 Plastics—Differential Scanning Calorimetry (DSC)—Part 3: Determination of Temperature and Enthalpy of Melting and Crystallization, 3rd ed. 2018. Available online: https://www.iso.org/standard/72460.html (accessed on 2 August 2022).
29. ISO/TC 61/SC 5 Physical-Chemical Properties. ISO 11357-6:2018 Plastics—Differential Scanning Calorimetry (DSC)—Part 6: Determination of Oxidation Induction Time (Isothermal OIT) and Oxidation Induction Temperature (Dynamic OIT), 3rd ed. 2018. Available online: https://www.iso.org/standard/72461.html (accessed on 2 August 2022).
30. Grabmayer, K.; Beißmann, S.; Wallner, G.M.; Nitsche, D.; Schnetzinger, K.; Buchberger, W.; Schobermayr, H.; Lang, R.W. Characterization of the influence of specimen thickness on the aging behavior of a polypropylene based model compound. *Polym. Degrad. Stab.* **2015**, *111*, 185–193. [CrossRef]
31. ISO/TC 61/SC 2 Mechanical Behavior. ISO 527-1:2019 Plastics—Determination of Tensile Properties—Part 1: General Principles, 3rd ed. 2019. Available online: https://www.iso.org/standard/75824.html (accessed on 2 August 2022).
32. ISO/TC 61/SC 2 Mechanical Behavior. ISO 527-2:2012 Plastics—Determination of Tensile Properties—Part 2: Test Conditions for Moulding and Extrusion Plastics, 2nd ed. 2012. Available online: https://www.iso.org/standard/56046.html (accessed on 2 August 2022).
33. Gall, M.; Freudenthaler, P.J.; Fischer, J.; Lang, R.W. Characterization of Composition and Structure–Property Relationships of Commercial Post-Consumer Polyethylene and Polypropylene Recyclates. *Polymers* **2021**, *13*, 1574. [CrossRef] [PubMed]
34. Messiha, M.; Frank, A.; Koch, T.; Arbeiter, F.; Pinter, G. Effect of polyethylene and polypropylene cross-contamination on slow crack growth resistance. *Int. J. Polym. Anal. Charact.* **2020**, *25*, 649–666. [CrossRef]
35. Kazemi, Y.; Ramezani Kakroodi, A.; Rodrigue, D. Compatibilization efficiency in post-consumer recycled polyethylene/polypropylene blends: Effect of contamination. *Polym. Eng. Sci.* **2015**, *55*, 2368–2376. [CrossRef]

36. Ehrenstein, G.W.; Riedel, G.; Trawiel, P. *Thermal Analysis of Plastics: Theory and Practice*; Hanser: Munich, Germany, 2004; ISBN 978-3-446-22673-9.
37. Baur, E.; Brinkmann, S.; Osswald, T.A.; Rudolph, N.; Schmachtenberg, E. *Saechtling Kunststoff Taschenbuch*, 31st ed.; Hanser: München, Germany, 2013; ISBN 978-3-446-43729-6.
38. Lanyi, F.J.; Wenzke, N.; Kaschta, J.; Schubert, D.W. On the Determination of the Enthalpy of Fusion of α-Crystalline Isotactic Polypropylene Using Differential Scanning Calorimetry, X-ray Diffraction, and Fourier-Transform Infrared Spectroscopy: An Old Story Revisited. *Adv. Eng. Mater.* **2020**, *22*, 1900796. [CrossRef]
39. Jansson, A.; Möller, K.; Gevert, T. Degradation of post-consumer polypropylene materials exposed to simulated recycling—Mechanical properties. *Polym. Degrad. Stab.* **2003**, *82*, 37–46. [CrossRef]
40. Traxler, I.; Fischer, J.; Marschik, C.; Laske, S.; Eckerstorfer, M.; Reckziegel, H. Structure-Property Relationships of Thermoformed Products based on Recyclates. *AIP Conf. Proc.* **2023**.
41. Lima, M.F.S.; Vasconcellos, M.A.Z.; Samios, D. Crystallinity changes in plastically deformed isotactic polypropylene evaluated by x-ray diffraction and differential scanning calorimetry methods. *J. Polym. Sci. B Polym. Phys.* **2002**, *40*, 896–903. [CrossRef]
42. Li, J.; Zhu, Z.; Li, T.; Peng, X.; Jiang, S.; Turng, L.-S. Quantification of the Young's modulus for polypropylene: Influence of initial crystallinity and service temperature. *J. Appl. Polym. Sci.* **2020**, *137*, 48581. [CrossRef]
43. Karger-Kocsis, J.; Bárány, T. (Eds.) *Polypropylene Handbook*; Springer International Publishing: Cham, Switzerland, 2019; ISBN 978-3-030-12902-6.
44. LyondellBasell Industries Holdings B.V. High-Performance Polypropylene for Rigid Packaging Applications. Available online: https://www.lyondellbasell.com/4a2134/globalassets/documents/polymers-technical-literature/product-information--high-performance-polypropylene-for-rigid-packaging-applicationss.pdf (accessed on 1 September 2022).
45. Ladhari, A.; Kucukpinar, E.; Stoll, H.; Sängerlaub, S. Comparison of Properties with Relevance for the Automotive Sector in Mechanically Recycled and Virgin Polypropylene. *Recycling* **2021**, *6*, 76. [CrossRef]
46. Gall, M.; Steinbichler, G.; Lang, R.W. Learnings about design from recycling by using post-consumer polypropylene as a core layer in a co-injection molded sandwich structure product. *Mater. Des.* **2021**, *202*, 109576. [CrossRef]

Disclaimer/Publisher's Note: The statements, opinions and data contained in all publications are solely those of the individual author(s) and contributor(s) and not of MDPI and/or the editor(s). MDPI and/or the editor(s) disclaim responsibility for any injury to people or property resulting from any ideas, methods, instructions or products referred to in the content.

Article

Compressive and Flexural Strengths of Mortars Containing ABS and WEEE Based Plastic Aggregates

Youssef El Bitouri [1] and Didier Perrin [2,*]

[1] LMGC, IMT Mines Ales, Univ Montpellier, CNRS, F-30100 Ales, France
[2] PCH, IMT Mines Ales, F-30100 Ales, France
* Correspondence: didier.perrin@mines-ales.fr

Abstract: The incorporation of plastic aggregates as a partial replacement of natural aggregates in cementitious materials is interesting in several ways. From a mechanical point of view, the partial substitution of sand with plastic aggregates could improve some properties (e.g., ductility, thermal insulation). This paper deals with the mechanical strength of mortars containing plastic aggregates as a partial replacement of sand. Part of the volume of sand in cement mortars is substituted with plastic aggregates which originate from WEEE (Waste from Electrical and Electronic Equipment) and consist of a mix of ABS (acrylonitrile-butadiene styrene), HIPS (high impact polystyrene) and PP (Polypropylene), or of monomaterial ABS from WEEE sorting. Three rates of replacement (by volume of sand) were tested: 10%, 15% and 30%. Mechanical tests were performed according to European standard EN196-1. The results show that compressive and flexural strength decrease with rate of replacement, but remain satisfactory for structural purposes. In addition, the density of mortar is reduced with the incorporation of plastic aggregates. The decrease of mechanical strength is mainly due to the weak bond between cement paste and plastic aggregates leading to the increase of porosity. Furthermore, it appears that mortars containing plastic aggregates could present a ductile rupture.

Keywords: plastic-aggregate; ABS; mortar; mechanical properties; flexural strength; rupture

Citation: El Bitouri, Y.; Perrin, D. Compressive and Flexural Strengths of Mortars Containing ABS and WEEE Based Plastic Aggregates. *Polymers* **2022**, *14*, 3914. https://doi.org/10.3390/polym14183914

Academic Editor: Enzo Martinelli

Received: 31 August 2022
Accepted: 15 September 2022
Published: 19 September 2022

Publisher's Note: MDPI stays neutral with regard to jurisdictional claims in published maps and institutional affiliations.

Copyright: © 2022 by the authors. Licensee MDPI, Basel, Switzerland. This article is an open access article distributed under the terms and conditions of the Creative Commons Attribution (CC BY) license (https://creativecommons.org/licenses/by/4.0/).

1. Introduction

The recycling and valorization of plastic wastes is a growing challenge from both an environmental and economic point of view. With more than 300 million tons produced each year around the world [1], recycling and valorization solutions must be proposed. Among the plastic wastes that could pose problems, we can cite the Wastes from Electrical and Electronic Equipment (WEEE). This type of polymer wastes contains a complex mixture of materials, some of which are hazardous [2]. Different types of thermoplastic polymers such as styrenic matrices (e.g., ABS for Acrylonitrile Butadiene Styrene, HIPS for High Impact Polystyrene, PP for Polypropylene) can be found in these wastes.

One of the interesting options for reusing these wastes is their incorporation into construction materials such as cementitious materials (mortars, concretes). This solution could present several possibilities of interests. In addition to the possibility of recovering large volumes and decreasing the landfilling, it would make it possible to confine the waste and limit its dissemination in the environment. The partial substitution of mineral aggregates by plastic aggregates would also make it possible to reduce the pressure on natural resources. Furthermore, the cementitious materials formulated with plastic aggregates could have interesting properties. In fact, thermal insulation performance can be improved by the addition of plastic aggregates, such as polyethylene (PET) [3], as well as the mechanical properties, especially ductility [4–6].

The incorporation of aggregates from plastic wastes is well documented in the scientific literature, and some critical reviews are available [7,8]. The data are quite dispersed in terms of the substituted component (fine or coarse aggregates), the type of plastic, the substitution rates and the pretreatments carried out on the plastic before incorporation into the cementitious materials [9–12]. Plastic wastes frequently studied are polyethylenes terephthalate (PET), high-density polyethylenes (HDPE), polyvinyl chlorides (PVC) and polypropylenes (PP). The pretreatments are sometimes used to improve the bond with the cement paste. Gamma radiation treatment of plastic polymers is the most widely used [10]. Other types of treatment based on atmospheric plasma were performed [9,11]. These treatments aim to physically modify the surface roughness while adjusting the nature of the functional groups of plastic polymers according to the surface energy of the cement pastes.

In general, the available data point out the deleterious effect of partial substitution of sand by plastic aggregates on the mechanical properties of concretes [13–16]. Here, we report only some studies performed on mechanical properties of mortars. Abed et al. [17] tested five different waste PET weight fractions of 0, 5, 15, 25 and 50% as a replacement of sand in cement mortar with constant cement content (525 kg/m^3) and water-to-cement ratio of 0.48. In general, it was found that compressive and flexural strengths were reduced as waste PET incorporation increased. The mortar containing 25% waste PET appears to be a lightweight mortar with satisfactory mechanical properties suitable for structural purposes. Ghernouti and Rabehi [18] reported a reduction of compressive and flexural strengths according to the increase in replacement rate of sand by plastic aggregates. These mechanical strengths remain almost close to those of the reference mortar for replacement rates below 20% by volume. According to Carneiro et al. [19], the substitution of sand by shredded PET in polymer mortar containing epoxy resin (12% by weight) as binder contributes to a decrease in both flexural and compressive strength. For flexural strength, the decrease is of about 48% for PET content of 20% (by weight), while compressive strength decreases by 59%. The authors explain this overall decrease by the poor bond between shredded PET waste and the epoxy matrix. Merlo et al. [20,21] studied the reuse of PVC deriving from WEEE as a partial substitute for sand to produce lightened mortars. The volume replacement rates of sand by plastic aggregates of 5, 10, 15 and 20% were used. The results obtained with ordinary Portland cement showed a strong decrease in both compressive and flexural strengths. The authors attribute this decrease to the mechanical characteristics of PVC and to the poor adherence between cement paste and plastic aggregates. Recently, the same authors published another study carried out with replacement rates of up to 90% (in volume) [21]. The decrease of the mechanical properties (compressive and flexural strengths) is of about 40% with a replacement rate of 15% vol. It should be noted that this decrease could be due not only to the poor adherence between cement paste and plastic aggregates, but also to the granular characteristics of the latter. In fact, the fines content of plastic aggregates was not sufficient to achieve the optimum packing leading to the minimum of porosity. The fineness modulus of plastic waste used was 2.5, while for standard sand, it was 3.0 [21].

Despite the large number of studies on the reuse of plastic wastes as aggregates in cementitious materials, only a few studies have focused on the styrenic materials encountered in WEEE, such as acrylonitrile-butadiene styrene (ABS) and high-impact polystyrene (HIPS) [4,22]. It can be noted that Makri et al. [22] investigated the physical and mechanical properties of cement mortars, partially replaced with ABS-based aggregates from plastic housing of LCD screens. The replacement percentages used were 2.5%, 5%, 7.5%, 10% and 12.5% (the authors do not specify whether the replacement rate is by mass or by volume), while the water-to-cement (w/c) ratio was maintained constant at 0.5. The obtained results show a decrease in compressive strength except for the replacement rate of 7.5% and 10%, which exhibited an increase by 15.4% and 7.8%, respectively. The elastic modulus (at 28 days) decreases by 39%, 48%, respectively, for the replacement rates of 2.5 and 5%, while only 19%, 24% of decrease is recorded for the replacement rate of 7.5% and 10%, respectively [19]. Another more recent article regarding physical properties and microstructure of WEEE-based plastic aggregate mortars made with acrylonitrile-butadiene-styrene (ABS), polycarbonate (PC), polyoxymethylene (POM), polyethylene (PET) and ABS/PC blend waste was produced by Kaur and Pavia [23]. The loss of compressive strength is significant with PET and POM-based aggregates, up to 42% with 20% (by volume) of replacement. At 5% of replacement (by volume), the PC, ABS and ABS/PC plastics increase the compressive strength by 6–15%. At 20% of replacement, the loss of compressive strength is of about 25%. The reduction in flexural strength is less pronounced than the compressive strength reduction. At 20% of sand replacement, the reduction of flexural strength is of about 21% for PET and 36% for PC and ABS/PC. Furthermore, the authors consider that the mechanical strengths (compressive and flexural strengths) met the standard strength requirements for masonry, rendering and plastering mortars. Moreover, the plastic aggregates increase the ability of the mortar to absorb deformation upon the application of stresses, retarding failure and transforming the brittle failure typical of cement mortars into a ductile failure. The authors notice that the particle size, shape and surface characteristics of the plastic particles have a significant impact on the properties of the resulting material. Finally, the results highlight the optimal performance from ABS scraps smaller than 2 mm, which hardly increases the hygric properties of mortars t, and shows high resistances and evidence of unbroken binding at the interfaces [23]. Therefore, by improving the compressive and flexural strengths, the use of recovered plastics from WEEE, such as styrenic thermoplastic materials mainly based on ABS, as recycled aggregates in partial substitution of sand in cement mortar could potentially prove a useful recycling alternative for plastic waste.

Therefore, it appears that the partial substitution of sand with plastic aggregates in cement mortars leads to a decrease in strength, especially with high replacement rates [14,18,20,21,23–25]. The loss of strength is often attributed to the weak cement paste/plastic aggregates bond induced by the hydrophobic nature of the plastic. However, the results in the literature are dispersed and depend not only on the nature of plastic aggregates, but on the particle size distribution of these aggregates. For instance, the maximum size of plastic particles is often different and sometimes exceeds that used in mortar sands (maximum 4 mm) [21], as well as the fineness modulus [14,18,20,21,25]. For the results to be comparable, it would be more appropriate to formulate mortars with plastic aggregates, whose the particle size distribution is close to that of sand according to the requirements of a standard such as the European standard EN 13139 or EN 196-1 [14,26,27]. In addition, only a few studies concern the styrenic materials encountered in WEEE (ABS).

In this work, the effect of a partial substitution of sand by plastic aggregates from styrenic materials originating from WEEE (Waste from Electrical and Electronic Equipment) on compressive and flexural strength of cement mortars is investigated. The plastic aggregates were grinded and sieved in order to obtain a range size of particles between 0 and 2 mm close to that of standard sand [26]. Furthermore, the quality of adherence between cement paste and these aggregates was investigated qualitatively through scanning electron microscopy (SEM) observations coupled with Energy-Dispersive X-Ray Analysis (EDX).

2. Materials and Methods

This study was focused on styrenic materials encountered in WEEE, which are principally represented by ABS (acrylonitrile-butadiene styrene), HIPS (high impact polystyrene) and PP (Polypropylene). Samples were provided by Suez Group (Berville-su-Seine, France) and consist of plastics from WEEE (mix of ABS, HIPS and PP, Figure 1a) and mono-material ABS from WEEE sorting (Figure 1b). These samples were cut with a Retsch® SM300 grinder (Haan, Germany) at 5000 rpm using a 2 mm grid. The obtained material is a mix of particles with different sizes, as shown in Figure 1. The waste absolute density was evaluated through a pycnometer (Micromcritics AccuPyc 1330, Micromeritics Instrument Corporation, Norcross, GA, USA). Density measurements were performed in triplicate (Table 1).

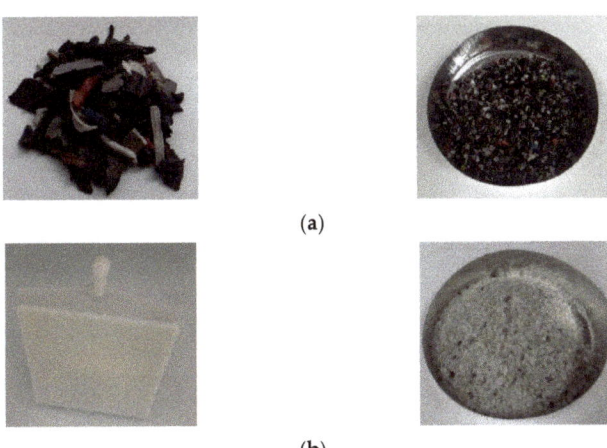

(a)

(b)

Figure 1. WEEE plastics (a) and ABS from WEEE sorting (b)-based wastes before and after grinding.

Table 1. Absolute density of materials.

	Cement	Sand	WEEE	ABS
Measured density (g/cm^3)	3.22	2.65	1.07	1.05

The CEN standard sand was used to prepare the mortars. It is a natural siliceous sand consisting of rounded particles and has a silica content of at least 98%. According to the European standard (EN-196-1), CEN standard sand has a specific particle size distribution ranging between 0.08 and 2.00 mm. The particle size distribution of CEN standard sand and plastic wastes was determined through vibratory sieving using sieves of 2 mm, 1.6 mm, 1 mm, 0.5 mm, 0.16 mm and 0.08 mm, and plastic particles larger than 2 mm were removed. These measurements were performed in triplicate. Particle size distributions of sand and plastic aggregates are shown in Figure 2.

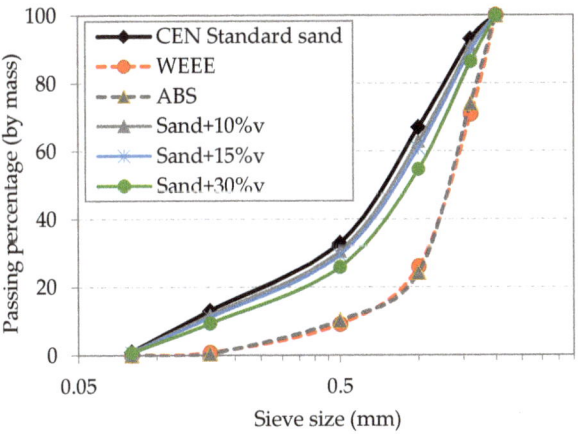

Figure 2. Particle size distribution of standard sand and plastic aggregates.

As shown in Figure 2, WEEE and ABS-based aggregates exhibit the same granulometry after grinding. The fines content of these aggregates is lower than standard sand.

An ordinary Portland cement (CEM I/42.5 N) was used to prepare the mortar mixtures.

The mortar mixtures were prepared according to the European standard (EN 196-1). The composition of the different mortars is provided in Table 2. Plastic aggregates were used as a partial replacement of sand volume with three replacement rates: 10, 15 and 30%.

Table 2. Composition of the different mortars.

Sample	Cement (g)	Water (g)	Sand (g)	Plastic (g)
Reference	450	225	1350	0
WEEE10	450	225	1215	54.5
WEEE15	450	225	1147.5	81.8
WEEE30	450	225	945	163.5
ABS10	450	225	1215	53.6
ABS15	450	225	1147.5	80.4
ABS30	450	225	945	160.8

The procedure of mixing was performed according to the following sequence: the water and the cement are placed into the bowl and mixed at low speed during 30 s. The sand is added steadily during the next 30 s, and the mixing is continued at high speed for additional 30 s. The mixer is then stopped for 90 s. During the first 30 s, the mixer walls are scraped to homogenize the mortar, and the mixing is continued for 60 s at high speed.

For each mortar composition, three specimens were prepared and placed in prismatic molds 40 × 40 × 160 mm^3. After 24 h, all samples were demolded and immersed in water at 20 °C for 28 days.

The chemical types of both WEEE and ABS plastic aggregates were checked by a Vertex 70 FT MIR spectrometer from Bruker (Billerica, MA, USA) with an ATR unit used (Figure 3). The used resolution was of 4 cm^{-1}, 32 scans for background acquisition and 32 scans for the sample spectrum. Spectra were acquired from 4000 to 400 cm^{-1} and analyzed using the OPUS software provided with the spectrometer. Most of the samples were directly analyzed on the crystal. Standard samples and dirty waste samples were previously cleaned with ethanol.

The FTIR spectroscopic analysis with ATR mode provides the identification as the family of styrenics through HIPS and ABS polymers. The most remarkable patterns of styrenics are the two aliphatic CH stretching signals at 2920 cm^{-1} and 2850 cm^{-1} as well as the 3–5 cm^{-1} aromatic CH stretching signals between 3000 and 3100 cm^{-1} and the two aromatic CH waging at ≈700 cm^{-1} and ≈750 cm^{-1}, the second one being thrice as small as the first one. The main difference between ABS and HIPS is the stretching of the carbon–nitrogen triple bond of acrylonitrile, which produces a fairly weak, but sharp signal at 2237 cm^{-1}, a very specific location. Moreover, the other signal specificities between the ABS and the HIPS styrenic family was found at 695 ± 2 cm^{-1} for PS-based samples, but at 700 ± 3 cm^{-1} for SAN-based polymers. It can be deduced that the presence of acrylonitrile in the polymer chains affect the aromatic CH vibrations.

Thus, styrenic-based polymers from WEEE plastics ((a) in Figure 1) were identified as HIPS and ABS, while (b) was checked as an ABS polymer after ABS WEEE sorting.

Flexural and compressive strength evaluations were performed with according to the European standard (EN 196-1) [26]. Flexural strength was determined using the three-point loading method. For each mortar composition, flexural strength was measured on three specimens. Then, compressive strength is measured on halves of the prism broken on flexural testing using 3R RP400E-425kN (Montauban, France). These measurements were performed on six specimens.

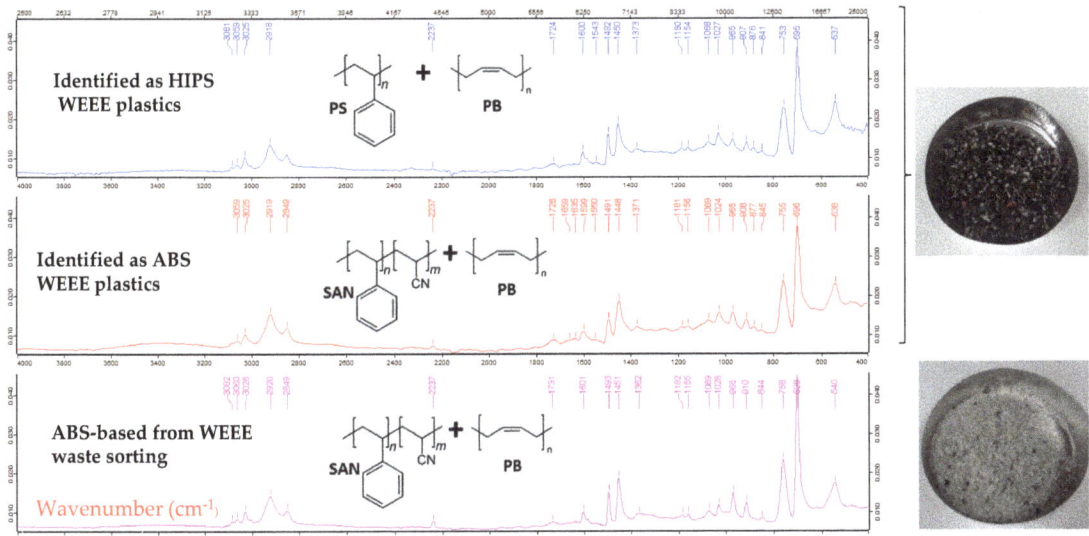

Figure 3. Identified FTIR with ATR mode as ABS-HIPS-based WEEE wastes (denoted (a) in Figure 1), and as ABS-based from WEEE waste sorting (denoted (b) in Figure 1).

In addition to mechanical tests, the polished sections of mortar were observed using Scanning Electron Microscopy (SEM Quanta 200 FEG (FEI Company, Hillsboro, OR, USA) with Energy-Dispersive X-Ray Analysis (EDX). Direct optical microscopy observations of the polished sections were performed with a Leica Laborlux 12 POL S optical microscope equipped with a 1600 × 1200 pixels mono-CDD Sony digital camera.

3. Results and Discussion

The aim of the mechanical characterization is to evaluate the effect of the partial substitution of sand by plastic aggregates (WEEE and ABS). The results are shown in Figure 4. The loss of compressive and flexural strength is represented in Figure 5.

These results show that the partial substitution of sand by plastic aggregates decreases both the compressive and the flexural strength. Furthermore, it appears that the compressive strength is more affected than the flexural strength, which is in accordance with literature [8,28–30]. For WEEE-based mortars, the compressive strength is decreased by about 11.7% for 10%v, 11.3% for 15%v, and 21% for 30%v. For ABS-based mortars, the compressive strength is reduced by 14.4% for 10%v, 22.8% for 15%v, and 23.5% for 30%v.

For ABS-based mortars, the compressive strength was slightly lower than that of WEEE-based mortars for replacement rates of 10%v and 30%v. Concerning the flexural strength, there is no significant difference between WEEE and ABS. It is interesting to note that the strength results of ABS-based samples were more dispersed than those of WEEE-based samples.

Concerning the flexural strength, the loss ranges between 8 and 17%. Mortars containing ABS-based aggregates show a flexural strength slightly better than that of WEEE-based aggregates. In addition, even if the displacement was not measured to estimate the fracture energy, it would seem that the ductility of ABS-based mortars is improved compared to that of other mortars, including the reference without plastic aggregates, as shown in Figure A1. In fact, the rupture is manifested by the propagation of a bending crack in the middle of the specimen without the specimen separating into two pieces, whereas for the other mortars, the rupture is sudden and without warning signs, which characterizes a brittle fracture under mode I. Further investigation with a complete mechanical characterization is required to examine this observation. In fact, if this increase of ductility is confirmed, the partial substitution of sand by ABS-based aggregates will be very interesting from a mechanical point of view (e.g., to limit the effect of drying shrinkage).

Furthermore, it has to be kept in mind that even if the partial substitution of sand by plastic aggregates leads to a decrease of compressive and flexural strength, the values of these strengths remain acceptable from a practical point of view. Moreover, a decrease in the self-weight of concrete could be obtained by the decrease of sand proportion. In fact, as shown in Figure 6, the decrease in density is of about 8% for WEEE and 11% for ABS at 30%v substitution.

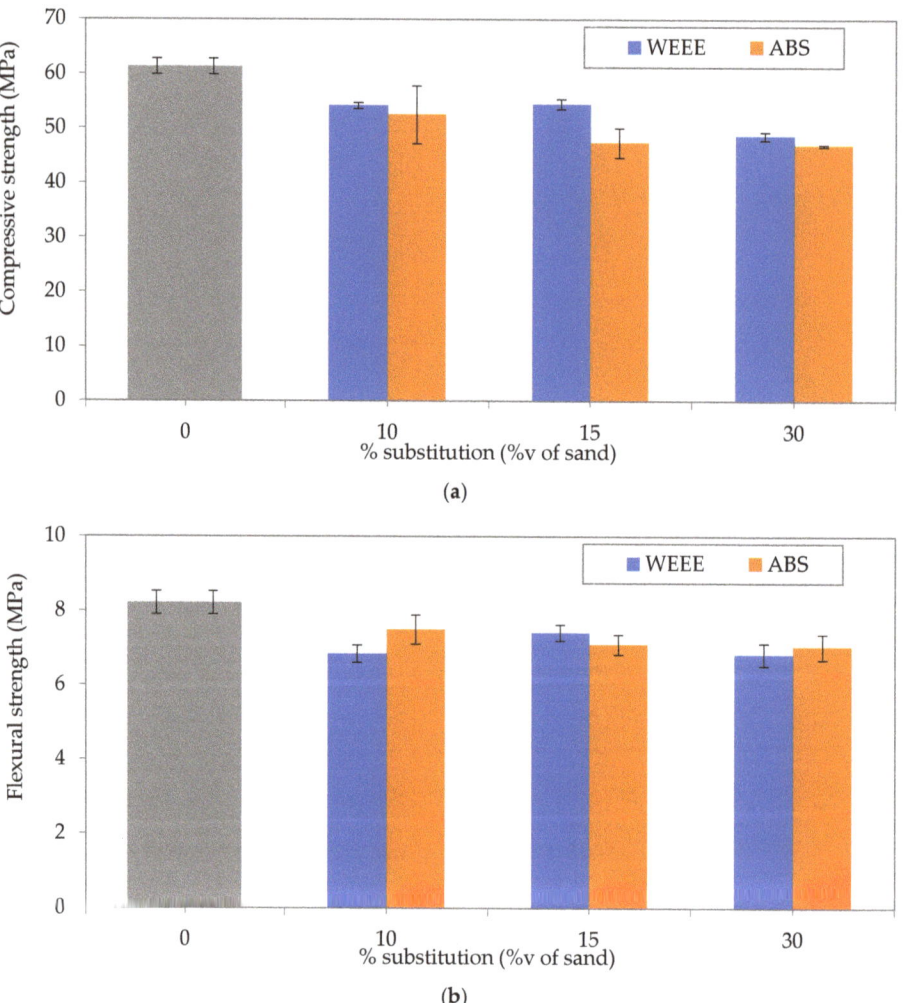

Figure 4. Compressive (**a**) and flexural (**b**) strength as a function of volume substitution of sand by plastic aggregates.

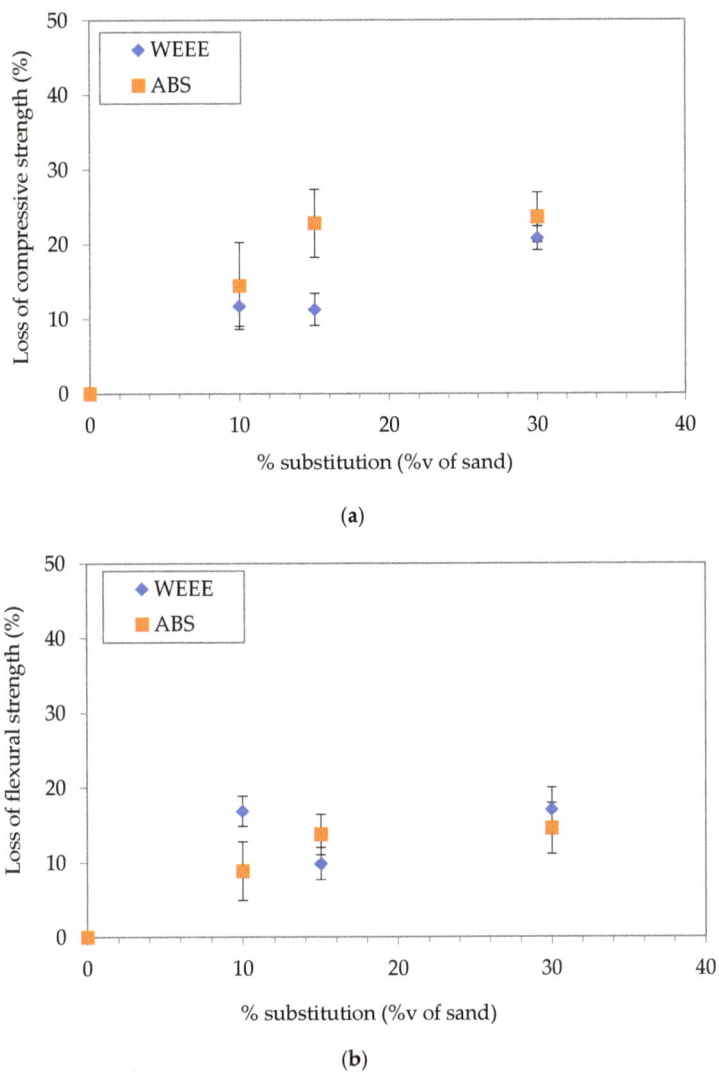

Figure 5. Loss of compressive (**a**) and flexural (**b**) strength compared to reference mortar without plastic aggregates.

The decrease in mechanical properties could be explained by different aspects. As shown in Figure 2, the particle size distributions certainly have slight differences, but they could lead to an increase in porosity in the hardened mortar. It should be noted that the initial compactness is the same for all the mixtures (of about 74.3%) since the substitution is performed by volume. In addition, the mechanical properties of plastic aggregates and sand are different. Finally, given the hydrophobic character of plastic aggregates, the quality of the cement paste/aggregate interface could be deteriorated, compared to that of sand, which can have epitaxial properties with the cement paste, i.e., the growth of cement hydration products on the aggregate surface and the chemical reaction between aggregate and cement. In order to evaluate the quality of this interface, SEM and optical observations were performed.

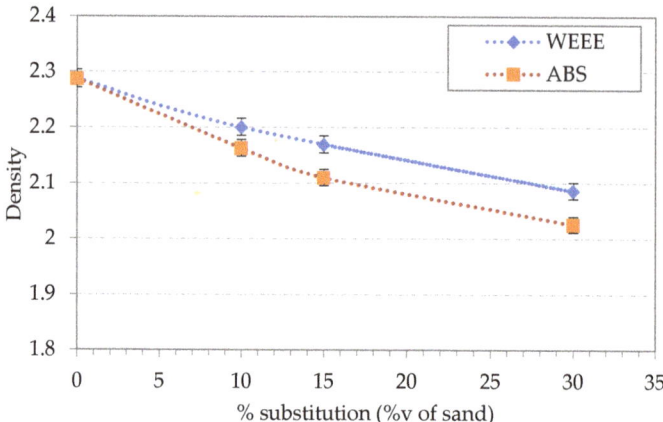

Figure 6. Density of mortars specimens as a function of volume substitution of sand by plastic aggregates.

As shown in Figure A2, cement matrix/plastic or sand aggregate interfaces display a decohesion zone. This decohesion is more visible with plastic aggregate. Using EDX analysis in different zones (Figures 7 and 8), the disturbed zones around plastic aggregates (WEEE and ABS) seem to be slightly more important than those around sand. The thickness of the disturbed zone is indicated by the slope of the signal drop of the marker element (C for plastic, Si for sand and Ca/Si for cement matrix) between the aggregate and the matrix (Figures 7 and 8).

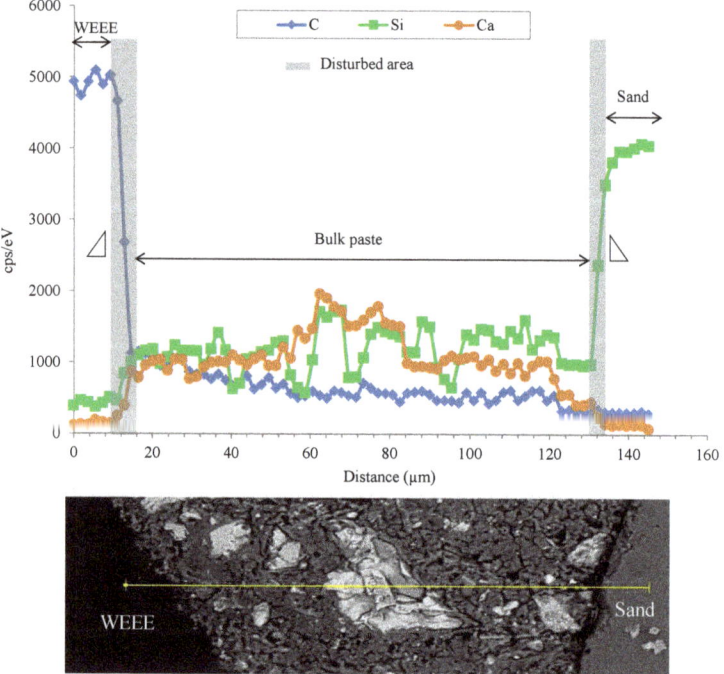

Figure 7. An example of Energy-Dispersive X-Ray analysis (EDX) of WEEE/cement matrix and sand/cement matrix interface.

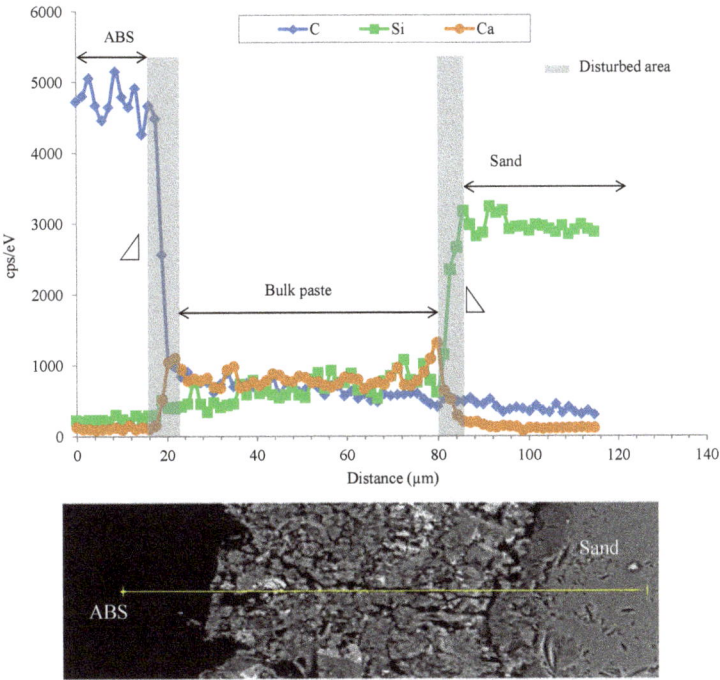

Figure 8. An example of Energy-Dispersive X-Ray analysis (EDX) of ABS/cement matrix and sand/cement matrix interface.

The difference between the particle size distribution of the reference mortar and the mortars containing plastic aggregates, and the quality of matrix/aggregate interface could lead to an increase in the porosity of mortar containing plastic aggregates (WEEE and ABS mortars). In fact, using the measured mechanical strengths and the extended Zheng model proposed by Chen et al. [31], the porosity of each mortar was assessed. As shown in (Figure 9), the porosity of the reference mortar is of about 5%, 10% for mortar with 10%v (ABS and WEEE), 13% and 9% for mortar with 15%v of ABS and WEEE, respectively, and 13% for mortars with 30%v of ABS and WEEE. It thus appears that porosity increases with increasing volume substitution. This increase of porosity is well correlated with the loss of strength.

The results obtained in this study are comparable to those found in the literature [14,17,20]. The compressive and flexural strength obtained with partial substitution of sand by plastic aggregates appear to be very interesting. In addition, the loss of mechanical strength observed on the mortar could be less important on the concrete.

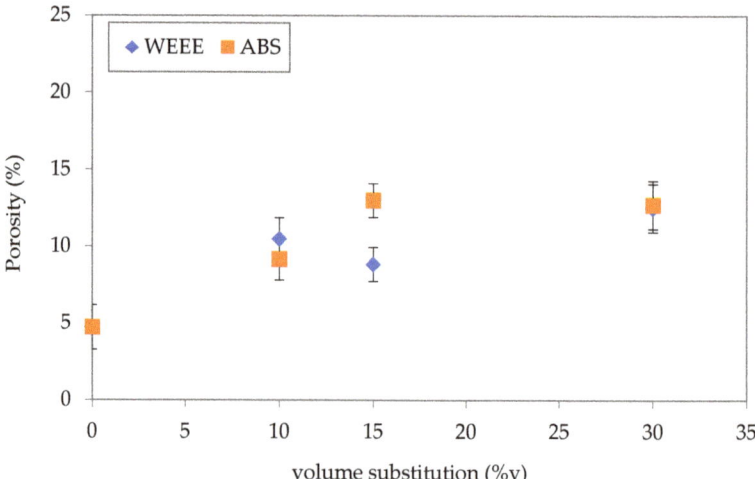

Figure 9. Porosity of mortar as a function of volume substitution.

4. Conclusions

The paper aimed to examine the effect of partial substitution of sand with plastic aggregates originated from WEEE on the mechanical strength of cement mortar. These plastic aggregates consist of a mix of ABS, HIPS and PP (WEEE-based aggregates) or only ABS from WEEE sorting (ABS-based aggregates). According to the obtained results, it appears that the partial substitution of sand leads to a decrease in strength.

For WEEE-based mortars, the compressive strength decreases by 11.7%, 11.3% and 20.8%, while the flexural strength decreases by 16.8%, 9.8% and 17%, respectively with 10%v, 15%v and 30%v of plastic aggregates. For ABS-based mortars, the loss of compressive strength is, respectively, of 14.4%, 22.8% and 23.6% with 10%v, 15%v and 30%v of plastic aggregates. The flexural strength of ABS-based mortars is reduced by 8.8% at 10%v, 13.7% at 15%v and 14.5% at 30%v. It can be noted that the flexural strength of cement mortar is less affected by the plastic aggregate replacement than the compressive strength, especially for ABS-based mortars.

The loss of strength can be explained by the increase of porosity induced mainly by the quality of plastic aggregate/matrix interface which is more porous than sand/matrix interface. Moreover, the density of the cement mortar is reduced by the substitution of sand by plastic aggregates.

Despite the loss of strength which remains acceptable from a practical point of view, the cement mortars containing plastic aggregates show a facies of rupture different from the reference mortar, suggesting an increase of ductility. Further investigations should be performed to examine this increase of ductility. In particular, a physical pre-treatment based on cold plasma that would attenuate the hydrophobic character of plastics could potentially lead to a better cohesion of the mixture, thus improving the properties of the final sample.

Author Contributions: Conceptualization, Y.E.B. and D.P.; methodology, Y.E.B. and D.P.; validation, Y.E.B. and D.P.; investigation, Y.E.B. and D.P.; writing—original draft preparation, Y.E.B. and D.P.; writing—review and editing, Y.E.B. and D.P.; visualization, Y.E.B. and D.P. All authors have read and agreed to the published version of the manuscript.

Funding: This research received no external funding.

Institutional Review Board Statement: "Not applicable" for studies not involving humans or animals.

Data Availability Statement: Data archived on a hard drive by the authors.

Conflicts of Interest: The authors declare no conflict of interest.

Appendix A

Figure A1. Facies of rupture: example of broken sample (30%v ABS).

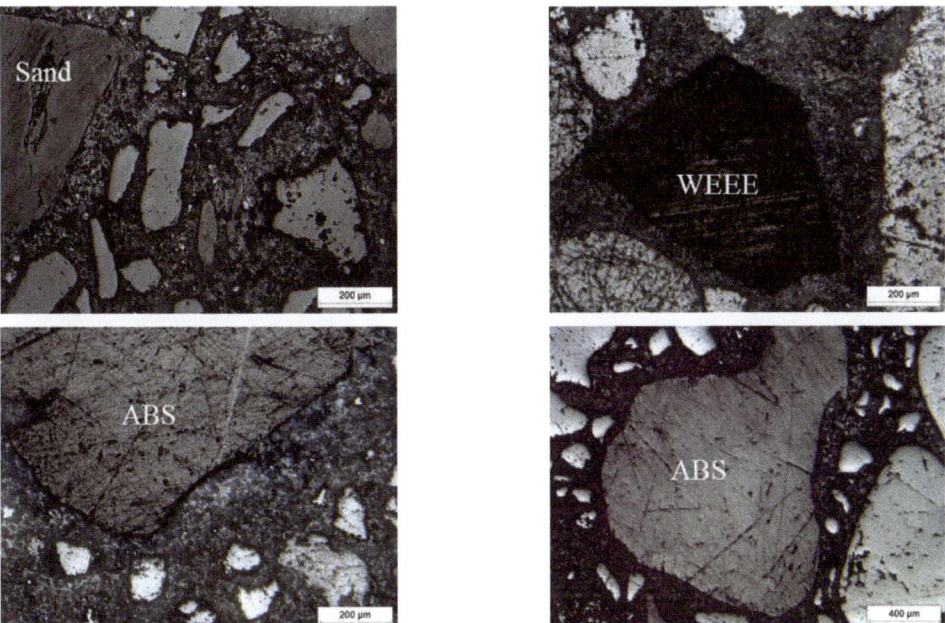

Figure A2. Example of optical observations.

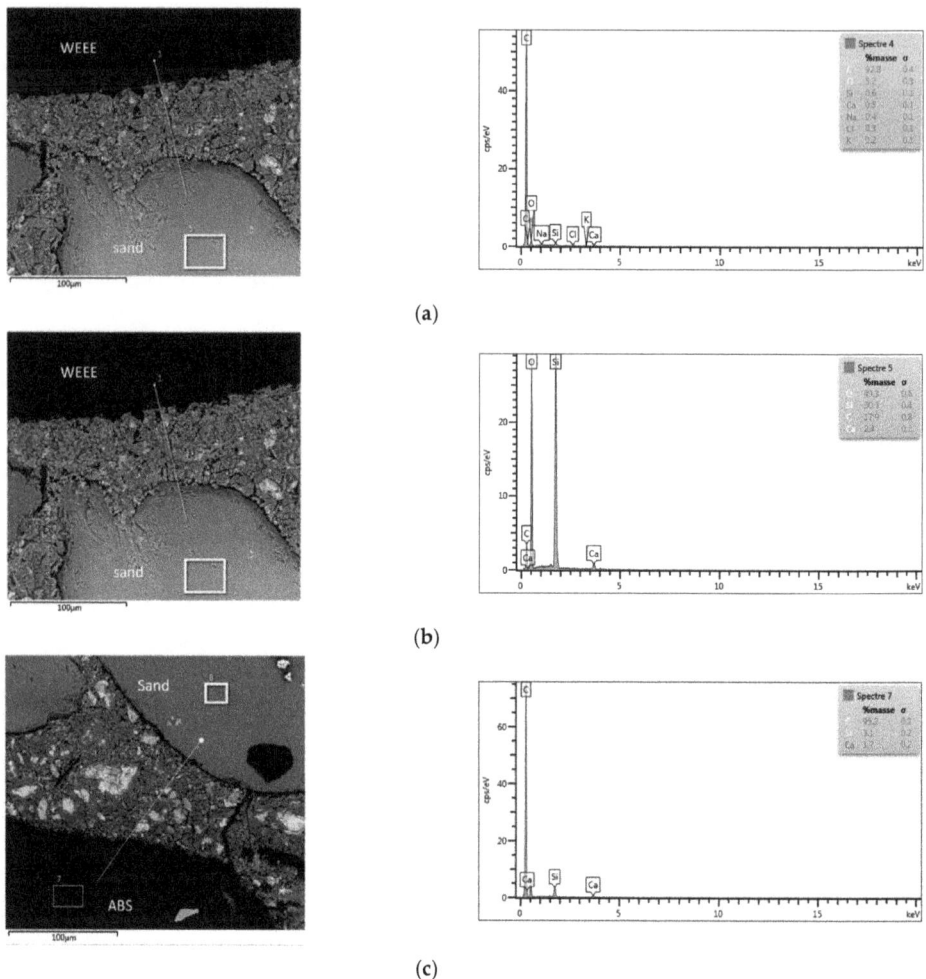

Figure A3. SEM/EDX observations. (**a**) EDX profile of WEEE. (**b**) EDX profile of sand. (**c**) EDX profile of ABS.

References

1. *Earth Day. Fact Sheet: Plastics in the Ocean.* 2018. Available online: https://www.earthday.org/fact-sheet-plastics-in-the-ocean/ (accessed on 20 June 2022).
2. Chandrappa, R.; Das, D.B. Waste From Electrical and Electronic Equipment. Environmental Science and Engineering, European Commission. 2012. Available online: https://ec.europa.eu/environment/topics/waste-and-recycling/waste-electrical-and-electronic-equipment-weee_en (accessed on 20 June 2022).
3. Yesilata, B.; Isiker, Y.; Turgut, P. Thermal insulation enhancement in concretes by adding waste PET and rubber pieces. *Constr. Build. Mater.* **2009**, *23*, 1878–1882. [CrossRef]
4. Palos, A.; D'Souza, N.A.; Snively, C.T.; Reidy, R.F. Modification of cement mortar with recycled ABS. *Cem. Concr. Res.* **2001**, *31*, 1003–1007. [CrossRef]
5. Gao, C.; Huang, L.; Yan, L.; Jin, R.; Kasal, B. Strength and ductility improvement of recycled aggregate concrete by polyester FRP-PVC tube confinement. *Compos. Part B Eng.* **2019**, *162*, 78–197. [CrossRef]
6. Babafemi, A.J.; Sirba, N.; Paul, S.C.; Miah, M.J. Mechanical and Durability Assessment of Recycled Waste Plastic (Resin8 & PET) Eco-Aggregate Concrete. *Sustainability* **2022**, *14*, 5725.
7. Sharma, R.; Bansal, P.P. Use of different forms of waste plastic in concrete—A review. *J. Clean. Prod.* **2016**, *112*, 473–482. [CrossRef]
8. Gu, L.; Ozbakkaloglu, T. Use of recycled plastics in concrete: A critical review. *Waste Manag.* **2016**, *51*, 19–42. [CrossRef]

9. Thibodeaux, N.; Guerrero, D.E.; Lopez, J.L.; Bandelt, M.J.; Adams, M.P. Effect of cold plasma treatment of polymer fibers on the mechanical behavior of fiber-reinforced cementitious composites. *Fibers* **2021**, *9*, 62. [CrossRef]
10. Liu, T.; Nafees, A.; Khan, S.; Javed, M.F.; Aslam, F.; Alabduljabbar, H.; Xiong, J.; Khan, M.I.; Malik, M.Y. Comparative study of mechanical properties between irradiated and regular plastic waste as a replacement of cement and fine aggregate for manufacturing of green concrete. *Ain Shams Eng. J.* **2022**, *13*, 101563. [CrossRef]
11. Nisticò, R.; Lavagna, L.; Boot, E.A.; Ivanchenko, P.; Lorusso, M.; Bosia, F.; Pugno, N.M.; D'Angelo, D.; Pavese, M. Improving rubber concrete strength and toughness by plasma-induced end-of-life tire rubber surface modification. *Plasma Process. Polym.* **2021**, *18*, 2100081. [CrossRef]
12. Lee, H.; Cheon, H.; Kang, Y.; Roh, S.; Kim, W. State-of-the-Art Modification of Plastic Aggregates Using Gamma Irradiation and Its Optimization for Application to Cementitious Composites. *Appl. Sci.* **2021**, *11*, 10340. [CrossRef]
13. Rahmani, E.; Dehestani, M.; Beygi, M.H.A.; Allahyari, H.; Nikbin, I.M. On the mechanical properties of concrete containing waste PET particles. *Constr. Build. Mater.* **2013**, *47*, 1302–1308. [CrossRef]
14. Senhadji, Y.; Escadeillas, G.; Benosman, A.S.; Mouli, M.; Khelafi, H.; Kaci, S.O. Effect of incorporating PVC waste as aggregate on the physical, mechanical, and chloride ion penetration behavior of concrete. *J. Adhes. Sci. Technol.* **2015**, *29*, 625–640. [CrossRef]
15. Kou, S.C.; Lee, G.; Poon, C.S.; Lai, W.L. Properties of lightweight aggregate concrete prepared with PVC granules derived from scraped PVC pipes. *Waste Manag.* **2009**, *29*, 621–628. [CrossRef] [PubMed]
16. Almeshal, I.; Tayeh, B.A.; Alyousef, R.; Alabduljabbar, H.; Mohamed, A.M. Eco-friendly concrete containing recycled plastic as partial replacement for sand. *J. Mater. Res. Technol.* **2020**, *9*, 4631–4643. [CrossRef]
17. Abed, J.M.; Khaleel, B.A.; Aldabagh, I.S.; Sor, N.H. The effect of recycled plastic waste polyethylene terephthalate (PET) on characteristics of cement mortar. *J. Phys. Conf. Ser.* **2021**, *1973*, 012121. [CrossRef]
18. Ghernouti, Y.; Rabehi, B. Strength and Durability of Mortar Made with Plastics Bag Waste (MPBW). *Int. J. Concr. Struct. Mater.* **2012**, *6*, 145–153. [CrossRef]
19. Carneiro, E.P.; Reis, J.M.L. Mechanical Properties of Polymer Mortar Containing PET Waste Aggregates. In Proceedings of the 21st Brazilian Congress of Mechanical Engineering, Natal, RN, Brazil, 24–28 October 2011.
20. Merlo, A.; Lavagna, L.; Suarez-Riera, D.; Pavese, M. Mechanical properties of mortar containing waste plastic (PVC) as aggregate partial replacement. *Case Stud. Constr. Mater.* **2020**, *13*, e00467. [CrossRef]
21. Merlo, A.; Lavagna, L.; Suarez-Riera, D.; Pavese, M. Recycling of weee plastics waste in mortar: The effects on mechanical properties. *Recycling* **2021**, *6*, 70. [CrossRef]
22. Makri, C.; Hahladakis, J.N.; Gidarakos, E. Use and assessment of "e-plastics" as recycled aggregates in cement mortar. *J. Hazard. Mater.* **2019**, *379*, 120776. [CrossRef]
23. Kaur, G.; Pavia, S. Physical properties and microstructure of plastic aggregate mortars made with acrylonitrile-butadiene-styrene (ABS), polycarbonate (PC), polyoxymethylene (POM) and ABS/PC blend waste. *J. Build. Eng.* **2020**, *31*, 101341. [CrossRef]
24. Ruiz-Herrero, J.L.; Nieto, D.V.; López-Gil, A.; Arranz, A.; Fernández, A.; Lorenzana, A.; Merino, S.; De Saja, J.A.; Rodríguez-Péreza, M.Á. Mechanical and thermal performance of concrete and mortar cellular materials containing plastic waste. *Constr. Build. Mater.* **2016**, *104*, 298–310. [CrossRef]
25. Senhadji, Y.; Siad, H.; Escadeillas, G.; Benosman, A.S.; Chihaoui, R.; Mouli, M.; Lachemi, M. Physical, mechanical and thermal properties of lightweight composite mortars containing recycled polyvinyl chloride. *Constr. Build. Mater.* **2019**, *195*, 198–207. [CrossRef]
26. EN 196-1:2016; Methods of Testing Cement—Part 1: Determination of Strength. European Commitee for Standardization: Brussels, Belgium, 2016.
27. BS EN 13139; Aggregates for Mortars. European Commitee for Standardization: Brussels, Belgium, 2013.
28. Saikia, N.; de Brito, J. Use of plastic waste as aggregate in cement mortar and concrete preparation: A review. *Constr. Build. Mater.* **2012**, *34*, 385–401. [CrossRef]
29. Alqahtani, F.K.; Khan, M.I.; Gurmel, G.; Dirar, S. Production of Recycled Plastic Aggregates and Its Utilization in Concrete. *J. Mater. Civ. Eng.* **2017**, *29*, 04016248. [CrossRef]
30. Hannawi, K.; Kamali-Bernard, S.; Prince, W. Physical and mechanical properties of mortars containing PET and PC waste aggregates. *Waste Manag.* **2010**, *30*, 2312–2320. [CrossRef]
31. Chen, X.; Wu, S.; Zhou, J. Influence of porosity on compressive and tensile strength of cement mortar. *Constr. Build. Mater.* **2013**, *40*, 869–874. [CrossRef]

Article

Influence of Biofillers on the Properties of Regrind Crystalline Poly(ethylene terephthalate) (CPET)

Victor S. Cecon [1,2], Greg W. Curtzwiler [1,2] and Keith L. Vorst [1,2,*]

[1] Polymer and Food Protection Consortium, Iowa State University, 536 Farm House Ln, Ames, IA 50011, USA
[2] Department of Food Science and Human Nutrition, Iowa State University, 536 Farm House Ln, Ames, IA 50011, USA
* Correspondence: kvorst@iastate.edu

Abstract: As the demand for plastics only increases, new methods are required to economically and sustainably increase plastic usage without landfill and environmental accumulation. In addition, the use of biofillers is encouraged as a way to reduce the cost of the final resin by incorporating agricultural and industrial waste by-products, such as rice hulls and coffee chaff to further reduce waste being sent to landfills. Crystalline poly(ethylene terephthalate) (CPET) is a resin commonly used for microwave and ovenable food packaging containers that have not been fully explored for recycling. In this article, we investigate how the incorporation of biofillers at 5% wt. and 10% wt. impacts critical polymer properties. The thermal and mechanical properties were not significantly altered with the presence of rice hulls or coffee chaff in the polymer matrix at 5% wt. loading, but some reduction in melt temperature, thermal stability, and maximum stress and strain was more noticed at 10% wt. The complex viscosity was also reduced with the introduction of biofillers. The levels of heavy metals of concern, such as Cd, Cr, and Pb, were below the regulatory limits applicable in the United States and Europe. Additional studies are suggested to improve the performance of CPET/biofiller blends by pre-treating the biofiller and using compatibilizers.

Keywords: crystalline poly(ethylene terephthalate); post-industrial recycling; polymer processing; biofillers; coffee chaff; rice hull

Citation: Cecon, V.S.; Curtzwiler, G.W.; Vorst, K.L. Influence of Biofillers on the Properties of Regrind Crystalline Poly(ethylene terephthalate) (CPET). *Polymers* **2022**, *14*, 3210. https://doi.org/10.3390/polym14153210

Academic Editors: Teofil Jesionowski and Łukasz Klapiszewski

Received: 1 July 2022
Accepted: 4 August 2022
Published: 6 August 2022

Publisher's Note: MDPI stays neutral with regard to jurisdictional claims in published maps and institutional affiliations.

Copyright: © 2022 by the authors. Licensee MDPI, Basel, Switzerland. This article is an open access article distributed under the terms and conditions of the Creative Commons Attribution (CC BY) license (https://creativecommons.org/licenses/by/4.0/).

1. Introduction

Plastics are global and found everywhere, from cars to kitchen utensils and especially in packaging materials. To meet the demands for these applications, worldwide plastic production has been increasing over the years, reaching 367 million tonnes in 2020 [1]. Packaging applications are the segment where plastics are mainly used, with more than 40% of plastic volume in Europe [1]. Food packaging has received a lot of attention from environmental and regulatory agencies due to the increased use and waste accumulation of single-use plastics (SUP) from the COVID-19 pandemic [2], which include takeout and ready-to-cook (RTC) food products that are commonly microwave heated. In addition to the basic concept of containing and protecting the food, microwavable packages are required to meet more sophisticated performance requirements such as high-temperature cooking for convenience products. These performance requirements create challenges and potential interaction with the food resulting in the migration of chemicals from the packaging and/or discoloration [3].

To meet the performance requirements of high heat convenience packaging the most common configuration consists of thermoformed trays made out of crystalline poly(ethylene terephthalate) (CPET). These trays are considered "dual ovenable" due to their ability to be used in both conventional and microwave ovens [4,5]. CPET is a semi-crystalline polymer of elevated crystallinity, a result of its linear macromolecular structure that is produced from PET with the addition of nucleants [6]. These crystallization agents can include ultrafine inorganic powders with concentrations up to 0.3% wt. that work as a nucleating agent and polyolefins, such as polypropylene (PP), high-density polyethylene (HDPE), or linear low-density polyethylene (PE-LLD) in concentrations close to 3% wt. acting as "crack stoppers" [5]. Despite the downside of losing transparency, the resulting trays made with thermoformed CPET have increased mechanical stability and light protection, and can be used between −40 and 220 °C, allowing its use from freezer storage to heating in a microwave oven [7].

Recently, natural fillers, or biofillers, have been gaining attention as a modifier in polymer composites, considering the low cost and great availability, as well as potentially increasing the performance of the polymers mixed with them, including rice hull (or rice husks) and coffee chaff (or coffee silverskin) [8–10]. Rice hull is a significant agricultural waste by-product from the rice milling that consists of up to 20% wt. of the rice grain, with annual production consisting of up to 151 million tonnes, based on the 756 million tonnes of rice harvested globally in 2020 [11,12]. Coffee chaff is a thin skin that covers green coffee beans and are produced as a by-product from the roasting process, representing approximately 4% wt. of the coffee bean [10,13,14], potentially resulting in annual production of almost 400,000 tonnes, considering that 9.9 million tonnes were harvested globally in 2020 [15]. The utilization of these biofillers have been reported in the literature for several polymer composites, with recent uses including both fossil fuel-based polymers as polypropylene (PP) [13,16,17], polyamides (PA) [12], polyethylene (PE) [18–20], and poly(ethylene terephthalate) (PET) [21], as well as biopolymers such as poly (lactic acid) (PLA) [22,23], poly (butylene succinate) (PBS) [24–28], and poly(butylene adipate-co-terephthalate/poly(3-hydroxybutyrate-co-3-hydroxyvalerate) (PBAT/PHBV) [29,30].

Considering the widespread utilization of CPET and the need to increase its sustainability appeal, biofillers are proposed to be mixed with recycled CPET trays. Not only reducing packaging costs, biofillers such as coffee chaff and rice hulls are potentially cheaper than PET resin. Therefore, substituting CPET polymers with very low cost fillers such as coffee chaff and rice hulls presents a sustainable message to brand owners and consumers while replacing and reducing the use of fossil fuel-based feedstocks and potentially improving the chemical recycling of the blends [31]. Coupled with biofillers, the use of regrind or post-industrial recycled (PIR) CPET also provides a low-waste approach by limiting the amount of CPET scraps bound to waste-to-energy (WTE) facilities or, even worse, landfills, considering that plastic production scraps could reach up to 40% of total production [32]. The recycling of post-consumer CPET must be considered as multiple large fast-moving consumer goods (FMCG) companies, such as Nestle, PepsiCo, and Unilever, have set ambitious targets of increasing post-consumer recycled content in their plastic packaging in the next few years [33]. In addition, the United States Environmental Protection Agency (U.S. EPA) has recently set a goal of increasing the national recycling rate to 50% by 2030 [34], which will require significant effort from all stakeholders in the section to achieve it, considering that only 8.7% of plastics were recycled from municipal solid waste in 2018 [35]. However, multiple challenges exist when considering the difficulty separating CPET from PET with the current recycling infrastructure and eventual contamination of PET feedstock streams.

In this work, coffee chaff and rice hulls were compounded with regrind CPET trays in two different ratios of 5 and 10% wt. The study's goal is to understand eventual changes that can be caused by the addition of the biofillers in the CPET properties, as few studies have investigated the reuse of CPET and there is greater attention towards the development of biocomposites [36,37]. The resulting blends were characterized and compared with the CPET without biofiller in terms of morphological, thermal, mechanical, and physical polymer properties. The presence of regulated heavy metals for food packaging was also investigated, as nineteen states in the United States limit the total incidental concentration of Cd, Cr^{6+}, Hg, and Pb to 100 ppm in any finished package or packaging component [38]. The results of this work suggest the addition of coffee chaff or rice hulls did not drastically affect the performance of CPET blends, but some slight reductions in thermal and mechanical properties were observed at the highest concentration (10% wt.) of biofiller tested. The use of coffee chaff and rice hulls can be a solution to reduce costs for packaging manufacturers while reducing the use of fossil fuel-based resins and diverting agricultural and industrial waste byproducts.

2. Materials and Methods

2.1. Preparation of Biofiller-CPET Blends

Crystalline poly(ethylene terephthalate) (CPET) was obtained from a packaging company in the form of black trays for food contact applications and then granulated into flakes using a commercial JS AO-10 granulator (Jumbo Steel Machinery, Taiwan), as shown in Figure 1. Coffee chaff (CC) and rice hulls (RH) were obtained from domestic sources from a coffee roaster and an agricultural company, respectively, as waste by-products. Blends containing 5% and 10% wt. of the biofillers were prepared using a Micro 18GL 18 mm twin-screen co-rotating extruder (Leistritz, Somerville, NJ, USA) and pelletized with a lab-scale pelletizer BT 25 (Bay Plastics Machinery, Bay City, MI, USA)., with the composition of each blend listed in Table 1. ASTM D638-14 injection-molded Type I dog bones were prepared using a production-grade HM90/350 90-ton horizontal injection molder (Wittmann Battenfeld, Torrington, CT, USA), as shown in Figure 2. Extrusion and injection molding feeds were flushed with N_2 to minimize CPET degradation due to atmospheric moisture. CPET feedstock was oven-dried at 120 °C for 24 h prior to processing to remove eventual moisture. Processing parameters are listed in Table 2.

(a) (b)

Figure 1. CPET (a) trays and (b) flakes used in CPET/biofiller blends.

(a) (b)

Figure 2. (a) Injection molder and (b) mold setup used in the production of Type I Dog Bone.

Table 1. CPET/biofiller blends formulation in terms of weight percentage.

Sample Name	CPET	Biofiller Type	Biofiller Content
CPET100	100	-	-
CPET95_CC5	95	Coffee Chaff	5%
CPET90_CC10	90	Coffee Chaff	10%
CPET95_RH5	95	Rice Hull	5%
CPET90_RH10	90	Rice Hull	10%

Table 2. Processing parameters of CPET/biofiller blends.

Extrusion	CPET100	CPET95 CC5	CPET90 CC10	CPET95 RH5	CPET90 RH10
Zone 1 Temperature (°C)	200	200	200	200	200
Zone 2 Temperature (°C)	265	250	250	265	250
Zone 3 Temperature (°C)	270	255	255	265	255
Zone 4 Temperature (°C)	270	255	255	270	255
Zone 5 Temperature (°C)	275	260	255	275	260
Zone 6 Temperature (°C)	285	260	255	280	260
Zone 7 Temperature (°C)	295	260	250	285	265
Die Temperature (°C)	300	255	245	285	265
Injection Molding					
Feed Zone Temperature (°C)			49		
Zone 1 Temperature (°C)			265		
Zone 2 Temperature (°C)			260		
Zone 3 Temperature (°C)			250		
Nozzle Temperature (°C)			285		

2.2. Microscope Imaging and Particle Size Analysis

Biofiller and injection molded (cryofractured) samples were analyzed using a 3D Surface Profiler VK-X1000 microscope (Keyence Corporation of America, Itasca, IL, USA). Biofiller images were captured at 5× magnification while injection molded samples were analyzed at 20× magnification. The particle size distribution of coffee chaff and rice hull were determined using a Mastersizer 2000 (Malvern Pananalytical, Worcestershire, UK), with the size range between 0.375 to 1000 µm.

2.3. Thermogravimetric Analysis

Thermal degradation properties such as the temperature of 5% mass loss and the ash residue of each blend of CEPT/biofiller were assessed through thermogravimetric analysis (TGA) using a Q5000IR thermogravimetric analyzer (TA Instruments, New Castle, DE, USA). For each CPET/biofiller blend, 5–10 mg specimens were weighted and loaded to a platinum pan, being heated at 10 °C/min up to 600 °C, under an N_2 atmosphere.

2.4. Differential Scanning Calorimetry

Thermal transitions were analyzed through a heat/cool/heat cycle between −30 °C and 310 °C at a rate of 10 °C/min under an N_2 atmosphere using a Q2000 differential scanning calorimeter (TA Instruments, New Castle, DE, USA). Each CPET/biofiller had specimens of 3–7 mg weighted and added into a hermetically sealed aluminum DSC pan that was crimped before the analysis.

2.5. Electromechanical Testing

Mechanical properties were assessed using an Autograph AGS-J (Shimadzu Corp., Kyoto, Japan) universal electromechanical tester with a 5 kN load cell and a manual non-shift wedge grip set MWG-5kNA (Shimadzu Corp., Kyoto, Japan) in the tensile mode. The specimens were evaluated for each CPET/biofiller blend following ASTM D638-14 with a 50 mm/min crosshead speed.

2.6. Fourier Transform Infrared Spectroscopy

Spectrometric analysis through Attenuated Total Reflectance Fourier Transform Infrared Spectroscopy (ATR-FTIR) was carried out using a Nicolet 6700 infrared spectrometer (Thermo Fisher, Waltham, MA, USA) at ambient temperature (22 °C) equipped with a DTGS detector. Each measurement had 32 scans and used a resolution of 2 cm^{-1}, with the ATR accessory cleaned with an isopropanol wipe after each run to prevent cross-contamination. Spectra were analyzed with OMINIC™ 8.3 software (Thermo Fisher, Waltham, MA, USA) to assess eventual interactions between the CPET and the biofillers.

2.7. Inductively Coupled Plasma—Optical Emission Spectroscopy

The presence of metals was evaluated using Inductively Coupled Plasma—Optical Emission Spectroscopy (ICP-OES). Specimens of 0.1500 ± 0.0005 g were weighted for each CPET/biofiller blend and digested via microwave-assisted digestion using an UltraWave digestion system (Milestone, Inc., Shelton, CT, USA) in 5 mL HNO$_3$ 67% v/v Trace Metal Grade (Fisher Scientific, Fair Lawn, NJ, USA) and 1 mL HCl 34% v/v Trace Metal Grade (Fisher Scientific, Fair Lawn, NJ, USA). An initial pressure of 40 bar (N$_2$) was applied, as well as a microwave power of 1350 W for 45 min (with a ramp of 5 min), from room temperature (22 °C) to 210 °C with a cooling time of 20 min. Initially, a pressure of 40 bar with N$_2$ was applied, followed by the use of a microwave power of 1500 W for 40 min, with a temperature increase from room temperature (22 °C) to 260 °C and a system pressure increase to 150 bar. A cooling period of 20 min was then applied. The method used is a slight modification of the modified Westerhoff digestion method described by Goodlaxson [39,40].

After the microwave digestion, the samples were diluted to 50 mL with ultra-pure, deionized water and analyzed using an iCap-7400 Duo ICP-OES (Thermo Scientific, Waltham, MA, USA). The analysis was carried out using the wavelength with the lowest limit of detection (LOD) for each metal. All measurements were performed in radial mode. Multi-element standard solutions containing Al (aluminum), Cd (cadmium), Cr (chromium), Fe (iron), Pb (lead), Sb (antimony), and Ti (titanium) were prepared for the concentrations of 0.1 µg/mL, 1 µg/mL, 5 µg/mL, 25 µg/mL, 50 µg/mL, and 100 µg/mL of each metal of interest, along with a 5 µg/mL yttrium internal standard. They were prepared using single standard solutions of 1000 µg/mL (Inorganic Ventures, Christiansburg, VA, USA) for each metal. Concurrent blanks (with no polymer added) were also run for each digestion batch for metals eventually present in the acids or leached from digestion vessels.

2.8. Parallel-Plate Oscillatory Melt Rheometry

Rheological analysis of the CPET/biofiller blends was carried out in oscillatory mode (frequency sweep) using a DHR-2 hybrid rheometer (TA Instruments, New Castle, UK). The rheometer was equipped with an environmental test chamber and a 25 mm parallel-plate geometry with a 1 mm gap between plates was used. All tested samples were dried in a convection oven at 140 °C overnight prior to analysis. The measurements were conducted within the linear viscoelastic region (previously tested) under 1% strain, at 270 °C, and air atmosphere, with the angular frequency ranging from 0.05 to 500 rad/s.

2.9. Statistical Analysis

Statistical analysis was conducted with a one-way ANOVA considering a 95% confidence level (α = 0.05) and grouped using Tukey's honestly significant difference (HSD) test. The JMP® 15 Pro (SAS Institute Inc., Cary, NC, USA) was utilized as the statistical analysis software. All analyses used five repeated measures unless otherwise noted.

3. Results and Discussion

3.1. Biofiller Characterization and Sample Imaging

The optical images of coffee chaff and rice hull are displayed in Figure 3, where a heterogenous particle size distribution and anisotropic particles can be observed. Coffee chaff particles are darker than rice hull particles and are wafer-like, while rice ones have a flake-like morphology. From Figure 4, it is shown that the biofillers are well dispersed in the matrix, with no agglomeration observed. The presence of the biofillers was noticed by the brown spots localized in the polymer matrix with different sizes, as the biofillers presented a large particle size distribution.

Figure 3. Optical images of the biofillers: (**a**) coffee chaff; (**b**) rice hull.

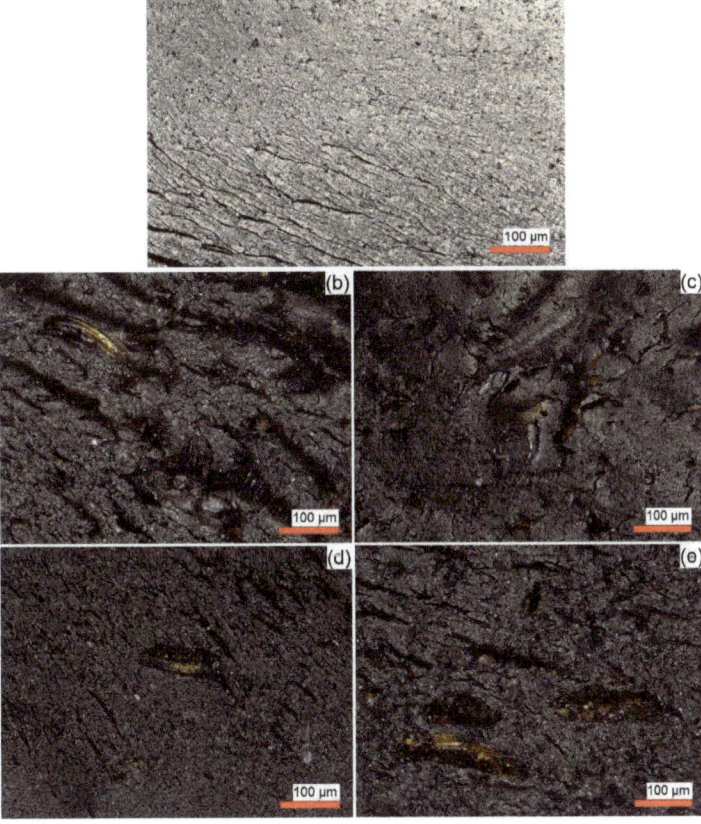

Figure 4. Optical images of cryofractured injection molded samples: (**a**) no biofiller; (**b**) 5% wt. coffee chaff; (**c**) 10% wt. coffee chaff; (**d**) 5% wt. rice hull; (**e**) 10% wt. rice hull.

The particle size analysis showed that coffee chaff, with a median size of 780.8 μm, had a larger particle size than rice hull, with a median size of 58.9 μm. Due to the equipment limitation on the maximum particle size range, sizes above 2000 μm were not recorded for coffee chaff, and the actual numbers would potentially indicate a larger median particle size. This result, illustrated by the particle size distribution in Figure 5, correlates with Figure 3, where it can be seen that coffee chaff has larger particles than rice hull and has a wider distribution.

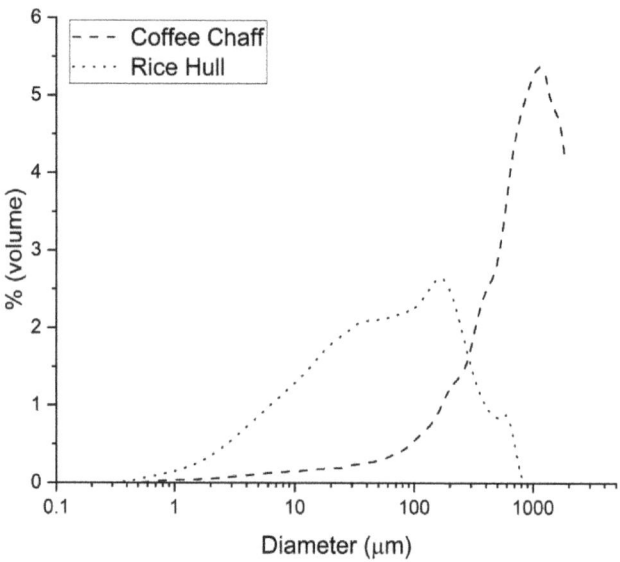

Figure 5. Particle size distribution for coffee chaff (dashed line) and rice hull (dotted line).

3.2. Impact of Biofillers on the Thermal Properties

Differential scanning calorimetry was performed to understand how the addition of biofiller influences the melt temperature and enthalpy of melting compared to CPET as it relates to thermal stability in rapid heating applications (microwave/oven). It was observed that adding both coffee chaff or rice hulls to CPET caused a statistically significant decrease ($p < 0.05$) in the melt temperature, as shown in Figure 6, although not enough to considerably change the polymer application. The increase from 5 to 10% wt. on the coffee chaff loading produced a significant decrease in temperature, whereas the two concentrations of rice hull tested did not yield a significant difference ($p < 0.05$). The enthalpy was consistent for all samples with the exception of the blend containing 10% wt. of coffee chaff, which displayed a statistically lower value. This result suggests that the biofillers do not act as a nucleant for CPET and additional studies are required to investigate the possible substitution of the inorganic nucleation agents by biofillers, considering that their particle size (not measured in this study) might have an important role. When taking into account only these two parameters, the blends containing rice hull displayed a lower deviation from the 100% CPET samples than coffee chaff.

The results from the thermogravimetric analysis in Figure 7 showed that at 5% wt., both biofillers did not significantly reduce the temperature at 5% mass loss, while the decrease was more noticeable when coffee chaff or rice hulls were mixed at 10% wt. The ash residue at 600 °C generally increased with more biofiller added to the blend, but there was no significant difference when coffee chaff was added at 5% wt. The analysis of the pure biofillers showed higher temperature and lower ash residue for coffee chaff when compared to rice hull, but the effect was not noticed in the CPET/biofiller blends analyzed. A review by Suhot, et al. [41] analyzed several publications discussing the use of rice husk in different polymers and found both positive and negative impacts on the composites' thermal properties made of polypropylene (PP) or low-density polyethylene (PE-LD), for example. A possible alternative to improve thermal stability is using halogen-free flame retardants to increase the fiber-matrix interaction, as observed in the work of Phan, et al. [42]. Aluminum diethylphosphinate and aluminum hydroxide were used as flame retardants in polyurethane composites and increased the temperature at 5% mass loss by 40 and 32 °C, respectively.

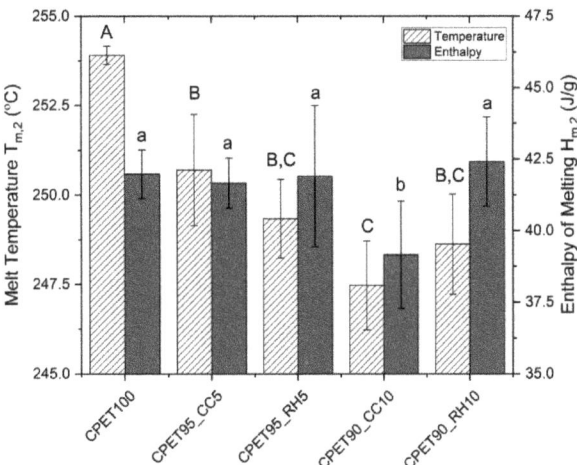

Figure 6. Melt temperature and enthalpy of melting from the second heat cycle. Data points associated with the same letter are statistically equivalent (α = 0.05); uppercase for melt temperature, lowercase for enthalpy.

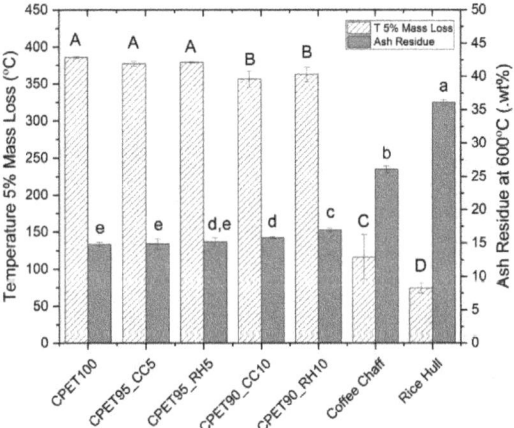

Figure 7. The temperature at 5% mass loss and ash residue at 600 °C. Data points associated with the same letter are statistically equivalent (α = 0.05); uppercase for temperature, lowercase for ash residue.

3.3. Impact of Biofillers on the Mechanical Properties

A tensile test was performed to evaluate the mechanical properties of injection-molded specimens made of CPET/biofiller blends. From Young's modulus results shown in Figure 8, it is possible to observe that there was no significant difference between all blends compared to the scenario without any biofiller, which can be correlated to the similar morphology observed in the cryofractured samples shown in Figure 4, despite the difference in biofiller particle size. However, from the maximum stress and strain results displayed in Figure 9, the blend containing 10% wt. of coffee chaff is the only blend that is significantly different from the 100% CPET samples for both stress and strain, with lower values, possibly due to larger particle size than rice hull. The blend with 10% wt. rice hulls also presented a strain value significantly different from the control scenario, being statistically equivalent to the 10% wt. coffee chaff blend. The samples containing 5% wt. of biofiller were statistically equivalent to 100% CPET, and the blends with 10% wt. biofiller.

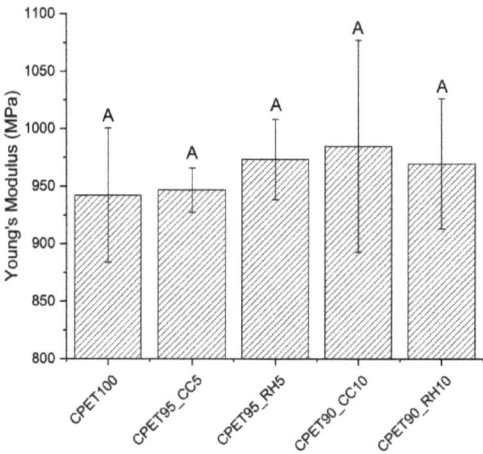

Figure 8. Young's Modulus for the CPET/biofiller blends. Data points associated with the same letter are statistically equivalent ($\alpha = 0.05$).

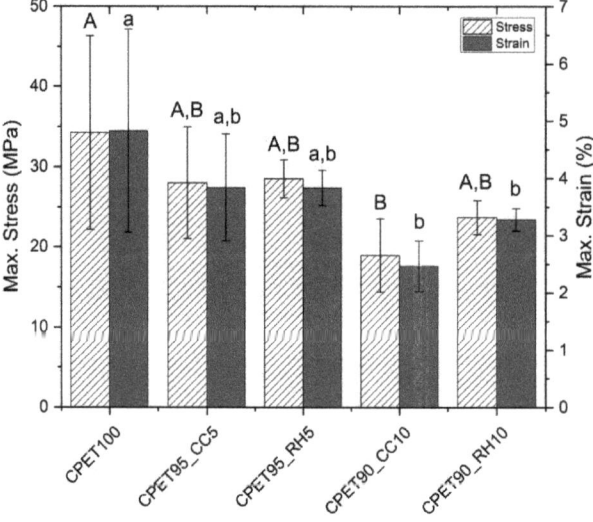

Figure 9. Maximum stress and strain for the CPET/biofiller blends. Data points associated with the same letter are statistically equivalent ($\alpha = 0.05$); uppercase for stress, lowercase for strain.

Testing rice husks concentrations between 10% wt. and 30% wt. with polypropylene (PP), Hidalgo-Salazar and Salinas [17] observed a significant increase in the tensile modulus as more biofiller was present in the blend. A study using recycled polyethylene (rPE) and poly(ethylene terephthalate) (rPET), blended at 75% rPE/25% rPET and compounded with rice husk in concentrations between 40% wt. and 80% wt. found that Young's modulus increased up to 70% wt. of rice husk added to the blend, with a sharp decrease at 80% wt [21]. Another blend containing 80% wt. PET and 20% wt. linear low-density polyethylene (PE-LLD) recovered from water bottles and used bale wrap, respectively, was mixed with a biocarbon, made of pyrolyzed spent coffee ground in concentrations between 5% wt. and 20% wt. Gupta, et al. [43] found increased tensile modulus with increased biofiller content; however, the ultimate tensile strength (UTS) and impact resistance presented an opposite response with decreased performance with more biocarbon. A direct comparison with these prior two studies cannot be made considering the different resins utilized and the use of compatibilizers. The pre-treatment of the rice hull and coffee chaff, as observed by Suhot, Hassan, Aziz, and Md Daud [41] for composites containing rice husks, and the use of a coupling agent would be beneficial for the mechanical performance of the CPET/biofiller blends. This is particularly useful when considering that CPET commonly has the presence of inorganic nucleation agents and "crack stoppers", as previously discussed.

3.4. Spectrometric Analysis of CPET/Biofiller Blends

The ATR-FTIR analysis did not find any new vibrational peak present with the addition of coffee chaff or rice hull, as shown in the spectra displayed in Figure 10. The main absorbance peaks observed are related to CPET, including at 1715 cm^{-1} for ν-C = O (stretching, ester, ring), 1240 cm^{-1} for ν-C(=O)-O (stretching), ν-C-C (ring, ester, stretching), and δ-C-H (ring, in-plane bending), 1095 cm^{-1} for ν-C-O (stretching), 1017 cm^{-1} for ν-C-O (stretching), 871 cm^{-1} for δ-C-H (ring, out-of-plane bending), δ-C-C (ring, ester, out-of-plane bending), and δ-C = O (out-of-plane bending, ring torsion), and 721 cm^{-1} for δ-C = O (out-of-plane bending), and δ-C-H (ring, out-of-plane bending) [44,45]. Additionally, no frequency shift or alteration in peak shape was noticed when the biofillers were introduced into the polymer blend, suggesting that there was no change in the intermolecular interactions. For example, it was reported in the literature that rice husks, which contain 29–35% of cellulose and 19–26% of lignin, could alter peaks at 3421 cm^{-1} (O-H bonds of lignin) and 1280 cm^{-1} (C-H bonds of crystalline parts of cellulose) [46]. Still, such an effect was not noticed in the CPET blends, possibly due to the low biofiller loading.

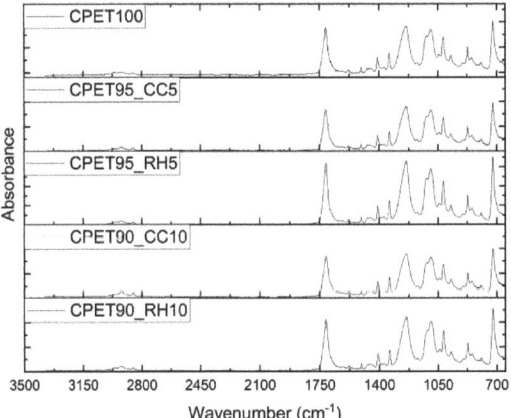

Figure 10. ATR-FTIR spectra for CPET/biofiller blends.

3.5. Rheological Analysis

The impact of biofillers on the sample viscosity was assessed using oscillatory melt rheology tests, as it could yield differences in polymer processing. It was noticed from the complex viscosity results, shown in Figure 11, that the addition of coffee chaff or rice hull can reduce the viscosity at higher frequencies for both biofillers. For coffee chaff, the viscosity increased at lower frequencies (<1 rad/s) with the addition of the biofiller, with 5% wt. displaying both the highest increase and the decrease at lower and higher frequencies, respectively. For rice hull, at 5% wt. and 10% wt. there was a reduction in viscosity, with the latter presenting a more noticeable decrease at lower frequencies while the 5% wt. blend had the smallest complex viscosity at higher frequencies. The results suggest that the incorporation of 5–10% wt. of coffee chaff or rice hull can reduce the viscosity of the final blend under certain processing conditions. In the blend of pyrolyzed coffee chaff with PET and PE-LLD, Gupta, Mohanty, and Misra [43] found that the complex viscosity increased when the biofiller was at 5% wt., followed by a decrease at higher concentrations. It was suggested that the biofiller could have prevented the entanglement and sliding of chains, which reduced the complex viscosity and could have happened at 5% wt. and 10% wt. in this study.

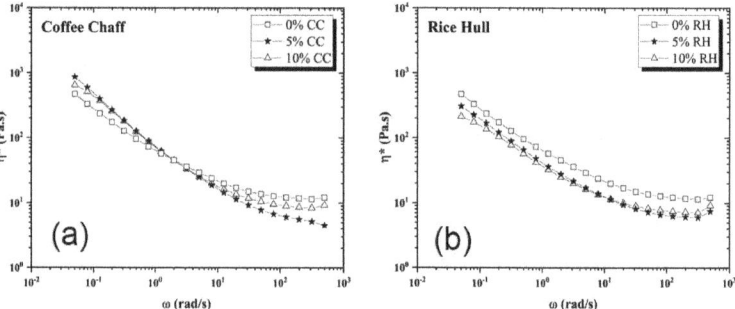

Figure 11. Complex viscosity plots of blends containing (**a**) coffee chaff and (**b**) rice hull at 270 °C.

We hypothesize that solid-state polymerization (SSP) and the use of a chain extender, as done for PET, might be beneficial to improve the melt behavior of the reprocessed CPET, as the thermal reprocessing of the material could lead to a molecular weight reduction and consequently, cause a reduction in mechanical and rheological properties [47,48]. Also, the addition of a compatibilizer might enhance the mixture between the biofillers and the CPET, having the potential to increase the availability of carboxylic or hydroxyl end groups from the biofillers to promote the esterification and transesterification reactions of SSP, and thus, increase the molecular weight of the reprocessed CPET.

3.6. Metal Analysis of CPET/Biofiller Blends

The metal analysis using ICP-OES was performed to assess the potential use of the CPET/biofiller blends for packaging applications, considering that multiple states in the United States have limits on the heavy metal content and main application of CPET for "dual-ovenable" use. The analysis did not find the presence of Cd and Pb (Hg was not tested but historically is not detected), and the amounts of Al, Cr, Fe, and Sb were summarized in Table 3. The samples would comply with the Toxics in Packaging legislation [38] in the United States and with Article 11 of the European Parliament and Council Direction 94/62/EC [49], limiting to 100 ppm the combined concentration of Cd, Cr^{6+}, Hg, and Pb, as only Cr was detected for the samples containing rice hulls, but at 2.58 and 5.32 ppm for 5% wt. and 10% wt. of the biofiller, respectively. The other metals found are commonly found in plastics due to its use as a catalyst during the polymerization process or as thermal and chemical stabilizers [50].

Table 3. Inorganic element composition of selected compounds (Al, Cr, Fe, Sb) in CPET and blends containing coffee chaff or rice hull as biofillers.

Sample Name	Al (ppm)	Cr (ppm)	Fe (ppm)	Sb (ppm)
CPET100	50.92	*b	43.41	168.61
CPET95_CC5	87.07	*b	53.61	159.13
CPET90_CC10	15.87	*b	51.12	96.75
CPET95_RH5	*a	2.58	33.55	162.41
CPET90_RH10	*b	5.32	86.26	160.53
LOD (ppm)	0.0101	0.0009	0.0015	0.0156
LOQ (ppm)	0.0337	0.0030	0.0050	0.0520

*a: value between LOD and LOQ; *b: value below LOD.

4. Conclusions

This work presented the impact on the polymer properties of adding coffee chaff and rice hulls to regrind CPET, being an alternative for plastic manufacturers to comply with consumer and government requests to reduce plastic use in packaging products while keeping costs down as fossil fuel-based plastic resins continue to rise based on oil prices and the low availability of recycled resins [51]. The analysis of the biofillers showed that coffee chaff presented higher particle size when compared to rice hull, but the images from the cryofractured blend samples did not show significant differences in morphology. Thermal analysis showed that blends containing 10% wt. of biofiller had a greater impact, with the reduction of melt temperature, enthalpy, and temperature at 5% mass loss being more significant when compared to the blend with only 5% wt. of coffee chaff or rice hull. When comparing both biofillers, the latter performed slightly better, with the enthalpy for the 10% wt. blend being statistically equivalent to the 5% wt. blend, which was not the case for coffee chaff.

There was no statistical difference between all samples for Young's modulus, although the maximum stress and strain results showed decreased performance for the blends containing 10% wt. of biofiller. The presence of coffee chaff or rice hull both at 5 and 10% wt. reduced the complex viscosity for the CPET blends, which could be potentially improved by the use of a compatibilizer. No noticeable difference was noticed in the spectra of the samples with or without biofiller, possibly due to the low loading of coffee chaff and rice hull. The metal analysis did not find levels of regulated metals (Cd, Cr, and Pb) above the limits established by legislation in the United States and Europe for packaging products.

Most of the literature reports the mixture of biofillers with polyolefins, with limited works involving the use of PET. Considering multiple applications of this resin for transparent products, biofillers could have an impact on these characteristics, which is not the case for CPET, a subtype of PET with also limited published research. The use of biofillers in recycled plastics helps to reduce landfill accumulation from both a traditional petroleum-based product (CPET) as well as industrial and agricultural by-product waste (coffee chaff and rice hull), contributing to a more circular economy and potential reduction of greenhouse gases from the production of virgin feedstocks as less virgin resin is used when the biofiller is incorporated. Despite the slight decrease in performance observed, this work expands the knowledge on CPET reuse, warranting additional studies to investigate the potential use of compatibilizers or the pre-treatment of the biofillers with the goal of improving the performance of blended CPET/biofiller package products. Future research should also look to understand the impacts of biofiller on product application. Using coffee chaff as an example, one could impart a characteristic odor to the final product made with the CPET/biofiller blend, and the utilization of techniques to improve molecular weight and viscosity of reprocessed CPET, as current literature is focused on the implications for PET.

Author Contributions: V.S.C.; conceptualization, investigation, visualization, writing—original draft. G.W.C.; supervision, writing—review & editing. K.L.V.; funding acquisition, project administration, conceptualization, supervision, writing—review & editing. All authors have read and agreed to the published version of the manuscript.

Funding: This research was funded by the Polymer and Food Protection Consortium, grant number GR-025895-00001 DD11523 and the Agricultural and Home Economics Experiment Station at Iowa State University through project No. IOW04202 by the Hatch Act, State of Iowa. The APC was partially funded by the Iowa State University Library–Publication Subvention Grants Program.

Data Availability Statement: The data in this study may be available upon request from the corresponding author.

Conflicts of Interest: The authors declare no conflict of interest.

References

1. PlasticsEurope. *Plastics—the Facts 2021*; An analysis of European plastics production, demand and waste data; PlasticsEurope: Brussels, Belgium, 2021.
2. Patrício Silva, A.L.; Prata, J.C.; Walker, T.R.; Duarte, A.C.; Ouyang, W.; Barceló, D.; Rocha-Santos, T. Increased plastic pollution due to COVID-19 pandemic: Challenges and recommendations. *Chem. Eng. J.* **2021**, *405*, 126683. [CrossRef] [PubMed]
3. Thanakkasaranee, S.; Sadeghi, K.; Seo, J. Packaging materials and technologies for microwave applications: A review. *Crit. Rev. Food Sci. Nutr.* **2022**, 1–20. [CrossRef] [PubMed]
4. Throne, J.L. Thermoforming crystallizing poly(ethylene terephthalate) (CPET). *Adv. Polym. Technol.* **1988**, *8*, 131–176. [CrossRef]
5. Throne, J.L. New Concepts in Thermoforming. *Polym.-Plast. Technol. Eng.* **1991**, *30*, 761–808. [CrossRef]
6. Galdi, M.R.; Olivieri, R.; Liguori, L.; Albanese, D.; Di Matteo, M.; Di Maio, L. PET based nanocomposite films for microwave packaging applications. *AIP Conf. Proc.* **2015**, *1695*, 020059.
7. Haldimann, M.; Blanc, A.; Dudler, V. Exposure to antimony from polyethylene terephthalate (PET) trays used in ready-to-eat meals. *Food Addit. Contam.* **2007**, *24*, 860–868. [CrossRef] [PubMed]
8. Matos Costa, A.R.; Lima, J.C.; Santos, R.; Barreto, L.S.; Henrique, M.A.; Carvalho, L.H.; Almeida, Y.M.B. Rheological, thermal and morphological properties of polyethylene terephthalate/polyamide 6/rice husk ash composites. *J. Appl. Polym. Sci.* **2021**, *138*, 50916. [CrossRef]
9. Arjmandi, R.; Hassan, A.; Majeed, K.; Zakaria, Z. Rice Husk Filled Polymer Composites. *Int. J. Polym. Sci.* **2015**, *2015*, 501471. [CrossRef]
10. Garcia, C.V.; Kim, Y.-T. Spent Coffee Grounds and Coffee Silverskin as Potential Materials for Packaging: A Review. *J. Polym. Environ.* **2021**, *29*, 2372–2384. [CrossRef]
11. FAO. Crops and Livestock Products. Available online: https://www.fao.org/faostat/en/#data/QCL/visualize (accessed on 17 July 2022).
12. Battegazzore, D.; Salvetti, O.; Frache, A.; Peduto, N.; De Sio, A.; Marino, F. Thermo-mechanical properties enhancement of bio-polyamides (PA10.10 and PA6.10) by using rice husk ash and nanoclay. *Compos. Part A Appl. Sci. Manuf.* **2016**, *81*, 193–201. [CrossRef]
13. Zarrinbakhsh, N.; Wang, T.; Rodriguez-Uribe, A.; Misra, M.; Mohanty, A.K. Characterization of Wastes and Coproducts from the Coffee Industry for Composite Material Production. *BioResources* **2016**, *11*, 7637–7653. [CrossRef]
14. Ballesteros, L.F.; Teixeira, J.A.; Mussatto, S.I. Chemical, Functional, and Structural Properties of Spent Coffee Grounds and Coffee Silverskin. *Food Bioprocess Technol.* **2014**, *7*, 3493–3503. [CrossRef]
15. International Coffee Organization. Coffee—Total Production by Crop Year. Available online: https://www.ico.org/historical/1990%20onwards/PDF/1a-total-production.pdf (accessed on 17 July 2022).
16. Morales, M.; Atencio Martinez, C.; Maranon, A.; Hernandez, C.; Michaud, V.; Porras, A. Development and Characterization of Rice Husk and Recycled Polypropylene Composite Filaments for 3D Printing. *Polymers* **2021**, *13*, 1067. [CrossRef] [PubMed]
17. Hidalgo-Salazar, M.A.; Salinas, E. Mechanical, thermal, viscoelastic performance and product application of PP-rice husk Colombian biocomposites. *Compos. Part B-Eng.* **2019**, *176*, 11. [CrossRef]
18. Hejna, A.; Barczewski, M.; Kosmela, P.; Mysiukiewicz, O.; Kuzmin, A. Coffee Silverskin as a Multifunctional Waste Filler for High-Density Polyethylene Green Composites. *J. Compos. Sci.* **2021**, *5*, 44. [CrossRef]
19. Wang, W.; Yang, X.; Bu, F.; Sui, S. Properties of rice husk-HDPE composites after exposure to thermo-treatment. *Polym. Compos.* **2014**, *35*, 2180–2186. [CrossRef]
20. Fávaro, S.L.; Lopes, M.S.; Vieira de Carvalho Neto, A.G.; Rogério de Santana, R.; Radovanovic, E. Chemical, morphological, and mechanical analysis of rice husk/post-consumer polyethylene composites. *Compos. Part A Appl. Sci. Manuf.* **2010**, *41*, 154–160. [CrossRef]
21. Chen, R.S.; Ahmad, S.; Gan, S. Rice husk bio-filler reinforced polymer blends of recycled HDPE/PET: Three-dimensional stability under water immersion and mechanical performance. *Polym. Compos.* **2018**, *39*, 2695–2704. [CrossRef]

22. Mort, R.; Peters, E.; Curtzwiler, G.; Jiang, S.; Vorst, K. Biofillers Improved Compression Modulus of Extruded PLA Foams. *Sustainability* **2022**, *14*, 5291. [CrossRef]
23. Tsou, C.-H.; Yao, W.-H.; Wu, C.-S.; Tsou, C.-Y.; Hung, W.-S.; Chen, J.-C.; Guo, J.; Yuan, S.; Wen, E.; Wang, R.-Y.; et al. Preparation and characterization of renewable composites from Polylactide and Rice husk for 3D printing applications. *J. Polym. Res.* **2019**, *26*, 227. [CrossRef]
24. Totaro, G.; Sisti, L.; Fiorini, M.; Lancellotti, I.; Andreola, F.N.; Saccani, A. Formulation of Green Particulate Composites from PLA and PBS Matrix and Wastes Deriving from the Coffee Production. *J. Polym. Environ.* **2019**, *27*, 1488–1496. [CrossRef]
25. Akindoyo, J.O.; Husney, N.A.A.B.; Ismail, N.H.; Mariatti, M. Structure and performance of poly(lactic acid)/poly(butylene succinate-co-L-lactate) blend reinforced with rice husk and coconut shell filler. *Polym. Polym. Compos.* **2021**, *29*, 992–1002. [CrossRef]
26. Ghazvini, A.K.A.; Ormondroyd, G.; Curling, S.; Saccani, A.; Sisti, L. An investigation on the possible use of coffee silverskin in PLA/PBS composites. *J. Appl. Polym. Sci.* **2022**, *139*, 52264. [CrossRef]
27. Sisti, L.; Totaro, G.; Rosato, A.; Bozzi Cionci, N.; Di Gioia, D.; Barbieri, L.; Saccani, A. Durability of biopolymeric composites formulated with fillers from a by-product of coffee roasting. *Polym. Compos.* **2022**, *43*, 1485–1493. [CrossRef]
28. Yap, S.Y.; Sreekantan, S.; Hassan, M.; Sudesh, K.; Ong, M.T. Characterization and Biodegradability of Rice Husk-Filled Polymer Composites. *Polymers* **2020**, *13*, 104. [CrossRef] [PubMed]
29. Sarasini, F.; Luzi, F.; Dominici, F.; Maffei, G.; Iannone, A.; Zuorro, A.; Lavecchia, R.; Torre, L.; Carbonell-Verdu, A.; Balart, R.; et al. Effect of Different Compatibilizers on Sustainable Composites Based on a PHBV/PBAT Matrix Filled with Coffee Silverskin. *Polymers* **2018**, *10*, 1256. [CrossRef]
30. Sarasini, F.; Tirillò, J.; Zuorro, A.; Maffei, G.; Lavecchia, R.; Puglia, D.; Dominici, F.; Luzi, F.; Valente, T.; Torre, L. Recycling coffee silverskin in sustainable composites based on a poly(butylene adipate-co-terephthalate)/poly(3-hydroxybutyrate-co-3-hydroxyvalerate) matrix. *Ind. Crops Prod.* **2018**, *118*, 311–320. [CrossRef]
31. Suriapparao, D.V.; Kumar, D.A.; Vinu, R. Microwave co-pyrolysis of PET bottle waste and rice husk: Effect of plastic waste loading on product formation. *Sustain. Energy Technol. Assess.* **2022**, *49*, 101781. [CrossRef]
32. Coles, R.; Kirwan, M.J. *Food and Beverage Packaging Technology*; Blackwell Publishing: Hoboken, NJ, USA, 2011; p. 342. [CrossRef]
33. Foundation, E.M. *The Global Commitment 2021 Progress Report*; Ellen McArthur Foundation: Cowes, UK, 2021.
34. EPA, U.S. The New National Recycling Goal. Available online: https://www.epa.gov/sites/default/files/2020-12/documents/final_one_pager_to_print_508.pdf (accessed on 9 June 2022).
35. EPA, U.S. Advancing Sustainable Materials Management: 2018 Tables and Figures. Available online: https://www.epa.gov/sites/default/files/2021-01/documents/2018_tables_and_figures_dec_2020_fnl_508.pdf (accessed on 9 June 2022).
36. Mohanty, A.K.; Vivekanandhan, S.; Pin, J.-M.; Misra, M. Composites from renewable and sustainable resources: Challenges and innovations. *Science* **2018**, *362*, 536–542. [CrossRef]
37. Vilaplana, F.; Strömberg, E.; Karlsson, S. Environmental and resource aspects of sustainable biocomposites. *Polym. Degrad. Stab.* **2010**, *95*, 2147–2161. [CrossRef]
38. TPCH. *Toxics in Packaging Clearinghouse—Fact Sheet*; TPCH: Brattleboro, VT, USA, 2021.
39. Goodlaxson, B.; Curtzwiler, G.; Vorst, K. Evaluation of methods for determining heavy metal content in polyethylene terephthalate food packaging. *J. Plast. Film. Sheeting* **2018**, *34*, 119–139. [CrossRef]
40. Westerhoff, P.; Prapaipong, P.; Shock, E.; Hillaireau, A. Antimony leaching from polyethylene terephthalate (PET) plastic used for bottled drinking water. *Water Res.* **2008**, *42*, 551–556. [CrossRef] [PubMed]
41. Suhot, M.A.; Hassan, M.Z.; Aziz, S.A.A.; Md Daud, M.Y. Recent Progress of Rice Husk Reinforced Polymer Composites: A Review. *Polymers* **2021**, *13*, 2391. [CrossRef] [PubMed]
42. Phan, H.T.Q.; Nguyen, B.T.; Pham, L.H.; Pham, C.T.; Do, T.V.V.; Hoang, C.N.; Nguyen, N.N.; Kim, J.; Hoang, D. Excellent Fireproof Characteristics and High Thermal Stability of Rice Husk-Filled Polyurethane with Halogen-Free Flame Retardant. *Polymers* **2019**, *11*, 1587. [CrossRef] [PubMed]
43. Gupta, A.; Mohanty, A.K.; Misra, M. Biocarbon from spent coffee ground and their sustainable biocomposites with recycled water bottle and bale wrap: A new life for waste plastics and waste food residues for industrial uses. *Compos. Part A Appl. Sci. Manuf.* **2022**, *154*, 106759. [CrossRef]
44. Bahl, S.K.; Cornell, D.D.; Boerio, F.J.; McGraw, G.E. Interpretation of the vibrational spectra of poly (ethylene terephthalate). *J. Polym. Sci. Polym. Lett. Ed.* **1974**, *12*, 13–19. [CrossRef]
45. Donelli, I.; Freddi, G.; Nierstrasz, V.A.; Taddei, P. Surface structure and properties of poly-(ethylene terephthalate) hydrolyzed by alkali and cutinase. *Polym. Degrad. Stab.* **2010**, *95*, 1542–1550. [CrossRef]
46. Sajith, S. Investigation on effect of chemical composition of bio-fillers on filler/matrix interaction and properties of particle reinforced composites using FTIR. *Compos. Part B Eng.* **2019**, *166*, 21–30. [CrossRef]
47. Cruz, S.A.; Zanin, M. PET recycling: Evaluation of the solid state polymerization process. *J. Appl. Polym. Sci.* **2006**, *99*, 2117–2123. [CrossRef]
48. Tavares, A.A.; Silva, D.F.A.; Lima, P.S.; Andrade, D.L.A.C.S.; Silva, S.M.L.; Canedo, E.L. Chain extension of virgin and recycled polyethylene terephthalate. *Polym. Test.* **2016**, *50*, 26–32. [CrossRef]

49. Union, E. European Parliament and Council Directive 94/62/EC of 20 December 1994 on Packaging and Packaging Waste. Available online: https://eur-lex.europa.eu/legal-content/EN/TXT/?uri=CELEX%3A01994L0062-20180704 (accessed on 24 May 2022).
50. Curtzwiler, G.; Vorst, K.; Danes, J.E.; Auras, R.; Singh, J. Effect of recycled poly(ethylene terephthalate) content on properties of extruded poly(ethylene terephthalate) sheets. *J. Plast. Film. Sheeting* **2011**, *27*, 65–86. [CrossRef]
51. Friedman, E. INSIGHT: 2022 to Be a Pivotal Year for US R-PET Market. Available online: https://www.icis.com/explore/cn/resources/news/2022/03/04/10740575/insight-2022-to-be-a-pivotal-year-for-us-r-pet-market (accessed on 15 July 2022).

Review

Possibility Routes for Textile Recycling Technology

Damayanti Damayanti [1,2], Latasya Adelia Wulandari [2], Adhanto Bagaskoro [2], Aditya Rianjanu [3] and Ho-Shing Wu [1,*]

[1] Department of Chemical Engineering and Materials Science, Yuan Ze University, 135 Yuan-Tung Road, Chung-Li, Taoyuan 32003, Taiwan; damayanti@tk.itera.ac.id
[2] Department of Chemical Engineering, Institut Teknologi Sumatera, Jl. Terusan Ryacudu, Way Huwi, Kec. Jati Agung, Lampung Selatan 35365, Indonesia; latasyadeliaw@gmail.com (L.A.W.); adhanto.118280076@student.itera.ac.id (A.B.)
[3] Department of Materials Engineering, Institut Teknologi Sumatera, Jl. Terusan Ryacudu, Way Huwi, Kec. Jati Agung, Lampung Selatan 35365, Indonesia; aditya.rianjanu@mt.itera.ac.id
* Correspondence: cehswu@saturn.yzu.edu.tw

Abstract: The fashion industry contributes to a significant environmental issue due to the increasing production and needs of the industry. The proactive efforts toward developing a more sustainable process via textile recycling has become the preferable solution. This urgent and important need to develop cheap and efficient recycling methods for textile waste has led to the research community's development of various recycling methods. The textile waste recycling process can be categorized into chemical and mechanical recycling methods. This paper provides an overview of the state of the art regarding different types of textile recycling technologies along with their current challenges and limitations. The critical parameters determining recycling performance are summarized and discussed and focus on the current challenges in mechanical and chemical recycling (pyrolysis, enzymatic hydrolysis, hydrothermal, ammonolysis, and glycolysis). Textile waste has been demonstrated to be re-spun into yarn (re-woven or knitted) by spinning carded yarn and mixed shoddy through mechanical recycling. On the other hand, it is difficult to recycle some textiles by means of enzymatic hydrolysis; high product yield has been shown under mild temperatures. Furthermore, the emergence of existing technology such as the internet of things (IoT) being implemented to enable efficient textile waste sorting and identification is also discussed. Moreover, we provide an outlook as to upcoming technological developments that will contribute to facilitating the circular economy, allowing for a more sustainable textile recycling process.

Keywords: textile recycling; mechanical recycling; pyrolysis; enzymatic hydrolysis; ammonolysis; glycolysis; IoT; sorting identification

1. Introduction

The improvement of living standards around the world due to economic development could be affected by the textile industry. The textile industry faces huge challenges related to environmental issues. In 2018, the total global textile production was 105 million metric tons [1–4]. In total, up to 64% of the textile fibers that are produced are produced from the petrochemical industry. The rest of the fibers, 36%, are shared by cotton, 24%, cellulosic fibers, 6%, and wool 1% and natural fibers. With its present fast fashion business model, which is defined by mass production, variety, agility, and affordability, the apparel industry has made a significant contribution to the amount and rate of trash generation [5]. The volume of textile waste that is primarily being disposed of in landfills or incinerators is continuing to significantly increase [6]. Less than 1% of the material used to produce clothing was recycled into new clothing [7].

Textile waste can be categorized into pre-consumer and post-consumer waste. Furthermore, there are several materials that can be used to produce fabric: (i) natural materials; the natural resources used to produce it, e.g., wool and silk (protein fibers); cotton, linen, and hemp (cellulosic); (ii) regenerated material: produced from natural polymers through processing, such as rayon, viscose, cuprammonium, and acetate (semi-synthesis fiber); and (iii) synthetic material: materials that come from the petrochemical industry, e.g., polyester and nylon [8]. Natural materials are environmentally friendly. Synthetic fibers are petrochemical-based products that are produced from materials that are not sustainable and that require enormous amounts of energy to be produced [9]. Textile waste can be recycled and utilized in secondhand stores to create value-added items. The circularity approach is currently being used in the textile reuse and recycling process, which is characterized by changing from the old linear paradigm of "take, make, and trash" to a circular model, where fiber, fabric, or garments can be used to their maximum potential [6,10]. Currently, recycling technologies have five problems: (i) lack of commercially viable recycling technologies for a low-grade textile fraction; (ii) lack of mainstreamed, up-scaled processes and know-how to separate fiber types from the mixed blends and composite structures; (iii) costly recovery process; (iv) the fact that the recycling end-market is dominated by low-quality materials and blends; (v) and costly logistics and the low availability of textile recycling plants on both local and regional levels [11]. In addition, textile recycling techniques should be pass the Life Cycle Assessment (LCA). The LCA suggests a critical area in the textile supply chain, specifically, it emphasizes recycling techniques and transportation distances between the textile recycling plant and waste collection areas due to the amount of energy that is consumed during transport [12–15].

Due to the enormous waste and pollution caused by the phenomena of fast fashion, the textile sector faces tremendous environmental and resource challenges [16]. The transition to a circular economy (CE) is already underway, and it offers an effective option for resource scarcity and waste disposal issues [17]. Sorting for textile material identification and separation takes place before textile waste is recycled due to the complex material in textile fibers [18]. Moreover, in the future, the large-scale separation and sorting of textile waste, regardless of the strategy that is selected, quick and reliable methods to accurately quantify the polyester content in cotton, polyester, and other textile material compound mixes are urgently needed [19]. Specific materials should be recycled, e.g., medical textiles, which are all fiber-based products and structures that are utilized for emergency treatment and clinical, surgical, and hygienic purposes [20]. The process of recycling textile waste can be easily achieved by Internet of Things (IoT) technology. IoT can assist in collecting the data provided by various sensors, including smart meters, in connecting stakeholders across the value chain [21]. Furthermore, IoT needs data collection due to the fact that smart sensor and RFID technologies require big data [22]. Fatimah has been introduced to modern technologies such as Indonesia's IoT-based sustainable CE approach; waste management has also benefited from IoT technology [23].

Identifying textile materials is the most challenging task in the textile recycling process because of the complicated structures of fabrics and textiles. Cellulose polyester mixtures are among the most popular multi-component garments on the marketplace and can be used to customize material attributes such as moisture absorption, wrinkle resistance, and wearing comfort [24]. There are some promising methods that can be used to analyze the complex components of different textiles. For example, applying near-infrared (NIR) spectroscopy and chemometrics for textile analysis was reported over two decades ago [25]. The cotton component, which comprises mixed cotton and polyester, was determined by genetic algorithms [25]. In addition, nuclear magnetic resonance (NMR) has been used to predict the proportion of cellulose and polyester in textile blends [24]. Nevertheless, solid-state NMR can show lower sample throughput and efficiency. Still, it allows for more precise measurements that are not influenced by factors such as particle size, dyes, or surface structure [26]. On the other hand, the nature of the sample surface, such as particle size, brightness, color, moisture content, and coating chemicals, has a significant impact on the NIR signals that are produced [27].

Polyester and cellulose performed very differently in terms of behavior, especially when considering differences such as thermal stability and hydrophilicity. The diversity of properties in these textiles allows them to be reprocessed and separated in different ways, such as through depolymerization, re-polymerization, and spinning [24]. Therefore, solving these problems requires a comprehensive analytical toolkit for the qualification and identification of textile materials [28]. Textile can be recycled through mechanical, thermal, and chemical methods. Textile recycling mainly entails the reprocessing of post- or pre-consumer waste fabric in order for it to be reused in new textile or non-textile products. The three types of textile recycling routes are the dissolution of natural fibers, the chemical degradation of polymeric fibers, mechanical treatment (pretreatment), and thermal treatment [29].

A research gap exists in the field of textile recycling technology, and it is possible for textile recycling technologies to be implemented based on the CE principle. Current technologies such as the IoT and big data can be used to enable the efficient collection of textile waste, identification, and recycling. Still, there are some challenges to recycling textile waste due to different materials that are added to fabric and textiles to make fashion items. Therefore, the strategies that can be used to recycle textile fibers, such as mechanical recycling, chemical recycling (pyrolysis, enzymatic hydrolysis, hydrothermal, ammonolysis, gasification, glycolysis), and decolorization technology in textile recyclizing, are reviewed to minimize the textile waste that ends up being incinerated or in landfills.

2. Circular Economy of Textile Recycling

A CE is a concept where material production can be redesigned. It utilizes resources to produce, use, and dispose of in favor of as much reuse and recycling as possible [8]. The CE approach has an environmental, industrial economic, and ecological basis [30–32]. There are some barriers impeding the implementation of CE in the fashion industry due to the various fashion styles, aesthetics, and the roles of shoppers. The regulations and policies regarding the environment and sustainability of CE need to be considered [19]. It has also been determined that the textile business also has a detrimental impact on society because it consumes enormous resources such as oil, carbon, and water [33,34]. Apart from that, it eliminates the over-generation of waste and obtains the total value of products by focusing on the CE through giving products a longer life and through reusing materials [8]. The CE is driven by three main methods and approaches (reduce, reuse, and recycle); all traditional waste managements are practiced [35].

Fabric waste reduction includes all of the manufacturing stages (including a minimum of feedstock) and various stages of use and consumption. The concept of the reuse stage is that textile materials are easily recyclable or repurposed for different types of applications. In addition, increasing reuse and repurposing will reduce the amount of textile materials that are needed for production [18,36]. Figure 1 shows this critical point in the textile CE. It is vital to have a CE because it emphasizes seeking scientific solutions to complete the loop. Waste is reduced at the source and is recycled back into the economy for reuse rather than production and consumption being stopped due to garbage disposal [37]. The CE cycles includes recycling, reuse, and remanufacturing. Textile reuse involves various methods for extending the useful life of textile items by redistributing them to a new buyer [38]. Nevertheless, the secondhand clothing markets will end up exporting these secondhand textiles to emerging market countries. On the other hand, CE emphasizes local consumer reuse, and the retailers must contribute to this circular system. This system can be applied to introduce new customer segments and new potential for reuse models [39].

Remanufacturing is used in the fashion and textile industry to describe the process of reconstructing worn clothes to create new garments by considering the quality of the textile product or the customer value [40,41]. In the early 21st century, the consideration of textile remanufacturing gained popularity because fashion designers and entrepreneurs were more aware of the sustainability of the post-consumer raw materials and textile waste [42,43]. The reasonable prices and the remanufacturing of the textile industry in some developing economies, such as in China, have been more competitive [44]. Furthermore, by using an eco-efficient value creation (EVC) model integrated with analysis cost, eco-cost, and retail value, remanufactured goods have a lower eco-cost due to their reduced materials and pollution. An EVC has a positive effect on the cost-benefit [45].

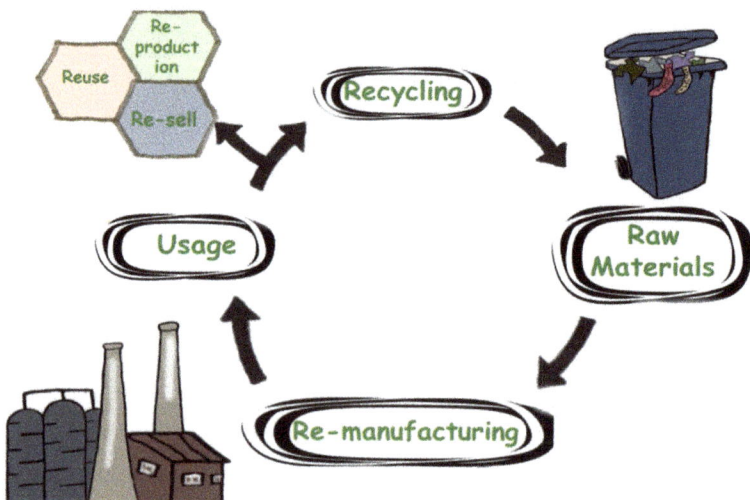

Figure 1. The critical point of CE.

Due to the textile sector, recycling is the principal environmental polluter, and textiles have complicated the manufacturing process. Furthermore, the microfibers in textile waste harm the oceans, and the textile industry contributes to this pollution. Globally, textile waste increases, and recycling could help reduce new material waste [14]. Mechanical and chemical recycling are two primary types of recycling technologies. Mechanical recycling via melt-extrusion is a technique that is used to extract fibers from waste materials, allowing them to then be spun into yarns. Therefore, to obtain great yarn strength and fineness, this technique must be applied to virgin fibers.

On the contrary, chemical recycling occurs when the polymers of textile waste are depolymerized into small monomers and/or are re-polymerized to make new fibers. The chemical recycling of cellulose fabric such as cotton or viscose also uses the dissolution approach. At the same time, organic solvents apply ionic liquids to dissolve cellulose polymers [45,46]. In addition, the valorization process offers another potential textile recycling method. Li et al. developed the term "textile waste valorization", which is related to the combination of "textile waste recycling and hydrolysis process". This process degrades the complex polymers found in textiles in order for them to become various value-added products that are supported by microorganisms such as enzymes. This kind of method is inspired by the CE approach, which aims provide multiple solutions for recycling textile materials [47].

Textile material recycling affects a variety of substances and contributes significantly to current social obligations. The recycling process allows businesses to make a more significant profit by avoiding textile waste being dumped in landfills. Meanwhile, the textile recycling contributes a positive impact on social and environmental issues, e.g., gives job opportunities to workers who are unemployed or underemployed, philanthropic commitments, and the expansion of reusable textile to the countries around the world that need proper cloth [48,49].

There are many challenges when transitioning the textile industry to a circular economy: (i) the market for recycled textile products is still limited; (ii) there is a lack profitable circular business models on the market; (iii) there are challenges in having supply chain partners work together to create innovations; and (iv) the quality of recycling of textiles of a relatively low grade is pricey in the short-term, with low economic benefits [50]. There are some fundamental strategies for recycling business models (RMBs); for instance, recycling process, partnership, and scale up the business models (BMs). Furthermore, approach RMBs are becoming more practicable. Collaboration between various stakeholders and across a variety of industry sectors is needed [51]. In addition, the participation of waste companies and educational institutions in investigating several options for closing material loops and identifying waste streams is a crucial change agent to facilitating the textile CE transformation [52].

3. Starting with Municipal Waste to Textile Trash in Applied IoT

Sustainability is also required in the waste management sector. It starts with the manufacturing process to the disposal phase. A loop feedback system for sustainable waste management comprises several parts, such as process activity and waste diversity [23,53]. The direct regulations of the European Union's CE action plan will prohibit the disposal of unsold textile products and a restriction on single-use products, and according to the Recycling and Waste Reduction Act 2020 with replaced the product stewardship Act 2011 in Australia, there are textile regulations that are related to the ambition of separate textile collection by 2025. A new French legislation has established a command-and-control prohibition on the destruction of various unsold products, including fashion items. This legislation requires that new products use a minimum amount of recycled material [54].

Furthermore, to support the current regulation and policy of waste textiles, digital innovations have been adopted. Recently, digital innovation and industry 4.0 have positively influenced CE transformation, such as data-driven product lifecycle analysis [55]. Industrial revolution 4.0 is achieving waste treatment sustainability, with waste treatment becoming more practicable, dependable, clear, efficient, and optimal due to the application of IoT, digitalization, and information and communication technology (ICT). ICT and IoT implementation can minimize the time and resources needed to improve waste management effectiveness and to move toward more sustainable and intelligent systems [56]. In addition, these methods will be selected as the best treatment technologies for the waste material in terms of technical, economic, and environmental feasibility. IoT and ICT can be combined to gather, analyze, and share data [23].

Digitalization can be improved in product/raw materials tracing and tracking by providing real-time data on product availability, location, and condition. At the same time, sensors and digital platforms can enhance the product life and can create CE solutions more effective and efficient [55]. IoT-enabled manufacturing is related to advanced concepts. Traditional manufacturing resources can be transformed into smart manufacturing objects (SMOs) that have the ability to interact, sense, and interconnect with one another to perform manufacturing logic [57]. The fundamental issue with these technologies is the sharing and collecting of real-time data, including wireless communication standards and radio frequency identification (RFID). Physical production flows, including material transportation, and associated information flows, such as the visibility and traceability of different manufacturing operations, can be easily connected utilizing RFID technology [58,59].

Cloud computing technology is a data management solution, and it can be used as a replacement for local servers [60]. It has resulted in a massive scale of data sharing among several users and has become the latest data management standard [61]. These processes and resources must be handled intelligently in cloud manufacturing technology because the product life cycle includes the start of the manufacturing process, and includes design, manufacture, and maintenance [62]. Resources can be controlled and managed automatically by applying various IoT technologies, making digital sharing easier. Underpinning technologies are typically service-oriented, and cloud manufacturing complements the conceptual framework of cloud manufacturing [63].

The design approach for the sorting and identification of textile waste is based on the IoT in Figure 2. Waste materials are sorted through a funnel and a single metallic sensor. The metallic sensor detects any metal compounds that are going through the funnel. The metal materials go into metal storage. If the object does not contain any non-metal materials, then it is placed on the sensing platform. The capacitive sensors sense the object while it is on the sensing platform based on its moisture content. Then, the materials that have moved through capacitive sensors are divided into two sections: plastic bottles and textiles. Furthermore, the sorted textile waste will undergo several sorting and identification processes due to the complex mixture of materials [64].

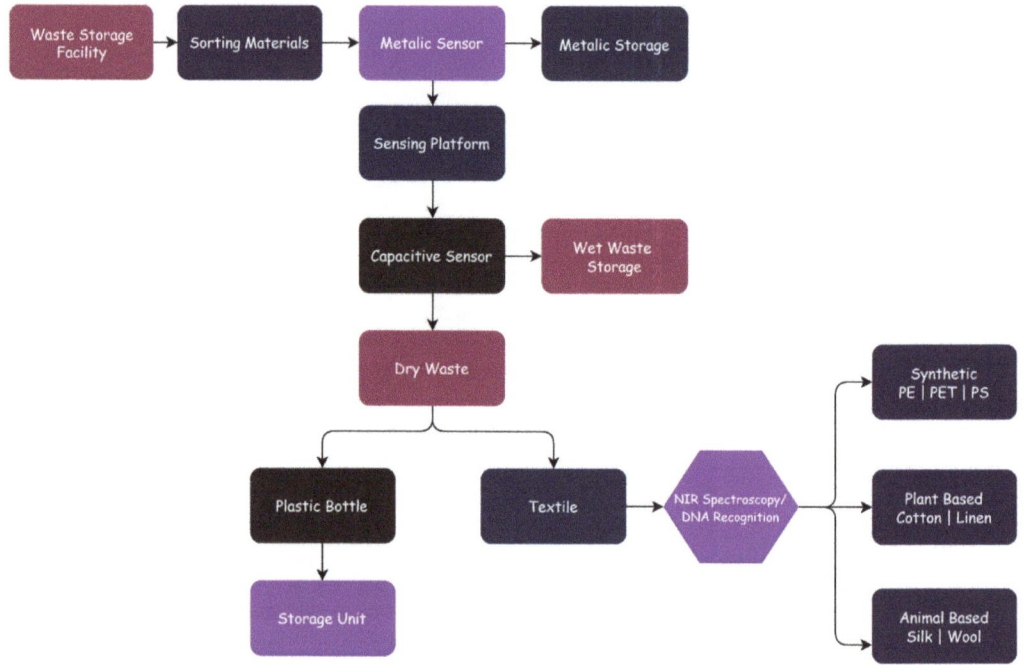

Figure 2. The sorting and identification of textiles using IoT.

On the other hand, manual sorting can be applied. Mechanical machines will transport the materials to a manual sorting platform by conveyors for preliminary sorting, which will sort the hazardous materials, bottles, large glasses, cotton, and textiles to be recycled. Each material will be separated based on its characteristics and will then be delivered to each storage bin. Each storage bin is equipped with a weight level sensor that sends the general information into the database. The database will receive and store all of the data related to the system's waste workflow. According to the performance of the trash that is collected and the existing processing technology, the database will suggest treatment options [23].

Yu et al. developed a hybridized intelligent framework (AIHIF) based on artificial intelligence (AI) for automated recycling to improve trash management. By combining machine learning and graph theory, the system enhances waste-collecting over a short distance. Furthermore, AI design technology can be supported through various approaches that can be adapted to specific interest groups, collecting their data and increasing ecological planning and urban management production, precision, and efficiency [65]. Enevo has already developed recycling systems and innovative waste treatment technologies with the Enevo One software. The software contains ultrasonic wireless sensors. Waste collection optimization has been performed not only by cloud service in the background but also by the web interface used for Enevo One and the sensors themselves. Enevo's sensors are compatible with more than 100 container types. The following waste types are currently supported: mixed, textile, glass, metal, and waste electrical and electronic equipment

(WEEE) [66]. In addition, Stack4things was created by the mobile and distributed systems lab at the University of Messina in Italy. The OpenStack add-on controls sensing and actuation resources, including remote control, virtualization, and network overlays. The main objective is applied through cloud computing to manage data and changes to a software as a service (SAaaS) perspective, in which developers or end-users are provided with actual or virtualized sensing and actuation resources [67].

4. Sorting and Identification of Textile Waste

The dilemma for textile recycling is sorting and identifying textile materials due to the complex structure of textile polymers. In addition, the majority of textile materials comes from fiber, and the main challenges are sorting and recycling due to the complicated fabric of the textile. The fabric waste is commonly mixed with other materials, such as buttons, zippers, or other decorations [68]. There are various ways to identify textile materials, including iso-standardized quantification methods based on distinct dissolution behavior (ISO 1833-1, for example) and morphological variations found using microscopy [69]. Figure 3 shows the source and mechanism of textile recycling, a common source of textile waste from consumers, and unsellable textile waste from the textile industry. In addition, the sorting and collecting of textile waste is not able to be achieved manually, which would improve material quality assessment [70].

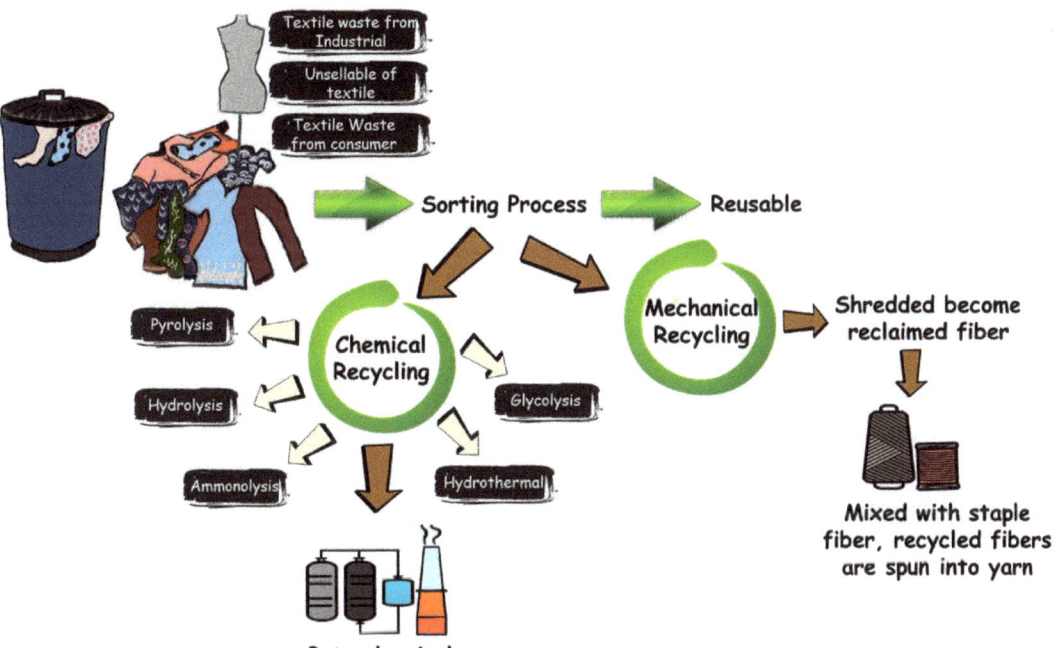

Figure 3. Mechanisms of possible textile waste recycling routes.

The thermal properties of textile materials are essential qualities of these materials and are regarded as the most crucial consideration when describing the comfort properties of apparel [71]. In addition, several different thermal properties have been demonstrated in numerous research studies, such as in textile fiber polymers, which are semi-crystalline, whose thermal conductivity is affected by the molecular structure, crystallization level, and molecular chain mobility of amorphous regions, etc. [72]. One of the methods that can be used to analyze the thermal properties of textile waste is by thermogravimetric analysis

(TGA). The data resulting from TGA are related to the thermal stability and degradation degree from the fibers. TGA is a simple method that can be used to obtain data on the thermal behavior of a textile before that textile waste is pyrolyzed in the industry [73]. Therefore, some real-time technology is required to develop the textile material recognition rate.

Table 1 indicates the recognition rate using different types of analyzers. A single type of textile waste polymer is easier to convert into small molecules with a higher value than when other types of waste polymers are presents. Recently, near-infrared (NIR) technology has been applied to sort textile waste. NIR spectroscopy can be implemented automatically by convolutional neural network machine learning (CNN) followed by wavelength recognition that has been transformed through image classification. Then, the surface area of the spectrum is converted and modified by Textile Recycling Net (Tr-Net), and the images are sorted using the softmax classifier, allowing the qualitative of textile compositions to be obtained [74,75].

Wilts et al. developed a new AI technology to sort textile waste from municipal waste. This robot with AI could be applied with a recognition ability of more than 95%. On the other hand, textile waste recovery was not achieved, with only 13% of textile waste being recovered due to the complicated structure of materials [76]. Xiong designed a cheap near-infrared device to sort the seven types of plastics using machine learning. Machine learning could identify the plastic types with 90–99% certainty [77]. Berghmans et al. studied infrared spectrometer techniques to identify the post-consumer and industrial carpet waste. The carpet waste materials came from polyamide-6, polyamide-66, polyester, polypropylene, and wool. The complex of the carpet materials can be determined by analyzing the radiation passing through the fibers. The range of wavelengths near and mid-infrared was (800–2500 mm) and (2500–25,000 mm), respectively [78].

Table 1. The recognition rate with different types of analyzers.

Textile	Analyzer	Mathematic Algorithms	Recognition Rate, %	Ref.
Pure (polyester slash; wool; cotton; polyester normal; nylon). Mix (polyester/nylon; polyester/wool; polyester/cotton slash; polyester/cotton).	NIR	SVM, MLP + CNN	92–98	[74]
Cotton, polyester, polyamide, acrylic, silk, wool.	NIR	SIMCA	93–97	[79]
Polypropylene, polyethylene terephthalate, polylactic acid, cashmere, Tencel, cotton, wool	NIR	PCA + SIMCA	100	[80]
Wool	NIR	PLS, ECR		[81]
Satin, twill, plain	ResNet-50	CNN	99.3	[82]
Satin, twill, plain	Epson Scan V330	TILT, HOG, FCM clustering, histogram equalization data	94.5	[83]
Non-woven fabric, plain weave, twill weave, double jersey, satin weave	Canon scanner 9950F, Self-organizing map (SOM) network	CIE + Co-occurrence matrix	92.6	[84]
Cotton, polyester, viscose	NIR	NA	76–100	[85]

CIE: C-Illuminant (light source C), FCM: fuzzy C-means, HOG: histogram of oriented gradients, PCA: principal component analysis, SIMCA: soft independent modeling of class analogy, SOM: self-organizing map, SVM: support vector machine, TILT: transform invariant low-rank textures. PLS: partial least squares. ECR: elastic component regression. NA: not available.

Riba et al. used an ATR-FTIR spectrometer (Attenuated Total Reflection–Fourier Transform Infrared) to sort and classify textiles based on the type of polymer materials in that textile. Mathematical modeling was applied to more accurately predict the data from the IR spectrum. The recognition rate percentage achieved by the mathematic algorithms (principal component analysis, canonical variate analysis, k-nearest neighbors algorithm (k-NN algorithm)) could detect synthetic and natural fibers, such as polyamide, viscose, polyester silk, linen, cotton, and wool, with an accuracy of up to 100% [86]. Furthermore, Peets et al. evaluated the capability of reflectance-FTIR (r-FTIR) to recognize textile fibers using an FTIR micro-spectrometer. In reflectance mode, the FTIR micro-spectrometer several reflectance spectra from textile fibers can be collected, and these spectra contain enough characteristics to create classification models and to identify unknown textile samples. Nevertheless, r-FTIR is a good approach for analyzing many textile samples quickly, easily, non-destructively, and non-invasively [87]. Morgado et al. identified PE, PET, PP, and PS using micro-ATR-FTIR spectra with known uncertainty [88].

The following three primary characteristics are reflected in the recognition and classification accuracy: (1) when physical properties, such as fabric color, yarn thickness, diameter, orientation, and uneven light, are considered, the model remains robust; (2) a model with fewer parameters should use the transfer learning technique, making it more computationally efficient; (3) the suggested model does not use handcrafted features, instead relying on a fully automated end-to-end architecture to extract and classify features [82].

5. Mechanical Recycling of Textile

Mechanical textile recycling has been developed since the industrial revolution began. Furthermore, mechanical recycling is one of the easiest recycling methods can be used with a low-cost budget. Textile waste is classified by the type of material and the color achieved through mechanical deconstruction; for instance, wool converts into useable yarn (re-woven or knitted) when carded yarn is spun with mixed shoddy [89,90]. Mechanical recycling can be categorized into several different methods based on the degree of breakdown that is undergone by the recovered material, such as fiber, fabric, polymer, and monomer recycling. The mechanical recycling applications in the wiper, fiber material, and prespun fiber industries could produce new fabrics for products [91].

On the other hand, there are few mechanical recycling options for post-consumer textile waste fabrics due to the variety and amount of fiber elements in one textile waste garment; mixed textiles cannot be mechanically recycled effectively. The shredding technique reduces the length of post-consumer textile waste garment fibers. The resulting short fibers are of a lower quality and strength than virgin cotton fibers, and they are frequently combined with virgin fibers to improve quality [11,13,92,93].

The metallic elements are removed when post-consumer textiles are converted to wipers. Textiles from the post- and pre-consumer markets can also be used as a raw material for filler products, including flocking, insulation, and nonwovens as well as shoddy fibers for recycled yarns [94]. A defector is developed with the mechanical ripping and carding processes during the production process. The first step of the defector process requires the feedstock to be shredded by the rotation crusher and sorted based on the type of fiber. Secondly, the fiber is crushed through correct fraction sizes and processes with the card oil. Furthermore, the defibering process is a three-phase mechanical ripping carding process. The materials are needled and folded to create a fiber blanket. In the needling process, other materials could be added to the product, e.g., plastic material. Then, the fiber blanket is trimmed to the correct size and is packed [95]. On the other hand, some textile materials cannot be recycled by mechanical process due to the low quality and difficulty of reprocessing these materials through the melting process to achieve high-value products, for instance, cotton or a mixture of synthesis fibers [90]. In addition, mechanical shredding harms natural fibers more than synthetic fibers in blended fabrics such as polyester–cotton. Chemical recycling becomes an alternative to recycle both fiber types [96].

Figure 4 shows an open and closed-loop garment recycling. The term "open-loop recycling" refers to methods in which fabrics or textiles are used to make new products [52]. Besides that, recycling wool through open-loop recycling can be used to mechanically extract the garments back into a fibrous form, and these materials can be used as raw materials. Non-woven fabrics are frequently made from these materials by garneting, carding, or air laying webs, which are then mechanically, thermally, or chemically bonded. Furthermore, hot water converges with the fibers when the melt-blown process is applied, followed by spun lace. The purpose of spun lace is to entangle fibers. The method utilizes extremely tiny high-velocity water jets rather than sharp needles when the textile is going through the drying process [97]. Besides that, the melt blowing process can be combined with polymer melt extrusion passed by the attenuation of the extrudate and orifices under extremely high-velocity air jets to generate nonwoven fibers in just one step [98].

Figure 4. The recycling of garments by open and closed-loop systems.

In addition, Tsai described a technology to develop an electret via saturating a nonwoven textile with a liquid and then removing the fluid. This technology contributed (i) a sheet made from a thermoplastic fiber nonwoven web; (ii) the ability to place the sheet underneath a roller: water jets were added in these steps to acquire a saturated sheet, and (iii) a suction process that is able to remove the water at the saturated sheet that then allows it to dry [99]. The applications of open-loop recycling include insulators for mattresses and sound absorption [97]. Nylon waste, such as fishing nets (nylon 6) and carpets, is the common raw material for that is used in mechanical recycling. Nylon materials are recycled via reshaping and melting to create new fibers with appropriate strength and length [100].

Closed-loop textile recycling refers to processes in which the recycled material is identical to the provided material. Both pre-and post-consumer mechanically recycled materials can be considered to be closed-loop recycled since the waste material or fiber re-enters the apparel generation chain. The fineness, length, polymer quality, and shading of the new fiber properties can determine the quality and to create the most appropriate final product [48]. A closed loop conducts the recycling of wool fibers. Using traditional techniques, recycled wool yarns can be turned into very high-value apparel products. Woolen materials are mainly formed from long fibers that have been carefully treated, allowing for mechanical recycling to covert the wool into fibers. Wool recycling follows the same methods as mechanical cotton recycling [101]. The Cardato brand coat from M&S Shops is manufactured using closed-loop recycled wool. The recycled wool waste should ideally be compatible with the same carding, spinning, and fabric production processes used to process raw wool during the production of woolen yarn [101–103]. The post-consumer wool recycling can be applied. This depends on effective strategies to reduce fiber breakage and to maximize residual fibers length following the mechanical pulling method being available [104].

Klaus-Nietrost et al. studied the recycling of cellulose-based textiles with a minimum of one synthetic plastic. The regeneration process was applied to acquire a cellulosic-molded body. In addition, the cellulosic-molded body was composed of a crucial substance in the form of lyocell—fibers or viscose—fibers. Chemical fibers, viscose fibers, and regenerated fibers are all terms that are used to describe different types of fibers. They can be produced using the viscose technique, which is a wet-spinning method [105,106]. The cellulosic material is soaked in N-methyl morpholine-N-oxide (NMMO) and is pressed with a spinneret. It coagulates and generates a new fiber [107]. When recycled fibers are applied in cellulose production, the main issue is the purity of the raw materials, such as old-fashioned textile materials, and more significantly, the contamination of synthetic polymer materials. Nevertheless, the synthetic polymer can be denoted as being combined with cellulose materials [105].

6. Chemical Recycling of Textile

Chemical treatment is seen as a promising textile recycling method. During this process, chemicals are added to degrade the complex textile polymer to create smaller polymer molecules. The "CRM", or chemical recycling to monomer, is a popular nowadays that can be used to degrade polymers such as those in textile waste to monomers. Furthermore, a solution-purified polymer, an oligomer, a monomer, or a crude feedstock in gaseous or liquid form are all examples of textile chemical recycling products. Re-polymerization can be used to renew polymers such as oligomers and monomers [90,108]. Table 2 lists the industrial scale that is applied when recycling textiles through various chemical processes. Typical chemical processes are glycolysis, hydrolysis, and alcoholysis. In addition, thermochemical methods consist of hydrocracking, pyrolysis, and gasification, which are typically conducted in the presence of catalysts at high temperatures [109]. Meanwhile, biochemical processes use enzymes or microorganisms in recycling processes such as enzymatic hydrolysis [106].

Table 2. Various chemical processes applied in textile recycling at the industrial scale.

Company	Material	Method	Country	Ref.
Ambercycle	Textile	Enzymatic treatment	Los Angeles	[110]
BlockTexx	Cotton and polyester (PET)	Fiber separation technology	Australia	[111]
FENC	PET and textile	Hydrolysis	Taiwan	[112]
Infinited Fiber	Cellulose and polyester (PET)	Alkaline extraction	Finland	[113,114]
Ioncell	Cotton and polyester (PET)	Ionic liquid solvent polymer dissolution	Finland	[115]
Lenzing	Pre-consumer cotton and post-consumer garments	Closed loops and chemical recovery	Austria	[116]
Tyton BioSciences	Cotton pulp, polyester (PET), poly-cotton blends	Subcritical water treatment	Danville	[117]
Worn Again Technologies	Cotton and polyester (PET)	Dissolution polymer Solvent separation	UK	[118]

The drawbacks and advantages of the various chemical textile recycling processes are shown in Table 3. Chemical polymer recycling for cellulose-based synthetics and mixtures of textile fibers has been studied. Textile fibers are destroyed by mechanical methods, such as shredding, and their destruction is followed by a chemical dissolution process. Hazardous solvents are often used; the polymer is retained while the fibers are regenerated and spun [119]. In addition, the dyes need to be removed from the textile waste, and they are removed using harmful chemicals such as bleach. Ionic liquid solvents have recently been used to reintroduce dyed post-consumer textile waste into a new life cycle and to reduce environmental effects [120]. Furthermore, the conversion of cotton to viscose is one of the most well-known examples of chemical textile recycling. Pure cotton fabric is depolymerized into a pulp, which is transformed to viscose. The production process is similar to that made from wood pulp. Industry leaders such as Lenzing and Birla Cellulose adapted the conversion of cotton to viscose technology, with the beginning of the process being the depolymerization of 100% cotton textile materials. Nevertheless, the yield of the synthesized polymer chains is lower than that of the pure wood pulp process due to decreased physical properties. Therefore the regenerated fibers must be combined with virgin viscose fibers to achieve a high yield [121].

Nylon 6 is a polyamide made from the seven-membered ring caprolactam, making it a valuable candidate for CRM. Since the early 1960s, researchers have been studying the depolymerization of nylon 6 back to a monomer, and CRM has been used for nylon 6 for decades [122]. The reactor is conducted by superheated steam and a catalyst to form a caprolactam distillate. Nylon 6 is made from crude caprolactam that has been distilled and depolymerized. In terms of purity, the caprolactam obtained is similar to virgin caprolactam. The repolymerized nylon 6 is spun into yarn and is tufted into carpet. The physical properties of the carpets produced using this technology are fairly similar to those of virgin caprolactam carpets [123–125]. In addition, textile waste can be used to produce geotextiles (thick ropes), which can be used to protect slopes from erosion and sliding. Natural fibers are dissolved with mixtures of 5% sodium hypochlorite (NaClO) and Schweizer's reagent for cellulosic fibers as solvents. Then, the geotextiles are installed in conditions where the ground is sloped and where there is vegetation—this method reduces soil movement and pedestalling. [126,127]. Another application for textile waste that has been processed by chemical recycling is for use as packaging. The cellulose nanofiber textile waste has been used to enhance the strength performance of biopolymers. The combination of polylactic acid and cellulose nano-fibrillated fiber (CNF) for packaging applications has shown excellent strength reinforcement ability, tensile modulus, and elongation [128,129].

The chemical recycling of textiles usually necessitates integrated material flow and energy within the chemical industry. Nevertheless, some sorted waste products from municipal solid waste have more potential for small-scale feedstock production [130]. Figure 5 shows the possible pathways to recycling textile using chemical methods. Furthermore, Yousef et al. studied a long-term approach for recovering cotton from textile waste and regenerating it to become a new substance. The primary processes used to recover the cotton were leaching nitric acid, dissolving dimethyl sulfoxide, and diluting hydrochloric acid for bleaching. This recovery rate, carbon footprint, and economic performance of this technology were determined to be 93%, -1534 CO_2 eq/ton, and 1466 USD/ton, respectively [73] if textile waste is used as a renewable resource for the production of thermal energies. Nunes et al. showed that cotton briquettes had a heating value of 16.80 MJ/kg and a cost of 0.006 EUR/kWh when used as a fuel [131].

Table 3. The drawbacks, advantages, and disadvantages of chemically recycling textiles.

Type of Technology	Advantages	Disadvantages	Ref.
Pyrolysis	<ul><li>Simple process</li><li>Many different kinds of raw materials can be used for the pyrolysis process.</li></ul>	<ul><li>High-temperature reaction process</li><li>High energy consumption</li></ul>	[132,133]
Enzymatic Hydrolysis	<ul><li>The novel microorganism enhances the production of biopolymers and conversion rate. It can effectively be spun into novel waste textile fibers</li><li>Under mild conditions</li><li>Low energy demand</li><li>Utilizes environmentally friendly solvents and chemicals</li></ul>	<ul><li>Only possible to recycle certain materials using enzymatic hydrolysis such as rayon, cotton, hemp, etc.</li><li>Needs large amounts of water</li></ul>	[120]
Hydrothermal	<ul><li>Low ash</li><li>High heating values</li><li>Relatively low temperatures than gasification and pyrolysis processes</li><li>Reduced oxygen content</li></ul>	<ul><li>Long reaction time</li><li>Lower purity and heterogeneity</li></ul>	[134–136]
Gasification	<ul><li>Potentially mixed textile waste</li></ul>	<ul><li>High energy consumption</li><li>High temperature is needed</li></ul>	[137]
Glycolysis	<ul><li>Low energy consumption</li></ul>	<ul><li>Low selectivity</li><li>Without a catalyst, slow processing</li></ul>	[138]
Ammonolysis	<ul><li>Various types of textile waste can be applied</li></ul>	<ul><li>Generates a mixture of primary, secondary amines</li><li>Applied toxic solvent (ammonia)</li><li>High pressure and temperature</li></ul>	[139]

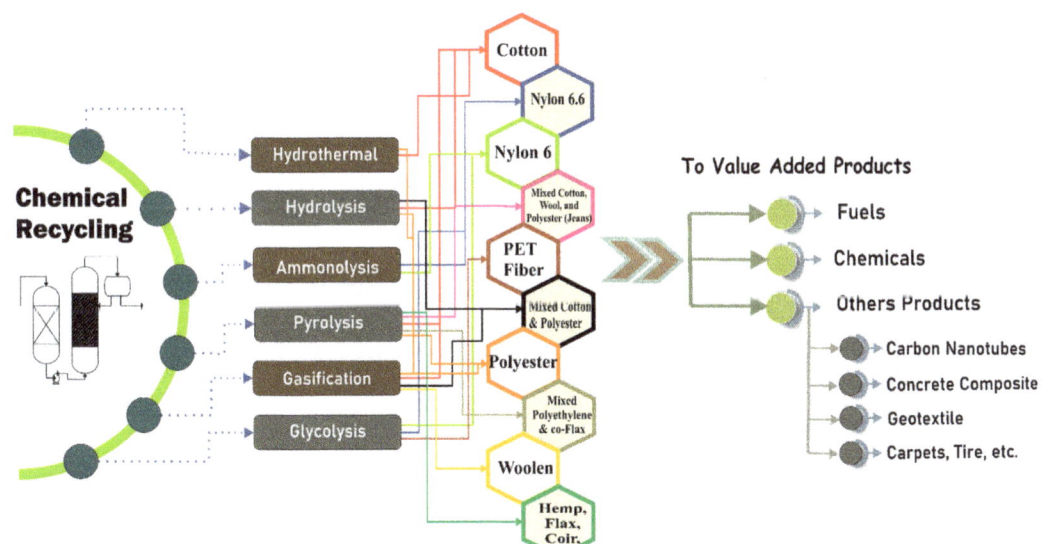

Figure 5. The possible pathways to recycling textiles using chemical methods.

6.1. Textile Recycling Using Pyrolysis

Textile recycling using pyrolysis is a promising technology that can be used to degrade the carbon-polymers of solid waste in order for them to become three pyrolysis products in the solid, liquid, and gas states [140–142]. Pyrolysis could be utilized in diverse textile materials that have not been sorted prior or in multi-material textile products that otherwise would have only been treated in waste-to-energy facilities [143]. Under oxygen-free circumstances, pyrolysis reallocates C/H/O elements from organic molecule waste into three-phase pyrolysates. Syngas (a combination of H_2 and CO) produced by thermochemical processes was utilized as a direct fuel or as a raw material to produce other hydrocarbons and alcohols [144–146].

The pyrolysis process does not require pretreatment, and it is a potential method for treating polluted waste. These factors make pyrolysis a more efficient process than chemical methods (including bio-chemicals), which require many chemicals and frequently produce a lot of trash that must be re-landfilled and takes longer to operate on a large scale [147–149]. Table 4 shows a summary of the diversity yields of the solid, liquid, and gas products produced via textile waste pyrolysis. Commonly, the liquid products produced from textile waste are mono-polyaromatic and oxygen compounds containing hydrocarbons, such as alcohols, aldehydes, ketones, and carboxylic acids [150]. The production of oil and tar formation from the pyrolysis process depends on raw materials and operation conditions [151].

The aromatic oxygen and hydrocarbon compounds were found at 500 °C. Furthermore, when the temperature reaction was as high as 600 °C, alkylphenols were produced, and at a temperature greater than 800 °C, the yield of the oxygen compounds began to decrease slowly. The condensable compounds at 800 °C were predominantly aromatics with no substituent groups, benzene, or naphthalene. On the other hand, if the temperature reaction was about 850 °C, the aromatic hydrocarbon compounds of 3 and 4 rings were produced [152].

Char derived via textile waste pyrolysis can be applied as a primary filler or a hybrid filler (carbon nanotubes/nanoball (CNTs/CB)) or as graphene oxide (GO/CB) in concrete composite applications. To synthesize CB of 10–20 nm, the process is conducted in a pyrolysis reactor followed by milling and chemical treatments to modify the cotton textile waste (CTW). Then, synthetic CB (0.05 wt%) and hybrid fillers (CNTs/CB and graphene/CB: 50/50) are applied to improve the cement paste characteristics [153]. On the other hand, Jagdale et al. studied applications of recycled and reused carbon produced from cotton waste by the pyrolysis process in active electrode battery materials. The cotton-based carbon fiber electrode demonstrated outstanding cycling behavior and provided a high discharge capacity, with a voltage range of 0.02–1.2 V [154].

Table 4. Yields of solid, liquid, and gas products produced via textile waste pyrolysis.

Raw Materials	Reactors	Catalyst	Reaction Time, Min	Temperature, °C	Yield (%) Solid	Yield (%) Liquid	Yield (%) Gas	Major Products	Ref.
Egyptian banknote ELCBs Cotton, 100%	Batch	NA	30	700	15.65	29.28	55.07	2-Propanone, toluene, 3-furaldehyde, 2-furalmethanol, benzaldehyde, phenol, acetophenone	[147]
Hemp, 100%	Batch	NA	120	750	24	48.5	27.50	Activated carbon	[155]
Flax, 100%	Fixed bed	$ZnCl_2$	220	450	44.80	NA	NA	Activated carbon	[156]

Table 4. Cont.

Raw Materials	Reactors	Catalyst	Reaction Time, Min	Temperature, °C	Yield (%)			Major Products	Ref.
					Solid	Liquid	Gas		
Egyptian banknote ELCBs Cotton, 100%	Batch	NA	30	600	17.26	30.28	52.46	1,3-Dioxolane-2-propanal, 2-methyl, furfural, 2-furanmethanol, 4,4-ethylenedioxy-1-pentylamine	[147]
Flax, 100%	Fixed bed	$ZnCl_2$	220	450	44.80	NA	NA	Activated carbon	[156]
Cotton, 70% Polyester, 30%	Batch	Dye-Originating Heavy Metals	35	500–700	17.79	37.59	44.62	Activated carbon	[147]
Egyptian banknote ELCBs Cotton, 100%	Batch	NA	30	500	20.74	39.75	39.51	Toluene, furfural, 2-furanmethanol, 2-furancarboxaldehyde, 5-methyl	[147]
Hemp, 100%	Fixed bed	$ZnCl_2$	220	450	41.60	NA	NA	Activated carbon	[156]
Cotton, 100%	Fixed bed	Na_2CO_3	120	600	16.25	29.49	54.26	Furans, ketones	[130]
Coir, 100%	Fixed bed	NA	240	800	37.40	47.40	18.20	Activated carbon	[157]
Abaca, 100%	Fixed bed	NA	240	800	28.60	48.10	23.60	Activated carbon	[157]
Cotton, 100%	Batch	NA	70	700	12.50	74.00	13.50	Double bond carboxyl and carbonyl liquid	[143]
Low Grade biomass fiber (Flax, 100%)	Batch	NA	120	900	20	55	25	Activated carbon	[155]
Flax, 100%	Fixed bed	H_3PO_4	220	450	39.20	NA	NA	Activated carbon	[156]
Biomass fiber waste Jute, 100%	Fixed bed	NA	240	800	24.60	59.60	15.90	Activated carbon	[157]

Furthermore, pyrolytic char derived from cotton waste can also be utilized as an adsorbent. A one-step low-temperature pyrolysis process produced char-based adsorbents made from CTW with different iron salts compounds. The Freundlich model fit the adsorption processes well when measuring Cr(VI) the adsorption performance of the chars, and char-$FeCl_3$ had the highest adsorptive capacity of 70.39 mg/g. Consequently, this low-temperature pyrolysis method was cost-effective and straightforward to implement, with a high adsorption capacity for Cr(VI) removal [158]. The catalytic pyrolysis of nylon 6 waste fishing nets (WFNs) over a ZSM-5 zeolite catalyst was investigated by Eimontas et al. The activation energy using ZSM-5/WFNs was larger than 112 kJ/mol via the free kinetic analysis method. The major volatile organic compounds were alkyl C–H, carbonyl (C=O), and caprolactam [159].

Silva et al. studied colored cotton wastes to generate renewable aromatic hydrocarbons by the catalytic reforming of ash pyrolysis utilizing a pyrolysis raw material (colored cotton wastes). These materials have high energetic compactness, which is good for use thermochemical processes. Light oxygenated carbon chains with up to four carbons resulted from the ash pyrolysis. The partial deoxygenation and cracking reactions were increased by pyrolysis vapor reformation at 500 °C, which could be improved by catalytic reforming [160].

The catalytic pyrolysis over Ultrastable Y (USY) zeolite was investigated by Wang et al. This method produces benzene, toluene, and xylene (BTX) via Diels–Alder reactions by means of catalytic co-pyrolysis with specified plastic wastes and furanic substituents via the catalytic deconstruction of textile waste. The selectivity of co-pyrolysis to BTX for polyethylene and 20% co-flax waste was 81.6% and 80%, respectively [161]. In addition, the primary components of dryer lint were produced when the drying garments were cotton and polyester. The range of the activation energy (Ea) of lint pyrolysis was 167–204 kJ/mol, and the primary products for a heating rate of 25 °C/min were furan, 3-methyl- (4.78%), furfural (6.48%), isobutane (10.77%), and carbon dioxide (11.68%) [162].

6.2. Textile Recycling Using Enzymatic Hydrolysis

Hydrolysis is a chemical breakdown process that occurs due to a reaction with water. Generally, the hydrolysis process requires a pretreatment process using acids, alkaline, and ionic liquid. Pretreatment is crucial in hydrolysis because it can affect the yield percentage [163]. Damayanti and Wu have already described recycled PET hydrolysis [109]. Before the recycling pf PET fibers from textile waste takes place, the pretreatment process is required. Pretreatment methods reduce the complexity of the structures of the textile materials by eliminating unwanted contaminants and improving hydrophilicity. PET fibers can be converted into raw materials through the following processes or reactions [164]:

The pretreatment process for textile waste materials that uses bases such as sodium, potassium, calcium, and ammonium hydroxides is known as alkali pretreatment. Alkali pretreatment will improve lignin solubilization and decreases cellulose crystallinity by enhancing cellulose digestibility. As a result, alkali pretreatment produces a high glucose yield. The reduced formation of fermentation inhibitors is a primary benefit of alkali pretreatment [165,166]. Furthermore, pretreatment using ionic liquids is a promising method due to the use of molecules that are more environmentally friendly that can dissolve cellulose at moderate temperatures without degrading the cellulose or the solvent. Solvent recovery can be applied through a thermal treatment with lower pressure [167]. The ionic liquid from organic solvents such as N-methylmorpholine-N-oxide (NMMO) and 1-allyl- 3-methyl-imidazolium chloride ([AMIM]Cl) was effective in pretreating cotton waste under enzymatic hydrolysis [168].

On the other hand, acid pretreatments such as sulfuric and phosphoric acid have been examined by researchers, even in the presence of cellulose solvents [169,170]. By breaking the polymeric structures in hemicellulose, acid pretreatments can hydrolyze the polymeric structures into monomers by increasing the availability of cellulose, hence boosting biodegradability [171]. There are some challenges such as the formation of side products (furfural), the cost of acid being relatively expensive, and the requirement of corrosion-resistant equipment [172].

Cotton waste from the textile industry has been discovered to be a viable feedstock for cellulosic ethanol production. Cotton waste can pre-treated and acid hydrolyzed to convert cellulose to reducing sugars. These findings are promising for ethanol production [173]. Table 5 lists the diverse range of textiles that can be used to generate glucose via hydrolysis. Furthermore, Pd-Au/SiO$_2$ bimetallic catalysts are effective in oxidizing glucose generated from cotton waste. These experiments were conducted under the correct temperature and pressure operating conditions for acidic hydrolysis [174]. The recovery of polyester and sugar compounds from textile waste via enzymatic hydrolysis was investigated by Li et al. The temperature of the pretreatment process was −20 °C for 6 h, with the concentration of NaOH/urea of 7%/12%. The glucose recovery was 98.3%, with a cellulase dosage of 20 FPU/g [47].

Table 5. Conditions for converting textile fiber to glucose via enzymatic hydrolysis.

Textile Materials	Type of Pretreatment	Condition of Pretreatment		Enzyme	Condition of Hydrolysis		Glucose Yield, %	Ref.
		T (°C)	T (h)		T (°C)	T (h)		
Cotton, 90% Wool, 5% Polyester, 5%	Protease (Enzymatic)	50	48	Cellic CTec3® and Savinase 12T®	50	70	95.0	[175]
Used Jeans	Phosphoric acid (Acid)	50	7	Cellulase *Trichoderma reesei* and *Aspergillus niger*	50	96	79.2	[176]
Cotton red T-shirt, 100% cotton black T-shirt, 100%	N-methyl-morpholine-N-oxide& Phosphoric Acid (Acid)	50	1	Cellulase AP3, *Aspergillus niger*	50	72	87.0–95.0	[177]
Cotton red T-shirt, 100% cotton black T-shirt, 100%	NaOH/Urea (Base)	50	1	Cellulase AP3, *Aspergillus niger*	50	72	48.0–55.0	[177]
Cotton T-shirts, 100%	1-allyl-3-methylimidazolium chloride (Ionic liquid)	110	1.5	*G. xylinus* (*Acetobacter aceti* subsp. *xylinus* or *A. xylinus*)	50	24	81.6	[168]
Towels, cellulose content 87.8%	Untreated Pretreatment	NA	NA	Cellulase	200	0.03	74.2	[169]
Waste blue jeans (polyester/cotton), 40%/60%	Sodium carbonate (Base)	150	2	Celluclast 1.5 L and β-glucosidase	45	72	81.71	[170]
Cotton, 100%	Sodium carbonate (Base)	150	2	Celluclast 1.5 L and β-glucosidase	45	72	88.0	[170]
Textile from End-of-life euro banknotes	NaOH/Urea (Base)	−20	6	Cellulase	50	382	96.0	[178]

Cellulases act as catalysts in hydrolysis reactions. In theory, these enzymes are unaffected by the reaction equilibrium and can be used again and again. Unfortunately, because these enzymes are massive organic compounds, some decomposition processes that occur over time must be considered [179]. Furthermore, these enzymes can break down material structures to change their physicochemical properties and biodistribution [180–182]. Enzymatic reactions involving various selections of materials can be initiated by certain enzymes [180,183]. As a result, the enzymatic breakdown of certain portions of these materials may be selective [184]. A commercial cellulase mixture typically contains three types of enzymes, each of which performs a different role in the catalytic process: (i) *Endoglucanases* degrade the cellulose chain at varying stages along the chain by increasing the number of accessible end parts, (ii) *Exoglucanases* degrade *cellobiose* units from both ends of the chain, and (iii) the depolymerization of the *disaccharide cellobiose* into monosaccharide units, e.g., glucose is carried out by beta-glucosidases. These three types of enzymes have generated excellent degradation performance in previously reported experiments [185].

Fermentation is an enzymatic degradation procedure for recovering textile waste such as cotton, polyester, nylon, and silk. For bioethanol production, five fermentation techniques can be implemented to degrade textile polymers: (i) Simultaneous saccharification and fermentation (SSF) converts sugars to ethanol [186]. Many researchers have used engineered microbes and have cultured them in biomass sugar solutions, such as glucose, xylose, mannose, galactose, and arabinose. The most commonly utilized microbes for ethanol production include *Saccharomyces cerevisiae*, *Zymomonas mobilis*, and *Aspergillus niger* [187]; (ii) separated hydrolysis fermentation (SHF) is a process where hydrolysis and fermentation are separated to produce bioethanol. SHF is a typical approach in which hydrolysis takes place first, followed by the fermentation process. This method allows the hydrolysis process to generate monosaccharide sugar initially, ensuring that sugar is available when fermentation starts [186,188,189]; (iii) semi-simultaneous saccharification and fermentation (SSnF): the excess sugar that accumulates during hydrolysis reduces enzyme activity, slowing down the process [186,190,191]. The short pre-hydrolytic method is used before the SSF process in SSnF. In addition, the yield of bioethanol production is be slightly higher than it would be if traditional SSF procedures were used [191]; (iv) consolidated bioprocessing (CBP) is required to degrade refractory biomass materials into solubilized sugars. Because CBP combines three processes (enzyme synthesis, saccharification, and fermentation), it is a viable technique for successful biofuel production. This technique can lower the cost of the reactor and enzymes, two main roadblocks to low-cost biomass processing [192,193]; (v) *submerged fermentation (SMF)* is one of the fermentation techniques that focuses on enhancing cellulase production. It is a widely accepted fermentation process for the synthesis of industrial enzymes because it is easy to regulate all of the factors such as pH, temperature, and operational approaches [186,187].

Table 6 lists ethanol production from various textile wastes using enzymatic hydrolysis and microbe fermentation. Jeihanipour and Taherzadeh studied ethanol production from the cotton from waste fabric. The alkali pretreatment was conducted in the concentration of NaOH (12 wt%) at 0 °C for 3 h, which was then fermented to ethanol via *Saccharomyces cerevisiae*, with an ethanol yield up to 99.1% [194]. Sanchis-Sebastiá et al. investigated green liquor as an alkaline solution in the processing of textiles to connect textile recycling with pulp mills. The textile recycling efficiency was predicted be 70% when using *Cellic CTec 2* as a hydrolysis enzyme. Nevertheless, the sodium reduction in the pulping process would be less than 15% [195]. On the other hand, Sebastia et al. studied the cotton fiber recycling process when it was conducted at 130 °C via acid hydrolysis. Sulfuric acid could depolymerize the structure of cotton to obtain glucose. The yield was over 90% when a two-step method was used [196].

Table 6. Ethanol production after enzymatic hydrolysis and microorganism fermentation of textile wastes.

Textile l	Type of Pretreatment	Hydrolysis		Enzyme for Hydrolysis	Fermentation Process		Microorganism	Type of Fermentation	EthanolYield, %	Ref.
		T (°C)	T (h)		T (°C)	T (h)				
Polyester/cotton (50%/50%)	N-methyl morpholine-N-oxide (Cellulose solvent)	NA	48	Cellulase and β-glucosidase	30	24	*Saccharomyces cerevisiae*	SHF	91.91	[197]
Polyester/cotton (40%/60%)	NaOH/Urea (Alkali)	45	72	30 FPU C ellulase and 60 IU β-glucosidase	36	72	*Saccharomyces cerevisiae*	SSF	70.00	[163]
Cotton, 100%	$Na_2S_2O_4$ and Na_2CO_3 (Alkali)	50	48	Cellulase AP3	37	48	*Zymomonas mobilis*	SSF	90.00	[198]
Viscose/polyester (60%/40%)	N-methylmorpholine-N-oxide (Cellulose solvent)	NA	48	Cellulase and β-glucosidase	30	24	*Saccharomyces cerevisiae*	SHF	94.99	[197]
Cotton, 100%	Na_2CO_3 (Alkali)	45	72	Cellulase and β-glucosidase	32	24	*Saccharomyces cerevisiae CCUG 53310*	NA	69.40	[170]
Cotton ginning trash, 100%	Sulfuric acid (Acid)	50	96	Cellic CTec 2 cellulase	30	NA	*Saccharomyces cerevisiae*	SSF	70.00	[199]

SHF: separated hydrolysis dermentation, SSF: simultaneous saccharification and fermentation, NA: not available.

6.3. Textile Recycling Using the Hydrothermal Method

The hydrothermal method is a decomposition process using chemical crystallization engineering techniques at high temperatures and pressures, with water as the main constituent of the reaction. The hydrothermal method is one of the most promising alternative methods for degrading carbon-polymer waste and organic components into three product phases: liquid, solid, and gas phases. It is conducted in an autoclave reactor in which organic acids can catalyze [200]. The hydrothermal process does not require pretreatment and is a method that utilizes water and can be classified into five groups based on the temperature range that is used: (i) hot water extraction, (ii) pressurized hot water extraction, (iii) hot liquid water treatment, (iv) hydrothermal carbonization, and (v) hydrothermal liquefaction. The method requires temperatures that are less than 280 °C [201]. Table 7 lists a summary of hydrothermal products produced via textile recycling. The hydrothermal conditions control the powder formation and particle size. The drawbacks of the hydrothermal process are the high temperatures, high pressure, and long reaction time that are required for the material manufacturing process [202].

Hongthong et al. studied the conversion of nylon 6 fishing net waste in the hydrothermal liquefaction of the macroalgae *Fucus serratus*. The hydrothermal liquefaction of macroalgae was performed for 10 min with a temperature reaction of up to 350 °C. The scope of nylon polymers included nylon 6 and nylon 6,6. A bio-crude yield of up to 17 wt% could be with a combination of 50% nylon 6 and *F serratus*. Furthermore, nylon 6 fully degraded in the process that produced the molecule caprolactam [203]. On the other hand, Xu et al. determined that CTW was catalyzed using surfactants during hydrothermal carbonization. The application of Span 80 and sodium dodecyl benzenesulfonate enhanced the conversion of CTW into bio-oil. Nevertheless, Span 80 was better for the formation of pseudo-lignin, which gave the produced hydro-chars a higher energy density and improved their fuel quality and combustion performance. The majority of bio-oil yields are aromatic chemicals, acids, hydrocarbons, ketones, and esters, with yields of up to 12.79%, 11.3%, 25.4%, 23.09%, and 14.88%, respectively. In addition, cyclopentenone substances (3-methyl-2-hydroxy-2-cyclopentene-1-one and 3-ethyl-2-hydroxy-2-cyclopentene-1-one) were also found [135]. H_2SO_4 successfully degraded CTW through hydrolysis [204].

Table 7. Products and conditions of textile recycling using hydrothermal methods.

Feedstock	Reactor	Catalyst	Time, h	Temperature, °C	Major Product	Yield, %	Ref.
A blue cotton/polyester,	Batch	Hydrochloric acid	3	150	Cellulose powder	49.3	[205]
Waste cotton fabrics	Batch	Hydrochloric acid	1.6	150	Microcrystalline cellulose	85.5	[206]
Cotton/synthetic fibers,	Autoclave	NA	90	280	Volatile compound (CH_4, C_2H_4)	98.0	[207]
Waste cottons	Hydrothermal reactor	β-cyclodextrin	3	250	Activated carbon	NA	[208]
Cotton textile waste	Tubular	Ferric chloride	1	700	Nanopowder	32.6	[209]
Waste cotton woven	Tubular	Ferric chloride	1.5	700	Activated carbon	NA	[210]
Light blue and white uniform with polyester/cotton,	Hydrothermal reactor	Citric acid	1	225	5'-hydroxymethylfurfural	12.5	[211]
Cotton	Semi-batch	Formic acid	1	250	Glucose	83.8	[212]

When recycling cotton-polyester textile waste with an organic acid catalyst, the hydrothermal process includes two steps: (i) splitting the mixture of polyester–cotton fiber waste into pieces and (ii) dispersing the mixture an organic acid catalyst in an aqueous phase to produce products. The reaction temperature was up to 140 °C under high pressure. The recycling yield of the obtained polyester fiber aggregate was 99%, and the recycling yield of the cotton fiber fragments was 81% [213]. The cotton was dissolved easily in a citric acid solution, and it was degraded in a strong acid solution. On the other hand, the cotton was unaffected by strong alkaline solutions [211]. Qi et al. used $FeCl_3$ as a catalyst to increase cotton textile waste conversion to hydrochars, achieving highly efficient hydrophobicity by lowering the hydrothermal carbonization temperature. In addition, the simultaneous impacts of $FeCl_3$ and hydrothermal carbonization produced a side reaction pathway by allowing more furfural compounds as derivatives [134].

6.4. Textile Recycling Using Ammonolysis

The ammonolysis process is the primary depolymerization method of nylon 6,6 carpet waste. Mckinney studied the reactions of nylon 6,6 and nylon 6,6/nylon 6 mixtures with ammonia or ammonium phosphate as a catalyst. The temperature and pressure of this process were 300–350 °C and 68 atm, respectively. The major products resulting from the depolymerization of nylon 6,6 were 5-cyanovaleramide, hexamethylenediamine, and adiponitrile. Furthermore, 6-aminocapronitrile, ε-caprolactam, and 6-aminocaproamide were the primary products of nylon 6 [214]. The depolymerization mechanism of the nylon 6,6 and nylon 6 mixture was described by Kalfas et al. The amine chain would have broken down, which would then be followed by the dehydration process and the ring addition and ring-opening reactions for the presence of cyclic lactams in nylon 6 [139].

6.5. Textile Recycling Using Gasification

Textile recycling by means of gasification processes is conducted under temperatures up to 400–1000 °C. The reactions occur under low-oxygen or oxygen-free conditions [152,215]. Compared to pyrolysis, gasification is more complicated since chemical reactions occur on the substance [216]. The gasification approach generates gases with high carbon and hydrogen concentrations, similar to the pyrolysis process. Nevertheless, the gaseous fraction in gasification is dramatically higher than that in pyrolysis [216]. The increasing pressure can lead to higher yield, calorific value, CO, and H_2 concentrations when using gasification [217]. In contrast, the syngas output increases when the operating temperatures rise, whereas the char yield lowers [218]. Gasifiers can be clarified based on the reactor bed and flow type. Fixed bed and fluidized bed gasifiers are the most common gasification process. On the contrary, the spouted bed gasifier is a unique case of a fluidization type. Low gas flow rates can manage coarser solid particles, although segregation is hampered by a unique hydraulic structure [219].

Synthesis gas, often known as syngas, is the desired end product in gasification, and includes compounds such as hydrogen, carbon dioxide, carbon monoxide, and methane (CH_4). The by-products of ethylene and ash can be formed [151]. Vela et al. studied the steam gasification of mixed textile waste using a fluidized bed bench-scale reactor at the reaction temperature up to 850 °C. The yields of CH_4 with pure polyester, cotton, and a mixed (50% polyester and 50% cotton) were 7.0%, 3.4%, and 10.7%, respectively [220]. Furthermore, the textile wastes that were used were cotton, wool, and polyester fibers with activation energies of 144.49, 79.37, and 178.15 kJ mol^{-1}, respectively. Char gasification took place in the temperature range of 800–1000 °C, with the primary product being CO_2 [221]. On the other hand, textile waste activation energies (cotton, woolen, polyester fiber) up to 89.0 kJ/mol were achieved with a 5 wt% Fe_2O_3 catalyst [222].

6.6. Textile Recycling Using Glycolysis

Glycolysis is a degradation process that can be used to convert the large molecules of a textile to small molecules. This process is widely used to recycle textile materials such as PET fiber and polyurethane. PET fiber is the most widely used textile fabric in the textile industry due to its outstanding mechanical characteristics and inexpensiveness. Consequently, the generation of PET waste is also very large [223,224]. The large volume of PET fiber waste has wreaked havoc on the ecosystem. It is essential to find a sustainable and economical technique for recycling PET fiber waste to protect our resources and the environment. The glycolysis process has many advantages: it has a short reaction time and uses minimal energy. Nevertheless, PET fiber of glycolysis is slowly processed without a catalyst [225,226].

For degradation without a catalyst, it is challenging to glycolyze PET waste. A lot of effort focuses on developing a high-performance and eco-friendly catalyst to degrade the macromolecules of textile fibers to bis(2-hydroxyethyl) terephthalate (BHET) monomers and the other chemicals [227]. The recycling of polyester textiles with Mg-Al double oxides as a catalyst was studied by Guo et al. The yield of BHET was over 82 mol% using a Mg-Al double oxide catalyst with wet mixing or co-precipitation. Furthermore, the recycled PET (r-PET) fibers that were formed from the re-polymerization process had similar spinnability and mechanical characteristics to virgin PET fibers [228].

The depolymerization process of nylon 6 via the glycolysis process was investigated by Hommez et al. [229]. The glycolysis process was conducted at 250 °C in the presence of phosphoric acid. The free carboxylic acid and carboxylic acid were esterified with ethylene glycol; the major products were caprolactam, N-(5-hydroxy-3-oxa-pentyl)-caprolactam, N,N'-ethylene-di(caprolactam), and N-(2-hydroxyethyl)-caprolactam, and linear oligomers were reported. Huczkowski et al. investigated the glycolysis of nylon 6 in boiling ethylene glycol with and without a catalyst, resulting in oligoamides containing amino- and hydroxyl end-groups [230,231]. Holland and Hay presented the thermal degradation of nylon 6 and nylon 6,6 to produce a crosslinking formation and non-volatile char [232]. Kim et al. studied the glycolysis of nylon 6,6 with ethylene glycol to obtain major products from the bis(β-hydroxyethyl)adipate and β-hydroxyethylester group [233].

The glycolysis of polyester fibers is traditionally conducted in a boiling ethylene glycol solution under atmospheric pressure with several metal catalysts, e.g., zinc acetate $(Zn(Ac)_2)$. Zinc acetate can be used as a catalyst in the glycolysis process. The operation conditions for this process are 1 h and 96 °C. The PET fiber conversion and monomer yield of BHET achieved 100% and 80%, respectively. The oligomers were broken down quickly into dimers, which caused an instability process, resulting in a low molecular weight [234]. On the other hand, the glycolysis process of PET fibers to produce BHET with acetic acid, lithium hydroxide, sodium sulfate, and potassium sulfate resulted in yields of up to 64.42, 63.50, 65.72, and 64.42%, respectively [235].

6.7. Decolorization Technology in Textile Recycling

Decolorization technology is a critical problem in the high-quality chemical recycling and recovery of textile wastes. Only 1% of textile wastes, mainly white-colored wastes, are recycled, so the final color quality of regenerated fibers is uncontrollable. Color removal is required for the large-scale circulation of non-thermoplastic fibers Technologies for color removal from textile wastes are dye destruction or extraction, which take place during the pre-recycling process. Dye destruction methods, such as oxidation and photodegradation, may damage polymers and change the dyeability of the regenerated textiles [236]. To lower the chemical potential of the dyes in solution, harsh extraction conditions such as ultra-high temperature (>150 °C), lengthy time, and the use of excessive corrosive solutions can lead to structural damages to the polymers. Dye extraction focuses on finding solvents with high dye solubility [237], but dye extraction fails to remove all of the dye from textiles. Mu and Yang demonstrated the minimization of fiber density by solvents and temperatures, completely removing dispersed dyes, acid dyes, and direct dyes from PET, nylon, and cotton fibers, respectively [238].

When color removal takes place after textile degradation, the final colors are not predictable due to the uncontrollable mixture of colors in the collected textiles, for example, when cotton with multiple shades is dissolved in ionic liquids to spin regenerated cellulose fibers [239]. Li et al. used nitric acid-modified activated carbon (AC-HNO$_3$) as an adsorbent for the removal of C.I. Disperse Red 60 (DR60) from the glycolysis products of waste PET fabrics and for the rapid decolorization of textile dyes using ceria and tin-doped ZnO nanoparticles [240,241]. Huang et al. presented ion-exchange resin (D201) to efficiently remove colorants after PET glycolysis [226]. The removal rate of the colorant and the retention efficiency of BHET was over 99% and 95%, respectively. Huang et al. presented a modified recrystallization process using ethyl acetate as the solvent needed to remove colored impurities from BHET. The process showed a decoloring rate of over 97.5% for the model-colored impurities, performing better than the reported physical adsorption processes [242].

7. Conclusions

The fashion industry faces some challenges in implementing a CE; it must consider environmental laws, policies, and the CE's long-term viability. Digital innovation and industry 4.0 have significantly impacted CE transformation with data-driven product lifecycle analysis. By implementing IoT and RFID technology, physical production processes, such as material transportation, and the associated information flows, such as the visibility and traceability of numerous manufacturing operations, can be easily connected. Furthermore, the recognition rate of textile waste can be achieved with up to 100% accuracy using NIR technology. Mechanical and chemical recycling are common methods that take place during the textile recycling process. Textile materials such as cotton and wool can be converted into useable yarn (re-woven or knitted) by spinning carded yarn and mixed shoddy. In addition, garnering, carding, or air laying webs, which are subsequently mechanically, thermally, or chemically bonded, can be widely used to make non-woven fabrics from these materials. On the other hand, chemical treatment is seen as a promising method for textile recycling. Chemical textile recycling can be applied to degrade textile waste into feedstock monomers. Cotton waste from the textile industry has been identified as a suitable feedstock to produce cellulosic ethanol with a yield of more than 90% by means of enzymatic hydrolysis. It may be worthwhile for future experimental studies to investigate chemical recycling methods that use enzymatic hydrolysis with an ionic liquid to obtain high-value fibers. Nevertheless, the toxicity of the dissolving solvents is still a major concern; the amount of energy and water that are consumed are affected by environmental issues. Furthermore, real-time sensors to identify various types of textile materials are needed.

Author Contributions: Conceptualization, H.-S.W. and D.D.; validation, H.-S.W., D.D. and A.R.; investigation, D.D., L.A.W., A.B. and A.R.; writing—original draft preparation, D.D., L.A.W. and A.B.; writing—review and editing, D.D.; visualization, D.D.; supervision, H.-S.W. All authors have read and agreed to the published version of the manuscript.

Funding: This research received no external funding.

Data Availability Statement: No applicable.

Acknowledgments: We thank the information obtained from the Far Eastern New Century Corporation Research & Development Center, Taiwan.

Conflicts of Interest: The authors declare no conflict of interest.

Abbreviations

AI	Artificial intelligence
BHET	Bis(2-hydroxyethyl) terephthalate
CBP	Consolidated bioprocessing
CE	Circular economy
CIE	C-Illuminant (light source C)
CNN	Convolutional neural network
CTW	Cotton textile waste
Ea	Activation energy
EG	Ethylene Glycol
FCM	Fuzzy C-means
HOG	Histogram of oriented gradients
ICT	Information and communication technology
IoT	Internet of Things
k-NN	k-nearest neighbors algorithm
LCA	Life cycle assessment
MLP	Multi-layer perceptron
NIR	Near-infrared spectroscopy
NMR	Nuclear magnetic resonance
PCA	Principal component analysis
PET	Polyethylene terephthalate
r-PET	Recycled polyethylene terephthalate
RMBs	Recycling business models
SAaaS	Software as a service

SHF	Separated hydrolysis fermentation
SIMCA	Soft independent modeling of class analogy
SMF	Submerged fermentation
SMOs	Smart manufacturing objects
SOM	Self-organizing map
SSF	Simultaneous saccharification and fermentation
SSnF	Semi-simultaneous saccharification and fermentation
SVM	Support vector machine
TILT	Transform invariant low-rank textures
WEEE	Waste electrical and electronic equipment

References

1. Stone, C.; Windsor, F.M.; Munday, M.; Durance, I. Natural or synthetic–how global trends in textile usage threaten freshwater environments. *Sci. Total Environ.* **2020**, *718*, 134689. [CrossRef] [PubMed]
2. Gardetti, M.Á. Sustainability in the textile and fashion industries: Animal ethics and welfare. In *Textiles and Clothing Sustainability*; Muthu, S., Ed.; Springer: Singapore, 2017; pp. 47–73.
3. Chapagain, A.K.; Hoekstra, A.Y.; Savenije, H.H.; Gautam, R. The water footprint of cotton consumption: An assessment of the impact of worldwide consumption of cotton products on the water resources in the cotton producing countries. *Ecol. Econ.* **2006**, *60*, 186–203. [CrossRef]
4. Pfister, S.; Bayer, P.; Koehler, A.; Hellweg, S. Environmental impacts of water use in global crop production: Hotspots and trade-offs with land use. *Environ. Sci. Technol.* **2011**, *45*, 5761–5768. [CrossRef] [PubMed]
5. Navone, L.; Moffitt, K.; Hansen, K.-A.; Blinco, J.; Payne, A.; Speight, R. Closing the textile loop: Enzymatic fibre separation and recycling of wool/polyester fabric blends. *J. Waste Manag.* **2020**, *102*, 149–160. [CrossRef]
6. Mäkelä, M.; Rissanen, M.; Sixta, H. Machine vision estimates the polyester content in recyclable waste textiles. *Resour. Conserv. Recycl.* **2020**, *161*, 105007. [CrossRef]
7. MacArthur, E. *A New Textiles Economy: Redesigning Fashion's Future*; Ellen MacArthur Foundation: Cowes, UK, 2017.
8. Shirvanimoghaddam, K.; Motamed, B.; Ramakrishna, S.; Naebe, M. Death by waste: Fashion and textile circular economy case. *Sci. Total Environ.* **2020**, *718*, 137317. [CrossRef]
9. Payne, A. Open-and closed-loop recycling of textile and apparel products. In *Handbook of Life Cycle Assessment (LCA) of Textiles and Clothing*; Elsevier: Amsterdam, The Netherlands, 2015; pp. 103–123.
10. Heikkilä, P.; Cura, K.; Heikkilä, J.; Hinkka, V.; Ikonen, T.; Kamppuri, T.; Knuutila, H.; Kokko, M.; Lankiniemi, S.; Lehtinen, L. *Telaketju: Towards Circularity of Textiles*; VTT Research Report; No. VTT-R-00062-19; VTT Technical Research Centre of Finland: Espoo, Finland, 2019; 117p.
11. Koszewska, M. Circular Economy—Challenges for the Textile and Clothing Industry. *Autex Res. J.* **2018**, *18*, 337–347. [CrossRef]
12. Park, S.H.; Kim, S.H. Poly (ethylene terephthalate) recycling for high value added textiles. *Fash. Text.* **2014**, *1*, 1–17. [CrossRef]
13. Sandin, G.; Peters, G.M. Environmental impact of textile reuse and recycling–A review. *J. Clean. Prod.* **2018**, *184*, 353–365. [CrossRef]
14. Dahlbo, H.; Aalto, K.; Eskelinen, H.; Salmenperä, H. Increasing textile circulation—consequences and requirements. *Sustain. Prod. Consum.* **2017**, *9*, 44–57. [CrossRef]
15. Muthu, S.S.; Li, Y.; Hu, J.Y.; Ze, L. Carbon footprint reduction in the textile process chain: Recycling of textile materials. *Fibers Polym.* **2012**, *13*, 1065–1070. [CrossRef]
16. Franco, M.A. Circular economy at the micro level: A dynamic view of incumbents' struggles and challenges in the textile industry. *J. Clean. Prod.* **2017**, *168*, 833–845. [CrossRef]
17. Galvão, G.D.A.; de Nadae, J.; Clemente, D.H.; Chinen, G.; de Carvalho, M.M. Circular economy: Overview of barriers. *Procedia CIRP* **2018**, *73*, 79–85. [CrossRef]
18. Hussain, A.; Kamboj, N.; Podgurski, V.; Antonov, M.; Goliandin, D. Circular economy approach to recycling technologies of postconsumer textile waste in Estonia: A review. *Proc. Est. Acad. Sci.* **2021**, *70*, 82–92.
19. Manglani, H.; Hodge, G.L.; Oxenham, W. Application of the internet of things in the textile industry. *Text. Prog.* **2019**, *51*, 225–297. [CrossRef]
20. Das, S.K.; Chinnappan, A.; Jayathilaka, W.A.D.M.; Gosh, R.; Baskar, C.; Ramakrishna, S. Challenges and potential solutions for 100% recycling of medical textiles. *Mater. Circ. Econ.* **2021**, *3*, 1–12. [CrossRef]
21. Pagoropoulos, A.; Pigosso, D.C.; McAloone, T.C. The emergent role of digital technologies in the circular economy: A review. *Procedia CIRP* **2017**, *64*, 19–24. [CrossRef]
22. Zhou, J.; Yao, X.; Zhang, J. Big data in wisdom manufacturing for industry 4.0. In Proceedings of the 2017 5th International Conference on Enterprise Systems (ES), Beijing, China, 22–24 September 2017; pp. 107–112.
23. Fatimah, Y.A.; Govindan, K.; Murniningsih, R.; Setiawan, A. Industry 4.0 based sustainable circular economy approach for smart waste management system to achieve sustainable development goals: A case study of Indonesia. *J. Clean. Prod.* **2020**, *269*, 122263. [CrossRef]

24. Haslinger, S.; Hietala, S.; Hummel, M.; Maunu, S.L.; Sixta, H. Solid-state NMR method for the quantification of cellulose and polyester in textile blends. *Carbohydr. Polym.* **2019**, *207*, 11–16. [CrossRef]
25. Ruckebusch, C.; Orhan, F.; Durand, A.; Boubellouta, T.; Huvenne, J. Quantitative analysis of cotton–polyester textile blends from near-infrared spectra. *Appl. Spectrosc.* **2006**, *60*, 539–544. [CrossRef]
26. Park, S.; Johnson, D.K.; Ishizawa, C.I.; Parilla, P.A.; Davis, M.F. Measuring the crystallinity index of cellulose by solid state 13 C nuclear magnetic resonance. *Cellulose* **2009**, *16*, 641–647. [CrossRef]
27. Rodgers, J.; Beck, K. NIR characterization and measurement of the cotton content of dyed blend fabrics. *Text. Res. J.* **2009**, *79*, 675–686. [CrossRef]
28. Lv, F.; Wang, C.; Zhu, P.; Zhang, C. Isolation and recovery of cellulose from waste nylon/cotton blended fabrics by 1-allyl-3-methylimidazolium chloride. *Carbohydr. Polym.* **2015**, *123*, 424–431. [CrossRef]
29. Spathas, T. The Environmental Performance of High Value Recycling for the Fashion Industry. Master's Thesis, Chalmers University of Technology, Gothenburg, Sweden, 2017.
30. Rosa, P.; Sassanelli, C.; Terzi, S. Towards Circular Business Models: A systematic literature review on classification frameworks and archetypes. *J. Clean. Prod.* **2019**, *236*, 117696. [CrossRef]
31. Winans, K.; Kendall, A.; Deng, H. The history and current applications of the circular economy concept. *Renew. Sust. Energ. Rev.* **2017**, *68*, 825–833. [CrossRef]
32. Geissdoerfer, M.; Savaget, P.; Bocken, N.M.; Hultink, E.J. The Circular Economy–A new sustainability paradigm? *J. Clean. Prod.* **2017**, *143*, 757–768. [CrossRef]
33. Kazancoglu, I.; Kazancoglu, Y.; Yarimoglu, E.; Kahraman, A. A conceptual framework for barriers of circular supply chains for sustainability in the textile industry. *J. Sustain. Dev.* **2020**, *28*, 1477–1492. [CrossRef]
34. Allwood, J.M.; Laursen, S.E.; de Rodriguez, C.M.; Bocken, N.M. Well dressed?: The present and future sustainability of clothing and textiles in the United Kingdom. *J. Home Econ. Inst. Aust.* **2015**, *22*, 42.
35. Manickam, P.; Duraisamy, G. 3Rs and circular economy. In *Circular Economy in Textiles and Apparel*; Elsevier: Amsterdam, The Netherlands, 2019; pp. 77–93.
36. Chen, X.; Memon, H.A.; Wang, Y.; Marriam, I.; Tebyetekerwa, M. Circular Economy and sustainability of the clothing and textile Industry. *Mater. Circ. Econ.* **2021**, *3*, 1–9. [CrossRef]
37. Balanay, R.; Halog, A. Tools for circular economy: Review and some potential applications for the Philippine textile industry. *Circ. Econ. Text. Appar.* **2019**, 49–75. [CrossRef]
38. Fortuna, L.M.; Diyamandoglu, V. Optimization of greenhouse gas emissions in second-hand consumer product recovery through reuse platforms. *J. Waste Manag.* **2017**, *66*, 178–189. [CrossRef] [PubMed]
39. Lüdeke-Freund, F.; Gold, S.; Bocken, N.M. A review and typology of circular economy business model patterns. *J. Ind. Ecol.* **2019**, *23*, 36–61. [CrossRef]
40. Pal, R.; Samie, Y.; Chizaryfard, A. Demystifying process-level scalability challenges in fashion remanufacturing: An interdependence perspective. *J. Clean. Prod.* **2021**, *286*, 125498. [CrossRef]
41. Dissanayake, G.; Sinha, P. An examination of the product development process for fashion remanufacturing. *Resour. Conserv. Recycl.* **2015**, *104*, 94–102. [CrossRef]
42. Niinimäki, K.; Hassi, L. Emerging design strategies in sustainable production and consumption of textiles and clothing. *J. Clean. Prod.* **2011**, *19*, 1876–1883. [CrossRef]
43. Gwilt, A.; Rissanen, T. *Shaping Sustainable Fashion: Changing the Way We Make and Use Clothes*; Routledge: England, UK, 2012.
44. Wen-hui, X.; Dian-yan, J.; Yu-ying, H. The remanufacturing reverse logistics management based on closed-loop supply chain management processes. *Procedia Environ. Sci.* **2011**, *11*, 351–354. [CrossRef]
45. Vogtlander, J.G.; Scheepens, A.E.; Bocken, N.M.; Peck, D. Combined analyses of costs, market value and eco-costs in circular business models: Eco-efficient value creation in remanufacturing. *J. Remanufacturing* **2017**, *7*, 1–17. [CrossRef]
46. De Silva, R.; Wang, X.; Byrne, N. Recycling textiles: The use of ionic liquids in the separation of cotton polyester blends. *RSC Adv.* **2014**, *4*, 29094–29098. [CrossRef]
47. Li, Y.; Hu, Y.; Du, C.; Lin, C.S.K. Recovery of glucose and polyester from textile waste by enzymatic hydrolysis. *Waste Biomass Valorization* **2019**, *10*, 3763–3772. [CrossRef]
48. Kumar, P.S.; Yaashikaa, P. Recycled Fibres. In *Sustainable Innovations in Recycled Textiles*; Springer: Amsterdam, The Netherlands, 2018; pp. 1–7.
49. Yin, Y.; Yao, D.; Wang, C.; Wang, Y. Removal of spandex from nylon/spandex blended fabrics by selective polymer degradation. *Text. Res. J.* **2014**, *84*, 16–27. [CrossRef]
50. Huang, Y.-F.; Azevedo, S.G.; Lin, T.-J.; Cheng, C.-S.; Lin, C.-T. Exploring the decisive barriers to achieve circular economy: Strategies for the textile innovation in Taiwan. *Sustain. Prod. Consum.* **2021**, *27*, 1406–1423. [CrossRef]
51. Martina, R.A.; Oskam, I.F. Practical guidelines for designing recycling, collaborative, and scalable business models: A case study of reusing textile fibers into biocomposite products. *J. Clean. Prod.* **2021**, 128542. [CrossRef]
52. Christensen, T.B. Towards a circular economy in cities: Exploring local modes of governance in the transition towards a circular economy in construction and textile recycling. *J. Clean. Prod.* **2021**, *305*, 127058. [CrossRef]
53. Seadon, J.K. Sustainable waste management systems. *J. Clean. Prod.* **2010**, *18*, 1639–1651. [CrossRef]

54. Payne, A.; Nay, Z.; Maguire, R. Regulating a circular economy for textiles in Australia. In Proceedings of the 4th PLATE 2021 Virtual Conference, Limerick, Ireland, 26–28 May 2021.
55. Ghoreishi, M.; Happonen, A. The case of fabric and textile industry: The emerging role of digitalization, internet-of-Things and industry 4.0 for circularity. In Proceedings of the Sixth International Congress on Information and Communication Technology, London, UK, 25–26 February 2022; pp. 189–200.
56. Abdullah, N.; Alwesabi, O.A.; Abdullah, R. Iot-based smart waste management system in a smart city. In Proceedings of the International Conference of Reliable Information and Communication Technology, Kuala Lumpur, Malaysia, 23–24 June 2018; pp. 364–371.
57. Zhong, R.Y.; Dai, Q.; Qu, T.; Hu, G.; Huang, G.Q. RFID-enabled real-time manufacturing execution system for mass-customization production. *Robot. Comput. Integr. Manuf.* **2013**, *29*, 283–292. [CrossRef]
58. Lu, B.; Bateman, R.; Cheng, K. RFID enabled manufacturing: Fundamentals, methodology and applications. *Int. J. Agil. Syst. Manag.* **2006**, *1*, 73–92. [CrossRef]
59. Zhong, R.Y.; Li, Z.; Pang, L.; Pan, Y.; Qu, T.; Huang, G.Q. RFID-enabled real-time advanced planning and scheduling shell for production decision making. *Int. J. Comput. Integr. Manuf.* **2013**, *26*, 649–662. [CrossRef]
60. Mavropoulos, A.; Nilsen, A.W. *Industry 4.0 and Circular Economy: Towards a Wasteless Future or a Wasteful Planet?* John Wiley & Sons: New Jersey, NJ, USA, 2020.
61. Zhong, R.Y.; Xu, X.; Klotz, E.; Newman, S.T. Intelligent manufacturing in the context of industry 4.0: A review. *Engineering* **2017**, *3*, 616–630. [CrossRef]
62. Kumar, K.; Zindani, D.; Davim, J.P. Intelligent Manufacturing. In *Industry 4.0*; Springer: Amsterdam, The Netherlands, 2019; pp. 1–17.
63. Wu, D.; Greer, M.J.; Rosen, D.W.; Schaefer, D. Cloud manufacturing: Strategic vision and state-of-the-art. *J. Manuf. Syst.* **2013**, *32*, 564–579. [CrossRef]
64. Lopes, S.; Machado, S. IoT based automatic waste segregator. In Proceedings of the 2019 International Conference on Advances in Computing, Communication and Control (ICAC3), Mumbai, India, 20–21 December 2019; pp. 1–5.
65. Yu, K.H.; Zhang, Y.; Li, D.; Montenegro-Marin, C.E.; Kumar, P.M. Environmental planning based on reduce, reuse, recycle and recover using artificial intelligence. *Environ. Impact Assess. Rev.* **2021**, *86*, 106492. [CrossRef]
66. Giacobbe, M.; Puliafito, C.; Scarpa, M. The big bucket: An iot cloud solution for smart waste management in smart cities. In Proceedings of the European Conference on Service-Oriented and Cloud Computing, Vienna, Austria, 5–7 September 2016; pp. 43–58.
67. Merlino, G.; Bruneo, D.; Distefano, S.; Longo, F.; Puliafito, A. Stack4things: Integrating iot with openstack in a smart city context. In Proceedings of the 2014 International Conference on Smart Computing Workshops, Hongkong, China, 5 November 2014; pp. 21–28.
68. Marin Perez, M. Analysis of European Post-Consumer Textile Waste for Automated Sorting. Master's Thesis, Uppsala University, Stockholm, Sweden, 2021.
69. Bergfjord, C.; Holst, B. A procedure for identifying textile bast fibres using microscopy: Flax, nettle/ramie, hemp and jute. *Ultramicroscopy* **2010**, *110*, 1192–1197. [CrossRef]
70. Nørup, N.; Pihl, K.; Damgaard, A.; Scheutz, C. Development and testing of a sorting and quality assessment method for textile waste. *J. Waste Manag.* **2018**, *79*, 8–21. [CrossRef] [PubMed]
71. Choudhuri, P.K.; Majumdar, P.K.; Sarkar, B. Thermal behaviour of textiles: A review. *Man-Made Text. India.* **2013**, *41*, 3.
72. Sombatsompop, N.; Wood, A. Measurement of thermal conductivity of polymers using an improved Lee's disc apparatus. *Polym. Test.* **1997**, *16*, 203–223. [CrossRef]
73. Yousef, S.; Tatariants, M.; Tichonovas, M.; Sarwar, Z.; Jonuškienė, I.; Kliucininkas, L. A new strategy for using textile waste as a sustainable source of recovered cotton. *Resour. Conserv. Recycl.* **2019**, *145*, 359–369. [CrossRef]
74. Liu, Z.; Li, W.; Wei, Z. Qualitative classification of waste textiles based on near infrared spectroscopy and the convolutional network. *Text. Res. J.* **2020**, *90*, 1057–1066. [CrossRef]
75. Zitting, J. Optical Sorting Technology for Textile Waste: Development of an Identification Method with NIR Spectroscopy. Bechelor's Thesis, Lahti University, Lahti, Finland, 2017.
76. Wilts, H.; Garcia, B.R.; Garlito, R.G.; Gómez, L.S.; Prieto, E.G. Artificial intelligence in the sorting of municipal waste as an enabler of the circular economy. *Resources* **2021**, *10*, 28. [CrossRef]
77. Xiong, E. Autonomous sorting of plastic resin using near-infrared and machine learning. *CSFJ.* **2021**, *3*, 1–5.
78. Berghmans, A.C.; Huys, M.J.G. Progress for Identifying Objects Using an Optical Spectrometer and a Transport System. U.S. Patent 7,071,469, 4 July 2006.
79. Zhou, C.; Han, G.; Via, B.K.; Song, Y.; Gao, S.; Jiang, W. Rapid identification of fibers from different waste fabrics using the near-infrared spectroscopy technique. *Text. Res. J.* **2019**, *89*, 3610–3616. [CrossRef]
80. Zhou, J.; Yu, L.; Ding, Q.; Wang, R. Textile fiber identification using near-infrared spectroscopy and pattern recognition. *Autex Res. J.* **2019**, *19*, 201–209. [CrossRef]
81. Chen, H.; Tan, C.; Lin, Z. Quantitative determination of wool in textile by near-infrared spectroscopy and multivariate models. *Spectrochim. Acta A Mol. Biomol. Spectrosc.* **2018**, *201*, 229–235. [CrossRef] [PubMed]

82. Iqbal Hussain, M.A.; Khan, B.; Wang, Z.; Ding, S. Woven fabric pattern recognition and classification based on deep convolutional neural networks. *Electronics* **2020**, *9*, 1048. [CrossRef]
83. Xiao, Z.; Guo, Y.; Geng, L.; Wu, J.; Zhang, F.; Wang, W.; Liu, Y. Automatic recognition of woven fabric pattern based on TILT. *Math. Probl. Eng.* **2018**, *2018*, 9707104. [CrossRef]
84. Kuo, C.-F.J.; Kao, C.-Y. Self-organizing map network for automatically recognizing color texture fabric nature. *Fibers Polym.* **2007**, *8*, 174–180. [CrossRef]
85. Cura, K.; Rintala, N.; Kamppuri, T.; Saarimäki, E.; Heikkilä, P. Textile recognition and sorting for recycling at an automated line using near infrared spectroscopy. *Recycling* **2021**, *6*, 11. [CrossRef]
86. Riba, J.-R.; Cantero, R.; Canals, T.; Puig, R. Circular economy of post-consumer textile waste: Classification through infrared spectroscopy. *J. Clean. Prod.* **2020**, *272*, 123011. [CrossRef]
87. Peets, P.; Kaupmees, K.; Vahur, S.; Leito, I. Reflectance FT-IR spectroscopy as a viable option for textile fiber identification. *Herit. Sci.* **2019**, *7*, 1–10. [CrossRef]
88. Morgado, V.; Gomes, L.; Bettencourt da Silva, R.J.N.; Palma, C. Validated spreadsheet for the identification of PE, PET, PP and PS microplastics by micro-ATR-FTIR spectra with known uncertainty. *Talanta* **2021**, *234*, 122624. [CrossRef]
89. Gulich, B. Development of products made of reclaimed fibres. *Recycl. Text. Camb.* **2006**, *1*, 117–136.
90. Valerio, O.; Muthuraj, R.; Codou, A. Strategies for polymer to polymer recycling from waste: Current trends and opportunities for improving the circular economy of polymers in South America. *Curr. Opin. Green Sustain. Chem.* **2020**, *25*, 100381. [CrossRef]
91. Michaud, J.-C.; Farrant, L.; Jan, O.; Kjær, B.; Bakas, I. Environmental Benefits of Recycling—2010 Update. Material Change for a Better Environment. 2010. Available online: http://localhost:8080/xmlui/handle/123456789/174 (accessed on 22 August 2021).
92. Euratex. Circular Textiles, Prospering in the Circular Economy. Available online: https://euratex.eu/wp-content/uploads/EURATEX-Prospering-in-the-Circular-Economy-2020.pdf (accessed on 22 August 2021).
93. Pensupa, N.; Leu, S.-Y.; Hu, Y.; Du, C.; Liu, H.; Jing, H.; Wang, H.; Lin, C.S.K. Recent trends in sustainable textile waste recycling methods: Current situation and future prospects. *Chem. Chem. Technol. Waste Valoriz.* **2017**, 189–228. [CrossRef]
94. Bartlett, C.; McGill, I.; Willis, P. *Textiles Flow and Market Development Opportunities in the UK*; Waste & Resources Action Programme: Banbury, UK, 2013.
95. Korhonen, M.-R.; Dahlbo, H. *Reducing Greenhouse Gas Emissions by Recycling Plastics and Textiles into Products*; Finnish Environment Institute: Helsinki, Finland, 2007.
96. Johansson, L. On the Mechanical Recycling of Woven Fabrics: Improving the Reusable Fibre Yield of Mechanical Methods. Master's Thesis, Uppala University, Stockholm, Sweden, 2020.
97. Hutten, I.M. *Handbook of Nonwoven Filter Media*; Elsevier: Amsterdam, The Netherlands, 2007.
98. Jin, K.; Banerji, A.; Kitto, D.; Bates, F.S.; Ellison, C.J. Mechanically robust and recyclable cross-linked fibers from melt blown anthracene-functionalized commodity polymers. *ACS Appl. Mater. Interfaces* **2019**, *11*, 12863–12870. [CrossRef] [PubMed]
99. Tsai, P.P.-Y. Methods of Saturating Nonwoven Fabrics with Liquid and the Making of Electret Therof. U.S. Patent 17/056,276, 26 August 2021.
100. Mondragon, G.; Kortaberria, G.; Mendiburu, E.; González, N.; Arbelaiz, A.; Peña-Rodriguez, C. Thermomechanical recycling of polyamide 6 from fishing nets waste. *J. Appl. Polym. Sci.* **2020**, *137*, 48442. [CrossRef]
101. Russell, S.; Swan, P.; Trebowicz, M.; Ireland, A. Review of wool recycling and reuse. In *Natural Fibres: Advances in Science and Technology towards Industrial Applications*; Springer: Dordrecht, The Netherlands, 2016; pp. 415–428. [CrossRef]
102. Alexander, C.; Reno, J. *Economies of Recycling: The Global Transformation of Materials, Values and Social Relations*; Zed Books Ltd.: London, UK, 2012.
103. Klepp, I.G.; Tobiasson, T.S.; Laitala, K. Wool as an Heirloom: How Natural Fibres Can Reinvent Value in Terms of Money, Life-Span and Love. In *Natural Fibres: Advances in Science and Technology towards Industrial Applications*; Springer: Dordrecht, The Netherlands, 2016. [CrossRef]
104. Harmsen, P.; Scheffer, M.; Bos, H. Textiles for circular fashion: The logic behind recycling options. *Sustainability* **2021**, *13*, 9714. [CrossRef]
105. Klaus-nietrost, C.; Richard, H.; Weilach, C. Method of Reusing a Mixed Textile Comprising Cellulose and Synthetic Plastic. U.S. Patent 16/962,498, 18 March 2021.
106. Piribauer, B.; Bartl, A. Textile recycling processes, state of the art and current developments: A mini review. *Waste Manag. Res.* **2019**, *37*, 112–119. [CrossRef]
107. Liu, R.-G.; Shen, Y.-Y.; Shao, H.-L.; Wu, C.-X.; Hu, X.-C. An analysis of Lyocell fiber formation as a melt–spinning process. *Cellulose* **2001**, *8*, 13–21. [CrossRef]
108. Kaminsky, W.; Scheirs, J. *Feedstock Recycling and Pyrolysis of Waste Plastics: Converting Waste Plastics into Diesel and Other Fuels*; J. Wiley & Sons: New Jersey, NJ, USA, 2006.
109. Damayanti, D.; Wu, H.-S. Strategic possibility routes of recycled PET. *Polymers* **2021**, *13*, 1475. [CrossRef]
110. Ambercycle. Creating a Circular System. Available online: www.ambercycle.com (accessed on 29 July 2021).
111. BlockTexx. Textile Recovery Technology. Available online: https://www.blocktexx.com (accessed on 29 July 2021).
112. FENC. New Process Recycles any PET Waste. Available online: http://news.fenc.com/news_detail.aspx?lang=en&id=4557 (accessed on 4 March 2021).

113. Fiber, I. Infinited Fiber Company Brochure. Available online: https://infinitedfiber.com/wp-content/uploads/2019/06/IFC_Brochure_5-2019.pdf (accessed on 29 July 2021).
114. Asikainen, S.; Määttänen, M.; Harlin, A.; Valta, K.; Sivonen, E. Method of Producing Dissolving Pulp, Dissolving Pulp and Use of Method. U.S. Patent 14/428,377, 13 August 2015.
115. Ioncell. Ioncell®Process. Available online: https://ioncell.fi/research/ (accessed on 30 July 2021).
116. Lenzing. Closing the Loops. Available online: https://www.lenzing.com/sustainability/production/resources/chemicals (accessed on 4 September 2021).
117. BioSciences, T. Tyton BioSciences. Available online: www.tytonbio.com (accessed on 4 September 2021).
118. Technologies, W.A. Available online: https://wornagain.co.uk/ (accessed on 22 May 2021).
119. Sherwood, J. Closed-loop recycling of polymers using solvents. *Johnson Matthey Technol. Rev.* **2020**, 4–15. [CrossRef]
120. Ribul, M.; Lanot, A.; Tommencioni-Pisapia, C.; Purnell, P.; McQueen-Mason, S.J.; Baurley, S. Mechanical, chemical, biological: Moving towards closed-loop bio-based recycling in a circular economy of sustainable textiles. *J. Clean. Prod.* **2021**, 129325. [CrossRef]
121. Saha, S. Textile Recycling: The Chemical Recycling Process of Textiles. Available online: https://www.onlineclothingstudy.com/2020/08/textile-recycling-chemical-recycling.html (accessed on 24 October 2021).
122. Coates, G.W.; Getzler, Y.D. Chemical recycling to monomer for an ideal, circular polymer economy. *Nat. Rev. Mater.* **2020**, *5*, 501–516. [CrossRef]
123. Bruce, M.; Miller, W.D. Nylon Hydrolisis. U.S. Patent 2,840,606, 24 June 1958.
124. Bodrero, S.; Canivenc, E.; Cansell, F. *Chemical Recycling of Polyamide 6.6 and Polyamide 6 through a Two Step Ami-/Ammonolysis Process*; Georgia Institute of Technology: Atlanta, GA, USA, 1999.
125. Polk, M.B.; Leboeuf, L.L.; Shah, M.; Won, C.-Y.; Hu, X.; Ding, W. Nylon 66, nylon 46, and PET phase-transfer-catalyzed alkaline depolymerization at atmospheric pressure. *Polym. Plast. Technol. Eng.* **1999**, *38*, 459–470. [CrossRef]
126. Rickson, R. Controlling sediment at source: An evaluation of erosion control geotextiles. *Earth Surf. Process.Land. BGRG* **2006**, *31*, 550–560. [CrossRef]
127. Broda, J.; Przybyło, S.; Gawłowski, A.; Grzybowska-Pietras, J.; Sarna, E.; Rom, M.; Laszczak, R. Utilisation of textile wastes for the production of geotextiles designed for erosion protection. *J. Text. Inst.* **2019**, *110*, 435–444. [CrossRef]
128. Yang, Z.; Li, X.; Si, J.; Cui, Z.; Peng, K. Morphological, mechanical and thermal properties of poly (lactic acid)(PLA)/cellulose nanofibrils (CNF) composites nanofiber for tissue engineering. *J. Wuhan Univ. Technol. Mater. Sci. Ed.* **2019**, *34*, 207–215. [CrossRef]
129. Rizal, S.; Olaiya, F.G.; Saharudin, N.; Abdullah, C.; Ng, O.; Mohamad Haafiz, M.; Yahya, E.B.; Sabaruddin, F.; Khalil, H. Isolation of textile waste cellulose nanofibrillated fibre reinforced in polylactic acid-chitin biodegradable composite for green packaging application. *Polymers* **2021**, *13*, 325. [CrossRef]
130. Barisci, S.; Oncel, M.S. The disposal of combed cotton wastes by pyrolysis. *Int. J. Green Energy* **2014**, *11*, 255–266. [CrossRef]
131. Nunes, L.J.R.; Godina, R.; Matias, J.C.O.; Catalão, J.P.S. Economic and environmental benefits of using textile waste for the production of thermal energy. *J. Clean. Prod.* **2018**, *171*, 1353–1360. [CrossRef]
132. Czajczyńska, D.; Anguilano, L.; Ghazal, H.; Krzyżyńska, R.; Reynolds, A.; Spencer, N.; Jouhara, H. Potential of pyrolysis processes in the waste management sector. *Therm. Sci. Eng. Prog.* **2017**, *3*, 171–197. [CrossRef]
133. Czajczyńska, D.; Nannou, T.; Anguilano, L.; Krzyżyńska, R.; Ghazal, H.; Spencer, N.; Jouhara, H. Potentials of pyrolysis processes in the waste management sector. *Energy Procedia* **2017**, *123*, 387–394. [CrossRef]
134. Qi, R.; Xu, Z.; Zhou, Y.; Zhang, D.; Sun, Z.; Chen, W.; Xiong, M. Clean solid fuel produced from cotton textiles waste through hydrothermal carbonization with $FeCl_3$: Upgrading the fuel quality and combustion characteristics. *Energy* **2021**, *214*, 118926. [CrossRef]
135. Xu, Z.; Qi, R.; Xiong, M.; Zhang, D.; Gu, H.; Chen, W. Conversion of cotton textile waste to clean solid fuel via surfactant-assisted hydrothermal carbonization: Mechanisms and combustion behaviors. *Bioresour. Technol.* **2021**, *321*, 124450. [CrossRef]
136. Duman, G. Preparation of novel porous carbon from hydrothermal pretreated textile wastes: Effects of textile type and activation agent on structural and adsorptive properties. *J. Water Process. Eng.* **2021**, *43*, 102286. [CrossRef]
137. Yasin, S.; Massimo, C.; Rovero, G.; Behary, N.; Perwuelz, A.; Giraud, S.; Migliavacca, G.; Chen, G.; Guan, J. An alternative for the end-of-life phase of flame retardant textile products: Degradation of flame retardant and preliminary settings of energy valorization by gasification. *BioResources* **2017**, *12*, 5196–5211. Available online: https://ojs.cnr.ncsu.edu/index.php/BioRes/article/viewFile/BioRes_12_3_5196_Yasin_Alternative_Flame_Retardant_Textile/5343 (accessed on 29 July 2021).
138. Sert, E.; Yılmaz, E.; Atalay, F.S. Chemical recycling of polyethylene terephthalate by glycolysis using deep eutectic solvents. *J. Polym. Environ.* **2019**, *27*, 2956–2962. [CrossRef]
139. Kalfas, G.A. Mathematical modeling of the depolymerization of polyamide mixtures-Part I: Kinetic mechanism and parametric studies in batch reactors. *Polym. React. Eng.* **1998**, *6*, 41–67. [CrossRef]
140. Kwon, E.E.; Lee, T.; Ok, Y.S.; Tsang, D.C.; Park, C.; Lee, J. Effects of calcium carbonate on pyrolysis of sewage sludge. *Energy* **2018**, *153*, 726–731. [CrossRef]
141. Lee, T.; Nam, I.-H.; Jung, S.; Park, Y.-K.; Kwon, E.E. Synthesis of nickel/biochar composite from pyrolysis of Microcystis aeruginosa and its practical use for syngas production. *Bioresour. Technol.* **2020**, *300*, 122712. [CrossRef]
142. Damayanti, D.; Wu, H.S. Pyrolysis kinetic of alkaline and dealkaline lignin using catalyst. *J. Polym. Res.* **2018**, *25*, 7. [CrossRef]

143. Miranda, R.; Sosa_Blanco, C.; Bustos-Martinez, D.; Vasile, C. Pyrolysis of textile wastes: I. Kinetics and yields. *J. Anal. Appl. Pyrolysis* **2007**, *80*, 489–495. [CrossRef]
144. Lee, J.; Kim, K.-H.; Kwon, E.E. Biochar as a catalyst. *Renew. Sust. Energ. Rev.* **2017**, *77*, 70–79. [CrossRef]
145. You, S.; Ok, Y.S.; Chen, S.S.; Tsang, D.C.; Kwon, E.E.; Lee, J.; Wang, C.-H. A critical review on sustainable biochar system through gasification: Energy and environmental applications. *Bioresour. Technol.* **2017**, *246*, 242–253. [CrossRef]
146. Kwon, D.; Yi, S.; Jung, S.; Kwon, E.E. Valorization of synthetic textile waste using CO2 as a raw material in the catalytic pyrolysis process. *Environ. Pollut.* **2021**, *268*, 115916. [CrossRef]
147. Yousef, S.; Eimontas, J.; Striūgas, N.; Trofimov, E.; Hamdy, M.; Abdelnaby, M.A. Conversion of end-of-life cotton banknotes into liquid fuel using mini-pyrolysis plant. *J. Clean. Prod.* **2020**, *267*, 121612. [CrossRef]
148. Wang, S.; Dai, G.; Yang, H.; Luo, Z. Lignocellulosic biomass pyrolysis mechanism: A state-of-the-art review. *Prog. Energy Combust. Sci.* **2017**, *62*, 33–86. [CrossRef]
149. Damayanti, D.; Wulandari, Y.R.; Wu, H.-S. Product Distribution of Chemical Product Using Catalytic Depolymerization of Lignin. *Bull. Chem. React. Eng. Catal.* **2020**, *15*, 432–453. [CrossRef]
150. Molino, A.; Chianese, S.; Musmarra, D. Biomass gasification technology: The state of the art overview. *J. Energy Chem.* **2016**, *25*, 10–25. [CrossRef]
151. Devi, L.; Ptasinski, K.J.; Janssen, F.J. A review of the primary measures for tar elimination in biomass gasification processes. *Biomass Bioenerg.* **2003**, *24*, 125–140. [CrossRef]
152. Rittfors, J. *Thermochemical Textile Recycling: Investigation of Pyrolysis and Gasification of Cotton and Polyester*; Chalmers tekniska högskola: Göteborg, Sweden, 2020.
153. Yousef, S.; Kalpokaitė-Dičkuvienė, R.; Baltušnikas, A.; Pitak, I.; Lukošiūtė, S.-I. A new strategy for functionalization of char derived from pyrolysis of textile waste and its application as hybrid fillers (CNTs/char and graphene/char) in cement industry. *J. Clean. Prod.* **2021**, 128058. [CrossRef]
154. Jagdale, P.; Nair, J.R.; Khan, A.; Armandi, M.; Meligrana, G.; Hernandez, F.R.; Rusakova, I.; Piatti, E.; Rovere, M.; Tagliaferro, A. Waste to life: Low-cost, self-standing, 2D carbon fiber green Li-ion battery anode made from end-of-life cotton textile. *Electrochim. Acta* **2021**, *368*, 137644. [CrossRef]
155. Williams, P.T.; Reed, A.R. Pre-formed activated carbon matting derived from the pyrolysis of biomass natural fibre textile waste. *J. Anal. Appl. Pyrolysis* **2003**, *70*, 563–577. [CrossRef]
156. Williams, P.T.; Reed, A.R. High grade activated carbon matting derived from the chemical activation and pyrolysis of natural fibre textile waste. *J. Anal. Appl. Pyrolysis* **2004**, *71*, 971–986. [CrossRef]
157. Reed, A.R.; Williams, P.T. Thermal processing of biomass natural fibre wastes by pyrolysis. *Int. J. Energy Res.* **2004**, *28*, 131–145. [CrossRef]
158. Xu, Z.; Gu, S.; Sun, Z.; Zhang, D.; Zhou, Y.; Gao, Y.; Qi, R.; Chen, W. Synthesis of char-based adsorbents from cotton textile waste assisted by iron salts at low pyrolysis temperature for Cr (VI) removal. *Environ. Sci. Pollut. Res.* **2020**, 1–14. [CrossRef] [PubMed]
159. Eimontas, J.; Yousef, S.; Striūgas, N.; Abdelnaby, M.A. Catalytic pyrolysis kinetic behaviour and TG-FTIR-GC–MS analysis of waste fishing nets over ZSM-5 zeolite catalyst for caprolactam recovery. *Renew. Energy* **2021**, *179*, 1385–1403. [CrossRef]
160. da Silva, J.E.; Calixto, G.Q.; de Araújo Medeiros, R.L.B.; de Freitas Melo, M.A.; de Araújo Melo, D.M.; de Carvalho, L.P.; Braga, R.M. Colored cotton wastes valuation through thermal and catalytic reforming of pyrolysis vapors (Py-GC/MS). *Sci. Rep.* **2021**, *11*, 1–10. [CrossRef] [PubMed]
161. Wang, J.; Jiang, J.; Ding, J.; Wang, X.; Sun, Y.; Ruan, R.; Ragauskas, A.J.; Ok, Y.S.; Tsang, D.C. Promoting Diels-Alder reactions to produce bio-BTX: Co-aromatization of textile waste and plastic waste over USY zeolite. *J. Clean. Prod.* **2021**, 127966. [CrossRef]
162. Yousef, S.; Eimontas, J.; Striūgas, N.; Mohamed, A.; Abdelnaby, M.A. Morphology, compositions, thermal behavior and kinetics of pyrolysis of lint-microfibers generated from clothes dryer. *J. Anal. Appl. Pyrolysis* **2021**, *155*, 105037. [CrossRef]
163. Gholamzad, E.; Karimi, K.; Masoomi, M. Effective conversion of waste polyester–cotton textile to ethanol and recovery of polyester by alkaline pretreatment. *Chem. Eng. Sci.* **2014**, *253*, 40–45. [CrossRef]
164. Subramanian, K.; Sarkar, M.K.; Wang, H.; Qin, Z.-H.; Chopra, S.S.; Jin, M.; Kumar, V.; Chen, C.; Tsang, C.-W.; Lin, C.S.K. An overview of cotton and polyester, and their blended waste textile valorisation to value-added products: A circular economy approach–research trends, opportunities and challenges. *Crit. Rev. Environ. Sci. Technol.* **2021**, 1–22. [CrossRef]
165. Nikolić, S.; Lazić, V.; Veljović, Đ.; Mojović, L. Production of bioethanol from pre-treated cotton fabrics and waste cotton materials. *Carbohydr. Polym.* **2017**, *164*, 136–144. [CrossRef]
166. Yang, T.C.; Kumaran, J.; Amartey, S.; Maki, M.; Li, X.; Lu, F.; Qin, W. Biofuels and bioproducts produced through microbial conversion of biomass. *Bioenergy Res. Adv. Appl.* **2014**, 71–93. [CrossRef]
167. Elsayed, S.; Hellsten, S.; Guizani, C.; Witos, J.; Rissanen, M.; Rantamäki, A.H.; Varis, P.; Wiedmer, S.K.; Sixta, H. Recycling of superbase-based ionic liquid solvents for the production of textile-grade regenerated cellulose fibers in the lyocell process. *ACS Sustain. Chem. Eng.* **2020**, *8*, 14217–14227. [CrossRef]
168. Hong, F.; Guo, X.; Zhang, S.; Han, S.-f.; Yang, G.; Jönsson, L.J. Bacterial cellulose production from cotton-based waste textiles: Enzymatic saccharification enhanced by ionic liquid pretreatment. *Bioresour. Technol.* **2012**, *104*, 503–508. [CrossRef] [PubMed]
169. Sasaki, C.; Nakagawa, T.; Asada, C.; Nakamura, Y. Microwave-assisted hydrolysis of cotton waste to glucose in combination with the concentrated sulfuric acid impregnation method. *Waste Biomass Valoriz.* **2020**, *11*, 4279–4287. [CrossRef]

170. Hasanzadeh, E.; Mirmohamadsadeghi, S.; Karimi, K. Enhancing energy production from waste textile by hydrolysis of synthetic parts. *Fuel* **2018**, *218*, 41–48. [CrossRef]
171. Keskin, T.; Abubackar, H.N.; Arslan, K.; Azbar, N. Biohydrogen production from solid wastes. In *Biohydrogen*; Elsevier: Amsterdam, The Netherlands, 2019; pp. 321–346.
172. Dimos, K.; Paschos, T.; Louloudi, A.; Kalogiannis, K.G.; Lappas, A.A.; Papayannakos, N.; Kekos, D.; Mamma, D. Effect of various pretreatment methods on bioethanol production from cotton stalks. *Fermentation* **2019**, *5*, 5. [CrossRef]
173. Chandrashekhar, B.; Mishra, M.S.; Sharma, K.; Dubey, S. Bio-ethanol production from textile cotton waste via dilute acid hydrolysis and fermentation by Saccharomyces cerevisiae. *J. Ecobiotechnol.* **2011**, *3*, 6–9.
174. Binczarski, M.J.; Malinowska, J.; Stanishevsky, A.; Severino, C.J.; Yager, R.; Cieslak, M.; Witonska, I.A. A Model Procedure for Catalytic Conversion of Waste Cotton into Useful Chemicals. *Materials* **2021**, *14*, 1981. [CrossRef] [PubMed]
175. Quartinello, F.; Vecchiato, S.; Weinberger, S.; Kremenser, K.; Skopek, L.; Pellis, A.; Guebitz, G.M. Highly selective enzymatic recovery of building blocks from wool-cotton-polyester textile waste blends. *Polymers* **2018**, *10*, 1107. [CrossRef]
176. Shen, F.; Xiao, W.; Lin, L.; Yang, G.; Zhang, Y.; Deng, S. Enzymatic saccharification coupling with polyester recovery from cotton-based waste textiles by phosphoric acid pretreatment. *Bioresour. Technol.* **2013**, *130*, 248–255. [CrossRef]
177. Kuo, C.H.; Lin, P.J.; Lee, C.K. Enzymatic saccharification of dissolution pretreated waste cellulosic fabrics for bacterial cellulose production by Gluconacetobacter xylinus. *J. Chem. Technol. Biotechnol.* **2010**, *85*, 1346–1352. [CrossRef]
178. Yousef, S.; Kuliešienė, N.; Sakalauskaitė, S.; Nenartavičius, T.; Daugelavičius, R. Sustainable green strategy for recovery of glucose from end-of-life euro banknotes. *Waste Manag.* **2021**, *123*, 23–32. [CrossRef]
179. Piribauer, B.; Bartl, A.; Ipsmiller, W. Enzymatic textile recycling–best practices and outlook. *Waste Manag. Res.* **2021**, *39*, 1277–1290. [CrossRef] [PubMed]
180. Zhu, L.; Pelaz, B.; Chakraborty, I.; Parak, W.J. Investigating possible enzymatic degradation on polymer shells around inorganic nanoparticles. *Int. J. Mol. Sci.* **2019**, *20*, 935. [CrossRef] [PubMed]
181. Feliu, N.; Docter, D.; Heine, M.; Del Pino, P.; Ashraf, S.; Kolosnjaj-Tabi, J.; Macchiarini, P.; Nielsen, P.; Alloyeau, D.; Gazeau, F. In vivo degeneration and the fate of inorganic nanoparticles. *Chem. Soc. Rev.* **2016**, *45*, 2440–2457. [CrossRef] [PubMed]
182. Soenen, S.J.; Himmelreich, U.; Nuytten, N.; Pisanic, T.R.; Ferrari, A.; De Cuyper, M. Intracellular nanoparticle coating stability determines nanoparticle diagnostics efficacy and cell functionality. *Small* **2010**, *6*, 2136–2145. [CrossRef] [PubMed]
183. Spector, L.B. Covalent enzyme-substrate intermediates in transferase reactions. *Bioorg. Chem.* **1973**, *2*, 311–321. [CrossRef]
184. Chanana, M.; Rivera_Gil, P.; Correa-Duarte, M.A.; Liz-Marzán, L.M.; Parak, W.J. Physicochemical properties of protein-coated gold nanoparticles in biological fluids and cells before and after proteolytic digestion. *Angew. Chem. Int. Ed.* **2013**, *52*, 4179–4183. [CrossRef]
185. Duff, S.J.; Murray, W.D. Bioconversion of forest products industry waste cellulosics to fuel ethanol: A review. *Bioresour. Technol.* **1996**, *55*, 1–33. [CrossRef]
186. Olguin-Maciel, E.; Singh, A.; Chable-Villacis, R.; Tapia-Tussell, R.; Ruiz, H.A. Consolidated bioprocessing, an innovative strategy towards sustainability for biofuels production from crop residues: An overview. *Agronomy* **2020**, *10*, 1834. [CrossRef]
187. Damayanti, D.; Supriyadi, D.; Amelia, D.; Saputri, D.R.; Devi, Y.L.L.; Auriyani, W.A.; Wu, H.S. Conversion of Lignocellulose for Bioethanol Production, Applied in Bio-Polyethylene Terephthalate. *Polymers* **2021**, *13*, 2886. [CrossRef]
188. Dahnum, D.; Tasum, S.O.; Triwahyuni, E.; Nurdin, M.; Abimanyu, H. Comparison of SHF and SSF processes using enzyme and dry yeast for optimization of bioethanol production from empty fruit bunch. *Energy Procedia* **2015**, *68*, 107–116. [CrossRef]
189. Leong, Y.K.; Chew, K.W.; Chen, W.-H.; Chang, J.-S.; Show, P.L. Reuniting the Biogeochemistry of Algae for a Low-Carbon Circular Bioeconomy. *Trends Plant Sci.* **2021**. [CrossRef]
190. Parisutham, V.; Kim, T.H.; Lee, S.K. Feasibilities of consolidated bioprocessing microbes: From pretreatment to biofuel production. *Bioresour. Technol.* **2014**, *161*, 431–440. [CrossRef]
191. Hasunuma, T.; Kondo, A. Consolidated bioprocessing and simultaneous saccharification and fermentation of lignocellulose to ethanol with thermotolerant yeast strains. *Process. Biochem.* **2012**, *47*, 1287–1294. [CrossRef]
192. Okamoto, K.; Uchii, A.; Kanawaku, R.; Yanase, H. Bioconversion of xylose, hexoses and biomass to ethanol by a new isolate of the white rot basidiomycete Trametes versicolor. *Springerplus* **2014**, *3*, 1–9. [CrossRef] [PubMed]
193. Lynd, L.R.; Van Zyl, W.H.; McBride, J.E.; Laser, M. Consolidated bioprocessing of cellulosic biomass: An update. *Curr. Opin. Biotechnol.* **2005**, *16*, 577–583. [CrossRef] [PubMed]
194. Jeihanipour, A.; Taherzadeh, M.J. Ethanol production from cotton-based waste textiles. *Bioresour. Technol.* **2009**, *100*, 1007–1010. [CrossRef]
195. Sanchis-Sebastiá, M.; Novy, V.; Stigsson, L.; Galbe, M.; Wallberg, O. Towards circular fashion–transforming pulp mills into hubs for textile recycling. *RSC Advances* **2021**, *11*, 12321–12329. [CrossRef]
196. Sanchis-Sebastiá, M.; Ruuth, E.; Stigsson, L.; Galbe, M.; Wallberg, O. Novel sustainable alternatives for the fashion industry: A method of chemically recycling waste textiles via acid hydrolysis. *J. Waste Manag.* **2021**, *121*, 248–254. [CrossRef] [PubMed]
197. Jeihanipour, A.; Karimi, K.; Niklasson, C.; Taherzadeh, M.J. A novel process for ethanol or biogas production from cellulose in blended-fibers waste textiles. *J. Waste Manag.* **2010**, *30*, 2504–2509. [CrossRef] [PubMed]
198. Kuo, C.-H.; Lin, P.-J.; Wu, Y.-Q.; Ye, L.-Y.; Yang, D.-J.; Shieh, C.-J.; Lee, C.-K. Simultaneous saccharification and fermentation of waste textiles for ethanol production. *BioResources* **2014**, *9*, 2866–2875. [CrossRef]

199. McIntosh, S.; Vancov, T.; Palmer, J.; Morris, S. Ethanol production from cotton gin trash using optimised dilute acid pretreatment and whole slurry fermentation processes. *Bioresour. Technol.* **2014**, *173*, 42–51. [CrossRef]
200. Yildirir, E. Chemical Recycling of Waste Plastics via Hydrothermal Processing. Ph.D. Thesis, University of Leeds, Leeds, UK, 2015.
201. Wikberg, H.; Grönberg, V.; Jermakka, J.; Kemppainen, K.; Kleen, M.; Laine, C.; Paasikallio, V.; Oasmaa, A. Hydrothermal refining of biomass: An overview and future perspectives. *Tappi J.* **2015**, *14*, 195–207. [CrossRef]
202. Tang, K.; Zhang, Y.; Lin, D.; Han, Y.; Chen, C.-T.A.; Wang, D.; Lin, Y.-S.; Sun, J.; Zheng, Q.; Jiao, N. Cultivation-independent and cultivation-dependent analysis of microbes in the shallow-sea hydrothermal system off Kueishantao Island, Taiwan: Unmasking heterotrophic bacterial diversity and functional capacity. *Front. Microbiol.* **2018**, *9*, 279. [CrossRef]
203. Hongthong, S.; Leese, H.S.; Allen, M.J.; Chuck, C.J. Assessing the Conversion of Various Nylon Polymers in the Hydrothermal Liquefaction of Macroalgae. *Environments* **2021**, *8*, 34. [CrossRef]
204. Zheng, C.; Ma, X.; Yao, Z.; Chen, X. The properties and combustion behaviors of hydrochars derived from co-hydrothermal carbonization of sewage sludge and food waste. *Bioresour. Technol.* **2019**, *285*, 121347. [CrossRef] [PubMed]
205. Hou, W.; Ling, C.; Shi, S.; Yan, Z.; Zhang, M.; Zhang, B.; Dai, J. Separation and characterization of waste cotton/polyester blend fabric with hydrothermal method. *Fibers Polym.* **2018**, *19*, 742–750. [CrossRef]
206. Shi, S.; Zhang, M.; Ling, C.; Hou, W.; Yan, Z. Extraction and characterization of microcrystalline cellulose from waste cotton fabrics via hydrothermal method. *J. Waste Manag.* **2018**, *82*, 139–146. [CrossRef]
207. Lin, Y.; Ma, X.; Peng, X.; Yu, Z. A mechanism study on hydrothermal carbonization of waste textile. *Energy Fuels* **2016**, *30*, 7746–7754. [CrossRef]
208. Li, Y.; Shao, M.; Huang, M.; Sang, W.; Zheng, S.; Jiang, N.; Gao, Y. Enhanced remediation of heavy metals contaminated soils with EK-PRB using β-CD/hydrothermal biochar by waste cotton as reactive barrier. *Chemosphere* **2022**, *286*, 131470. [CrossRef]
209. Xu, Z.; Zhang, T.; Yuan, Z.; Zhang, D.; Sun, Z.; Huang, Y.; Chen, W.; Tian, D.; Deng, H.; Zhou, Y. Fabrication of cotton textile waste-based magnetic activated carbon using FeCl3 activation by the Box–Behnken design: Optimization and characteristics. *RSC Adv.* **2018**, *8*, 38081–38090. [CrossRef]
210. Xu, Z.; Zhou, Y.; Sun, Z.; Zhang, D.; Huang, Y.; Gu, S.; Chen, W. Understanding reactions and pore-forming mechanisms between waste cotton woven and FeCl3 during the synthesis of magnetic activated carbon. *Chemosphere* **2020**, *241*, 125120. [CrossRef]
211. Kawamura, K.; Sako, K.; Ogata, T.; Tanabe, K. Environmentally friendly, hydrothermal treatment of mixed fabric wastes containing polyester, cotton, and wool fibers: Application for HMF production. *Bioresour. Technol. Rep.* **2020**, *11*, 100478. [CrossRef]
212. Asaoka, Y.; Funazukuri, T. Hydrothermal saccharification of cotton cellulose in dilute aqueous formic acid solution. *Res. Chem. Intermed.* **2011**, *37*, 233–242. [CrossRef]
213. Keh, E.Y.M.; Yao, L.; Liao, X.; Liu, Y.; Cheuk, K.; Chan, A. Method for Separating and Recycling a Waste Polyester-Cotton Textile by Means of a Hydrothermal Reaction Catalyzed by an Organic Acid. U.S. Patent 2020/0262108 A1, 20 August 2020.
214. McKinney, R. Ammonolysis of Nylon. US 5302756A, 12 April 1994.
215. Molino, A.; Larocca, V.; Chianese, S.; Musmarra, D. Biofuels production by biomass gasification: A review. *Energies* **2018**, *11*, 811. [CrossRef]
216. Labaki, M.; Jeguirim, M. Thermochemical conversion of waste tyres—A review. *Environ. Sci. Pollut. Res.* **2017**, *24*, 9962–9992. [CrossRef]
217. Samolada, M.; Zabaniotou, A. Potential application of pyrolysis for the effective valorisation of the end of life tires in Greece. *Environ. Dev.* **2012**, *4*, 73–87. [CrossRef]
218. Portofino, S.; Donatelli, A.; Iovane, P.; Innella, C.; Civita, R.; Martino, M.; Matera, D.A.; Russo, A.; Cornacchia, G.; Galvagno, S. Steam gasification of waste tyre: Influence of process temperature on yield and product composition. *J. Waste Manag.* **2013**, *33*, 672–678. [CrossRef] [PubMed]
219. Yasin, S. Eco-Design for End-of-Life Phase of Flame Retardant Textiles. Ph.D. Thesis, Université Lille 1-Sciences et Technologies, Lille, France, 2017.
220. Cañete Vela, I.; Maric, J.; Seemann, M. Valorisation of textile waste via steam gasification in a fluidized bed reactor. In Proceedings of the 7th International Conference on Sustainable Solid Waste Management, Crete Island, Greece, 26–29 June 2019.
221. Wen, C.; Wu, Y.; Chen, X.; Jiang, G.; Liu, D. The pyrolysis and gasification performances of waste textile under carbon dioxide atmosphere. *J. Therm. Anal. Calorim.* **2017**, *128*, 581–591. [CrossRef]
222. Wu, Y.; Wen, C.; Chen, X.; Jiang, G.; Liu, D. Catalytic pyrolysis and gasification of waste textile under carbon dioxide atmosphere with composite Zn-Fe catalyst. *Fuel Process. Technol.* **2017**, *166*, 115–123. [CrossRef]
223. Ziyaei, M.D.; Barikani, M.; Honarkar, H. Recycling of Polyethylene Terephthalate (PET) via Glycolysis Method for Synthesis Waterborne Polyurethane. In Proceedings of the International Seminar on Polymer Science and Technology, Tehran, Iran, 10–22 November 2018; pp. 520–523.
224. Guo, Z.; Adolfsson, E.; Tam, P.L. Nanostructured micro particles as a low-cost and sustainable catalyst in the recycling of PET fiber waste by the glycolysis method. *J. Waste Manag.* **2021**, *126*, 559–566. [CrossRef] [PubMed]
225. Al-Sabagh, A.; Yehia, F.; Eshaq, G.; Rabie, A.; ElMetwally, A. Greener routes for recycling of polyethylene terephthalate. *Egypt. J. Pet.* **2016**, *25*, 53–64. [CrossRef]
226. Heiran, R.; Ghaderian, A.; Reghunadhan, A.; Sedaghati, F.; Thomas, S.; hossein Haghighi, A. Glycolysis: An efficient route for recycling of end of life polyurethane foams. *J. Polym. Res.* **2021**, *28*, 1–19. [CrossRef]

227. Al-Sabagh, A.M.; Yehia, F.Z.; Eissa, A.-M.M.; Moustafa, M.E.; Eshaq, G.; Rabie, A.-R.M.; ElMetwally, A.E. Glycolysis of poly (ethylene terephthalate) catalyzed by the Lewis base ionic liquid [Bmim][OAc]. *Ind. Eng. Chem. Res.* **2014**, *53*, 18443–18451. [CrossRef]
228. Guo, Z.; Eriksson, M.; de la Motte, H.; Adolfsson, E. Circular recycling of polyester textile waste using a sustainable catalyst. *J. Clean. Prod.* **2021**, *283*, 124579. [CrossRef]
229. Hommez, B.; Goethals, E.J. Degradation of nylon-6 by glycolysis. Part 1: Identification of degradation products. *J. Macromol. Sci.—Pure Appl. Chem.* **1998**, *35*, 1489–1505. [CrossRef]
230. Huczkowski, P.; Kapko, J.; Olesiak, R. Degradation of nylon-6 in ethylene glycol. *Polymer* **1978**, *19*, 77–80. [CrossRef]
231. Huczkowski, P.; Kapko, J. Degradation of nylon-6 in ethylene glycol: 2. Mathematical illustration of degradation. *Polymer* **1980**, *21*, 86–88. [CrossRef]
232. Holland, B.J.; Hay, J.N. Thermal degradation of nylon polymers. *Polym. Int.* **2000**, *49*, 943–948. [CrossRef]
233. Kim, K.J.; Dhevi, D.M.; Lee, J.S.; Dal Cho, Y.; Choe, E.K. Mechanism of glycolysis of nylon 6, 6 and its model compound by ethylene glycol. *Polym. Degrad. Stab.* **2006**, *91*, 1545–1555. [CrossRef]
234. Hu, Y.; Wang, Y.; Zhang, X.; Qian, J.; Xing, X.; Wang, X. Synthesis of poly (ethylene terephthalate) based on glycolysis of waste PET fiber. *J. Macromol. Sci. A* **2020**, *57*, 430–438. [CrossRef]
235. Shukla, S.; Harad, A.M. Glycolysis of polyethylene terephthalate waste fibers. *J. Appl. Polym. Sci.* **2005**, *97*, 513–517. [CrossRef]
236. Maryan, A.S.; Montazer, M.; Damerchely, R. Discoloration of denim garment with color free effluent using montmorillonite based nano clay and enzymes: Nano bio-treatment on denim garment. *J. Clean. Prod.* **2015**, *91*, 208–215. [CrossRef]
237. Groves, E.; Palenik, C.S.; Palenik, S. A survey of extraction solvents in the forensic analysis of textile dyes. *Forensic Sci. Int.* **2016**, *268*, 139–144. [CrossRef] [PubMed]
238. Mu, B.; Yang, Y. Complete separation of colorants from polymeric materials for cost-effective recycling of waste textiles. *Chem. Eng. Sci.* **2022**, *427*, 131570. [CrossRef]
239. Haslinger, S.; Wang, Y.; Rissanen, M.; Lossa, M.B.; Tanttu, M.; Ilen, E.; Määttänen, M.; Harlin, A.; Hummel, M.; Sixta, H. Recycling of vat and reactive dyed textile waste to new colored man-made cellulose fibers. *Green Chem.* **2019**, *21*, 5598–5610. [CrossRef]
240. Padmanaban, A.; Murugadoss, G.; Venkatesh, N.; Hazra, S.; Rajesh Kumar, M.; Tamilselvi, R.; Sakthivel, P. Electrochemical determination of harmful catechol and rapid decolorization of textile dyes using ceria and tin doped ZnO nanoparticles. *J. Environ. Chem. Eng.* **2021**, *9*, 105976. [CrossRef]
241. Li, Y.; Li, M.; Lu, J.; Li, X.; Ge, M. Decoloration of waste PET alcoholysis liquid by an electrochemical method. *Water Sci. Technol.* **2018**, *77*, 2463–2473. [CrossRef]
242. Huang, J.; Yan, D.; Dong, H.; Li, F.; Lu, X.; Xin, J. Removal of trace amount impurities in glycolytic monomer of polyethylene terephthalate by recrystallization. *J. Environ. Chem. Eng.* **2021**, 106277. [CrossRef]

Article

Polyethylene/Polyamide Blends Made of Waste with Compatibilizer: Processing, Morphology, Rheological and Thermo-Mechanical Behavior

Dorota Czarnecka-Komorowska [1,*], Jagoda Nowak-Grzebyta [2], Katarzyna Gawdzińska [3], Olga Mysiukiewicz [1] and Małgorzata Tomasik [4]

[1] Institute of Materials Technology, Poznan University of Technology, 60-965 Poznan, Poland; olga.mysiukiewicz@put.poznan.pl
[2] Faculty of Mechanical Engineering, Poznan University of Technology, 60-965 Poznan, Poland; jagoda.pa.nowak@doctorate.put.poznan.pl
[3] Department of Machines Construction and Materials, Maritime University of Szczecin, 71-650 Szczecin, Poland; k.gawdzinska@am.szczecin.pl
[4] Department of Interdisciplinary Dentistry, Pomeranian Medical University, 70-111 Szczecin, Poland; malgorzata.tomasik@pum.edu.pl
* Correspondence: Dorota.Czarnecka-Komorowska@put.poznan.pl

Abstract: The aim of this study was to develop a polyethylene/polyamide (R-PE/R-PA) regranulated product made from post-consumer wastes grafted with polyethylene-graft-maleic anhydride (PE-g-MAH) by reactive extrusion in a twin-screw extruder equipped with an external mixing zone. The compatibility effect of PE-g-MAH used as a modifier in R-PE/R-PA blends was evaluated by means of differential scanning calorimetry (DSC) and dynamic mechanical thermal analysis (DMTA), while the analysis of the chemical structure of this blend was carried out by Fourier transform infrared spectroscopy (FT-IR). The thermal properties, complex viscosity, and selected usage properties of R-PE/R-PA blends compatibilized with PE-g-MAH, i.e., density and water absorption, were evaluated. The morphology of the blends with and without the compatibilizer was observed by scanning electron microscopy. The R-PE/R-PA/MAH blend shows heterogenic structure, which is a result of the chemical reaction in reactive extrusion between functional groups of PE-g-MAH used as modifier and the end groups of R-PA6. The results show that the R-PE/R-PA blend with increased PE-g-MAH content showed increased hardness, stiffness, and ultimate tensile strength due to the increased degree of crystallinity. The increase in crystallinity is proportional to the improvement of the mechanical properties. Moreover, it is shown that 1 wt.% PE-g-MAH added to the R-PE/R-PA waste blend increases the interfacial interactions and compatibility between R-PE and R-PA, resulting in decreased polyamide particle size. Finally, the results show that it is possible to produce good quality regranulated products with advantageous properties and structure from immiscible polymer waste for industrial applications.

Keywords: recycling; plastics waste; extrusion; polyethylene/polyamide blends; compatibility; physical properties; structure

1. Introduction

Nowadays, polymer blends based on waste, from which new materials with tailor-made properties are made, are very popular among researchers and could be beneficial both from an environmental and economical point of view. Due to the change of the traditional model of the linear economy "take–make–dispose" on the circular model is considered [1]. The linear model represents the one-way consumption of plastic products, which results in the production of large amounts of waste materials [2]. Disposing of this waste in an unprocessed form results in an endless need to extract non-renewable raw materials for use in the production of polymer materials [3]. Consequently, this leads to the overexploitation of natural resources, increased greenhouse gas emissions, and environmental degradation [4,5]. An alternative solution is to introduce a circular economy [6], which assumes an increased level of raw material recovery from waste through material recycling, thus reducing the total greenhouse gas emissions [5]. The circular economy works to ensure that the value of materials, and products are preserved for as long as possible while minimizing waste [6]. The efficient processing of polymer waste remains a difficult challenge, and the recycling process is still the best way to manage wastes [7]. Many researchers are working on developing new technologies for recycling polymers [8–10]. They are also developing new recycled materials, which, when properly modified, can function as full-value and cheap "secondary" raw materials for plastic products that can be used in manufacturing processes either instead of or with virgin raw materials. The ideal materials that meet both economic and environmental criteria that can be used in mechanical engineering are recycled polymer regranulates of various engineering plastics, i.e., polyethylene (PE), polyamide (PA), polypropylene (PP), and polyoxymethylene (POM) [11,12]. Polymer regranulates, as a rule, have reduced mechanical properties and thermal resistance, and many procedures can be used to improve these properties, such as mixing with virgin polymers; adding fibrous fillers, i.e., glass fibers, and natural fibers; and modifying with powder fillers such as basalt [13], talc [14], polysilsesqioxanes (POSS) [15,16], or montmorillonite (MMT) [17].

Polyethylene is one of the commercial plastics widely used in packaging applications such as films and bottles, etc., which is equivalent to generating a lot of amounts of post-consumer waste [9]. Among the recycling methods, blending recycled polyethylene with other polymers is very attractive because of the ability to combine different material properties. For example, the use of polyamide can improve the stiffness and strength of materials. Unfortunately, but the above-mentioned PE/PA blending obtained by a simple blending technique shows reduced strength properties than polyethylene or polyamide due to poor compatibility between PE and PA. One way to improve the properties of blends can be to incorporate compatibilizers into immiscible polymers, such as polyethylene with polyamide in the molten state by extrusion [18].

In the case of waste polymer blends, the problem is much more serious, as they are often thermodynamically non-compatible blends with low or no adhesion [9]. This leads to the production of regranulates with unfavorable mechanical and processing properties. Improving the interfacial adhesion can be accomplished by chemical compatibilization by introducing a compatibilizer, an additional third component that is miscible with both phases [19,20] and copolymer whose one part is miscible with one phase and another with another phase [21] and or by reactive processing [20,21], i.e., reactive formation of graft, block, or lightly crosslinked copolymer, formation of ionically bonded; structures trans-reactions, and mechano-chemical blending [20]. The compatibilization must accomplish optimization of the interfacial tension, stabilize the morphology against high stresses during enhance adhesion between the phases in the solid state [20].

The most frequently used compatibilizers in the case of polymer mixtures are ethylene copolymers intended for the processing of waste polypropylene (PP) and polyethylene (PE) blends, lattices (S-E-B) for styrene blends, and maleic anhydride, especially in the case of polyolefins recyclates. For instance, Janik et al. [22] studied the improvement of mechanical properties such as tensile and flexural strength, and modulus, and elongation at break of polycarbonate/polypropylene mixtures using polypropylene-graft-maleic anhydride (PP-g-MA). Similarly, Mengual et al. [23] researched the use of PP-g-MA and polyethylene-graft-maleic anhydride (PE-g-MA) or styrene-ethylene/butylene-styrene-graft-maleic anhydride (SEBS-g-MA) for polymer modification. Three types of application were presented: improving the properties of the base compound, as a compatibilizer in the blending of two materials, and as an additional component in the mixing of various materials [23].

Moreover, in the last years, many papers have been done to improve the compatibility of PE/PA blends [24–29]. For instance, Jeziórska et al. [24] studied the influence of polyethylene grafted with maleic anhydride (PE-g-MAH) on the mechanical and structural properties of polyamide (PA) and textile wastes containing poly(ethylene terephthalate) (PET). The chemical reactions of PE-g-MAH with blended components were studied by Fourier transform infrared spectroscopy (FT-IR) [24]. Jeziórska at al. [25] also reported that the incorporation of ricinol-2-oxazoline methylmaleate (MRO) in PA6/LDPE blends leads to an increase in the elastic modulus and impact strength of the blends. The PA6/LDPE/MRO blends show heterogenic structure, which, as a result of chemical reactions going during the extrusion process, stabilizes at a microphase level probably of functional groups of PA with oxazoline groups grafted onto LDPE [25]. A similar effect was observed in case of polyamide 6 and polyolefins blends functionalized with acrylic acid (polyethylene—PE–AA and polypropylene—PP–AA) by Psarski et al. [26].

Wahab et al. [27] investigated the compatibilization effects of PE-g-MA on mechanical, thermal and swelling properties of high density polyethylene (HDPE)/natural rubber (NR)/thermoplastic tapioca starch (TPS) blends, and found the effectiveness of PE-g-MA as compatibilizer in improving the miscibility between HDPE/NR/TPS blends.

Majid et al. [28] studied the effects of PE-g-MA on the morphology, water absorption tensile and mechanical properties of LDPE/thermoplastic sago starch (TPSS) blends and found that PE-g-MA acted as an effective compatibilizer for the LDPE/TPSS blends. All these studies confirmed the beneficial effect of compatibilizers on the structure, especially the even distribution of components and selected functional properties of blends.

As described by Huitric et al. [29], the addition of a compatibilizer promotes good dispersion and structure stabilization by suppressing coalescence to increase interfacial adhesion. The compatibilizing elements can include diblock copolymers, mixed with both phases, or block or graft copolymers formed by the interface reaction during mixing. The copolymer (acting on the surface of the droplets) inhibits coalescence very well by lowering the interfacial tension, which facilitates disintegration. The main result of these studies is the essential size reduction of the dispersed phase due to the suppression of coalescence. This is achieved using steric repulsion between the block copolymers present at the interface and the Gibbs–Marangoni effect induced by the local concentration of the compatibilizer's gradient at the interface [29].

Therefore, the paper presents the development process of manufacturing R-PE/R-PA regranulates based on immiscible waste polymers. Regranulates from these materials, including polyethylene and polyamide, were modified with PE-g-MAH as compatibilizer. The obtained R-PE/R-PA/MAH regranulates are characterized by defined properties and structure, useful for industrial applications, including the production of blown films.

2. Experimental Sections

2.1. Materials and Procedures

The regranulate was prepared from two commercial plastics, low-density polyethylene and polyamide. Polyethylene and polyamide 6 from post-consumer waste and their blends, modified with polyethylene-graft-maleic anhydride (PE-g-MAH) copolymer as a compatibilizer were investigated in our research. PE-g-MA is a compatibilizer for polymer blends which serves as support for polar to nonpolar substances [23]. The compatibilizing agent PE-g-MAH (viscosity 1.700–4.500 cP at 140 °C, $T_{m\ (DSC)}$ = 105 °C, density 0.9 g/cm^3 at 25 °C) [30] was purchased from Sigma–Aldrich (St. Louis, MO, USA). Polyamide 6 (PA 6) grade wastes (trade name Tarnamid), supplied by Azoty Group S.A. (Tarnów, Poland), and low-density polyethylene (LDPE) wastes (trade name Malen), supplied by Basell Orlen Polyolefines (Płock, Poland), were obtained from a local Polish supplier. Sample codes and polymer blend compositions are detailed in Table 1.

Table 1. Sample codes and polymer blend compositions.

Sample Code	Composition R-PE/R-PA/PE g-MAH [wt.%]
R-PA	0/100/0
R-PE	100/0/0
R-PE/R-PA	80/20/0
R-PE/R-PA/MAH1	80/19/1
R-PE/R-PA/MAH3	80/17/3

The components were first dried at 80 °C for 24 h using an ULE 500 cabinet dryer (Memmert GmbH + Co. KG, Schwabach, Germany). The blends, containing 80 wt.% of R-PE with different contents of R-PA (20, 19, or 17 wt.%) and PE-g-MAH (1 and 3 wt.%), were manufactured by mixing in a molten state (Table 1). The mixing and reacting of the components was performed under controlled conditions [31] using an EH-16.2D co-rotating twin screw extruder (Zamak Mercator Sp. z o.o. Skawina, Poland) with a length/diameter ratio (L/D) of 40 mm, a screw diameter (D) of 16 mm, and a capillary die diameter (d) of 8 mm operating at 100 rpm and with the following temperatures: 250–240 °C, 235 °C, 235 °C, 230 °C, and 230 °C along the barrel and 240 °C for the die. The extruded rod was air-dried and pelletized. Next, the regranulates were dried for 24 h at 80 °C. Dumbbell-shaped specimens for tensile tests were injection-molded using an Eco e-mac 50 (Engel Austria GmbH, Schwertberg, Austria) with a temperature of 230–250 °C, mold temperature of 25 °C, and 35 s cooling time.

2.2. Methods

2.2.1. Scanning Electron Microscopy (SEM)

The morphology of the R-PE/R-PA grafted by PE-g-MAH blends was investigated with a MIRA3 scanning electron microscope (TESCAN Brno, s.r.o., Brno, Czech Republic) with high-resolution imaging. The specimens were fractured in liquid nitrogen and then coated with a thin layer of 20 μm carbon powder. The dispersion and dimension domains of R-PA particles in the R-PE matrix was investigated by scattered electron (SE) signal, with an accelerating voltage of 10 kV. A magnification of 5000× was used.

2.2.2. Density and Water Absorption

The solid mass of the R-PE, R-PA and their blends were measured by an electronic balance AXIS AD50-AD200 (AXIS, Gdansk, Poland). The density was measured based on PN-EN ISO 1183–1:2004 [32] standards using the hydrostatic method. Ethyl alcohol (analytical grade, supplied by Chempur, Piekary Śląskie, Poland) as an immersion liquid, was used, and measurements were made of 5 samples from each series.

Water absorption (WA) tests were carried out on sets of 5 samples for R-PE, R-PA, and each type of R-PE/R-PA blend in accordance with ISO 62:2008 [33]. In the test procedure, all specimens were dried in a cabinet dryer at 80 °C for 24 h, then their weights were measured. After 24 h immersion in water at 23 °C, samples were taken out of the water. The surface was dried with a cloth and weight was measured again. The percentage of water absorption by the different polymer blends was calculated with the following Equation (1):

$$WA(\%) = \frac{m_s - m_0}{m_0} \cdot 100 \qquad (1)$$

where m_0 is the weight of the dry sample (g) and m_s is the weight of the sample after 24 h water immersion (g).

2.2.3. Melt Mass-Flow Rate (MFR)

Melt flow properties provide significant insight for thermoplastic manufacturing. The MFR is an indicator of the flow properties of the material in molten state. The MFR of the blends with and without compatibilizer was measured using a Kayeness melt flow plastometer (model LMI 4003, Dynisco, Franklin, MA, USA), according to the standard test method ISO 1133-1:2011 [34]. The measurements were conducted at a temperature of 230 °C under a 2.16 kg load. The average of 10 readings was calculated to determine the MFR.

2.2.4. Melt Rheology

The rheological behavior of the R-PE/R-PA blends grafted by PE-g-MAH was evaluated using rotational rheometer in small-amplitude oscillation shearing mode. A MCR301 apparatus (Anton Paar GmbH, Graz, Austria) in plate–plate configuration with a 1 mm gap was used. The samples were dried overnight at 80 °C and tested in an angular frequency range of $1-500$ s^{-1} under strain of 1%, which was in the linear viscoelastic (LVE) range, as proven by a preliminary amplitude sweep. An evaluation of zero shear viscosity (η_0) was possible due to the rheological measurements in oscillatory mode and the calculations performed with Rheoplus 32 software V.3.40 (Anton Paar GmbH, Graz, Germany). A temperature of 230 °C was applied during the test to make sure that no polymer degradation would take place and violate the results.

2.2.5. Fourier Transform Infrared (FT-IR) Spectoscopy

Fourier transform infrared (FT-IR) spectroscopy was carried out using a FT/IR-4600 spectrometer (Jasco Europe S.R.L., Cremella, Italy). The spectra were collected at room temperature over a wavelength range of $4000-400$ cm^{-1}, with a resolution of 4 cm^{-1}, by averaging 40 scans. Spectroscopic data were evaluated using the software Spectra Manager (ver. 2, Jasco, Easton, MD, US).

2.2.6. Thermal Differential Scanning Calorimetry (DSC)

The melting and crystallization behavior of the R-PE/R-PA blends grafted by PE-g-MAH was studied using a DSC 204 F1 Phoenix (Netzsch GmbH, Selb, Germany), operated with a protective atmosphere (nitrogen flow of 20 mL/min). Samples of about 8 mg were first heated from room temperature to 270 °C at a rate of 10 °C/min and kept at that temperature for 5 min, followed by cooling to 20 °C at a rate of 10 °C/min to eliminate the thermal and mechanical history of polymers. The samples were subsequently reheated to 270 °C to determine the enthalpy of melting and cooled at the same rate. The thermal positions of the exothermic and endothermic peaks were taken as crystallization and melting temperatures (T_{cr} and T_m). The crystallization and melting enthalpy (ΔH_m) were evaluated from the exothermic and endothermic peak areas. The degree of crystallinity (X_c) of the polyethylene, a major component of the R-PE/R-PA blends, was determined using the following equation:

$$X_c(\%) = \frac{\Delta H_m}{\varphi_{PE} \Delta H_m^0} \cdot 100 \qquad (2)$$

where ΔH_m is the melting enthalpy of the sample (J/g), $\Delta H_m^0 = 293$ J/g is the melting enthalpy of 100% crystalline polyethylene [35], and φ_{PE} is the R-PE content in the samples.

2.2.7. Thermogravimetry (TGA)

To check the composition of the blends, thermogravimetric analysis (TGA) was carried out using a TG 209 F1 (Netzsch GmbH, Selb, Germany), operated with a protective nitrogen atmosphere with a flow rate of 60 mL/min. Samples (10 ± 0.2 mg) were placed in a ceramic crucible and heated from room temperature to 800 °C at a rate of 10 °C/min. The thermal properties of the blends were determined from TGA and DTG measurements as the temperature at which the mass loss was 5% ($T_{5\%}$) and the temperature of the maximum mass loss rate (T_{max}).

2.2.8. Mechanical Testing

The mechanical properties of the R-PE/R-PA blends were determined by tensile tests performed using a Zwick/Roell Z010 universal testing machine (Zwick GmbH & Co. KG, Ulm, Germany), together with the testXpert II program. The sample size was 110 mm × 6 mm × 4 mm, according to the ISO 527-2:2012 standard [36]. Tensile characteristics were measured at room temperature with a crosshead speed of 50 mm/min. Young's modulus (E), ultimate tensile strength (σ_M), tensile stress at break (σ_B), and elongation at break (ε_B) were evaluated from the tensile stress–strain curves. The reported data are averages of the results of 10 specimens. The hardness of samples was also measured using a Shore hardness tester (HBD 100–0, Sauter GmbH, Balingen, Germany) according to the PN-EN ISO 868:2005 [37]. The hardness was indicative of an average penetration value (Shore degrees on the D scale) based on 10 readings from tests.

3. Results

3.1. Morphological Aspects

The morphology and compatibility of R-PE/R-PA blends with and without PE-g-MAH were investigated by scanning electron microscopy (SEM). The results are given in Figure 1.

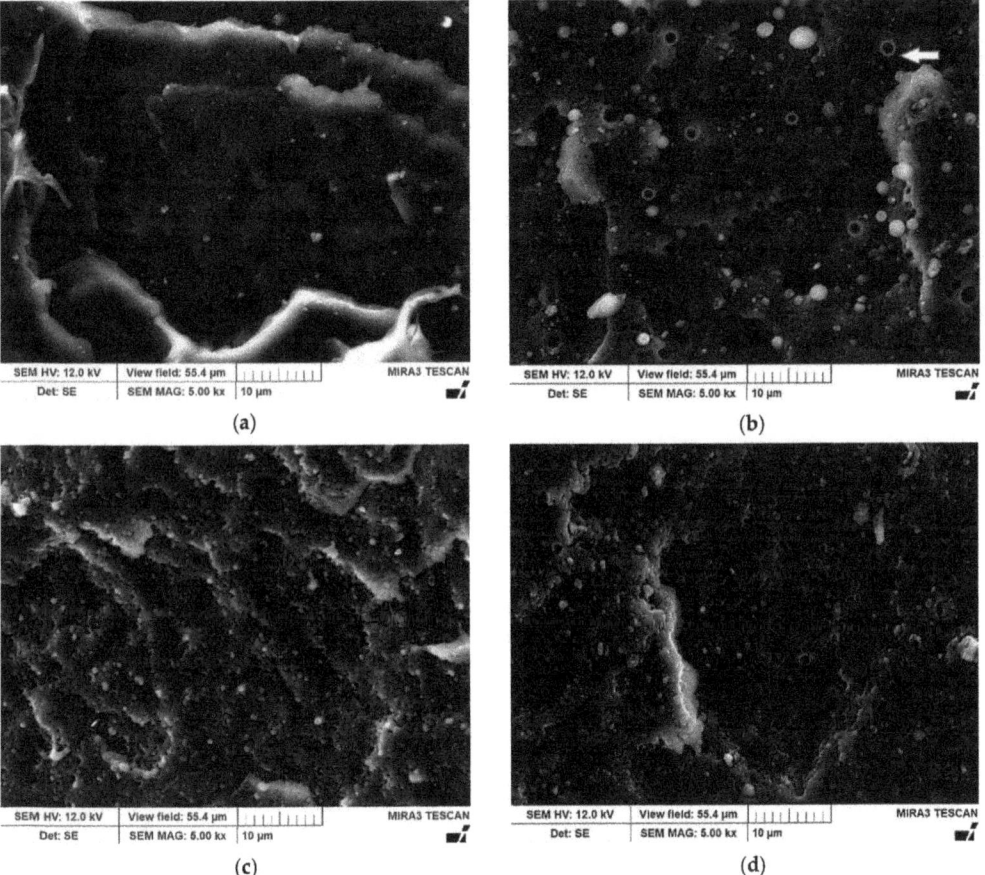

Figure 1. SEM micrographs of fractured surface of (**a**) R-PE and R-PE/R-PA (**b**) without and (**c**) with compatibilizer R-PE/R-PA/MAH1 and (**d**) R-PE/R-PA/MAH3; magnification 5000×.

Figure 1 shows SEM images of the fractured surfaces of R-PE, R-PE/R-PA binary blend, and R-PE/R-PA/MAH blends. This shows that PE and PA regranulates are incompatible by melt blending (Figure 1b). A two-phase morphology is observed on the fractured surface of the R-PE/R-PA blend. Polyamide disperses in a spherical shape in the continuous phase of polyethylene matrix. Figure 1b shows the "holes" (red circles) formed during fracture, which means that the weakly bound PA dispersed phase was "pulled out" from the polyethylene matrix. A similar effect in the PET/PA6 blend was observed by Lin et al. [38]. Because there is no adhesion between the R-PA phase and the R-PE polymer matrix, the cavities of the R-PA particles are clear and smooth. Adding a small quantity of compatibilizer to the blend resulted in changes in the morphology of the R-PE/R-PA blends (Figure 1c,d).

Figure 1c,d shows SEM micrographs of R-PE/R-PA blends after compatibilization by different PE-g-MAH content. By adding 1 wt.% PE-g-MAH, the size of the domain phase decreased, and the distribution became more uniform. It can be seen that the dimensions of dispersed PA domains decreased (from 2 ± 0.03 µm to less than 0.5 ± 0.05 µm). Based on the blend morphology, it can be seen that 1 wt.% PE-g-MAH added to the R-PE/R-PA waste blend increased the interfacial interactions and compatibility between polyethylene and polyamide 6, resulting in decreased polyamide particle size. Similar results were obtained for PA6/PE blends-clay nanocomposites by Scaffaro [39] and found that in the blends PA6/HDPE with an ethylene-co-acrylic acid copolymer (EAA) the dimension of the particles of the PA6 dispersed phase is smaller and the interfacial adhesion better [39].

3.2. Density and Melt Mass-Flow Rate (MFR) Analysis

The results of density and melt flow rate, the basic processing property of polymers, are presented in Table 2. MFR enables indirect assessment of the melt behavior of blends (viscosity), and the selection of technologies and parameters for further processing [10].

Table 2. Results of density and melt flow rate (MFR) of samples.

Sample	Density [g/cm^3]	MFR $_{(2.16\ kg,\ 230\ °C)}$ [g/10 min]
R-PA	1.181 ± 0.110	9.560 ± 0.174
R-PE	0.949 ± 0.006	1.030 ± 0.095
R-PE/R-PA	1.008 ± 0.014	1.535 ± 0.010
R-PE/R-PA/MAH1	1.012 ± 0.039	1.330 ± 0.062
R-PE/R-PA/MAH3	1.008 ± 0.019	1.237 ± 0.467

As shown in Table 2, for the R-PE/R-PA blend with 1 wt.% compatibilizer, the density increased, and the density did not change for the blend with 3 wt.% PE-g-MAH content. This may be due to the "better packing" of R-PA in a R-PE matrix as a result of better dispersion of the components involving compatibilizer. The compatibilizing agent improved the adhesion between the R-PA and R-PE phases, which decreased the gaps and free spaces at the interface, thus increasing the density of the blend.

The analysis of processing properties based on MFR shows that for the blend of R-PE/R-PA waste with 1 wt.% compatibilizer, a decrease in MFR occurred, i.e., an increase in the viscosity of the mixture, compared to polyethylene. This was due to the change in intermolecular interactions resulting from the formation of block polymers, which was a result of chemical reactions, as confirmed in the literature [19].

3.3. Water Absorption Analysis

The changes of water absorbency for polyamide, polyethylene, and their blends with grafted PE-g-MAH are shown in Table 3.

Table 3. Water absorption test results of samples.

Sample	Water Absorption [%]
R-PA	2.01 ± 0.11
R-PE	0.00 ± 0.00
R-PE/R-PA	0.56 ± 0.16
R-PE/R-PA/MAH1	0.14 ± 0.04
R-PE/R-PA/MAH3	0.22 ± 0.01

It can be seen from Table 3 that water absorptivity increased with addition of R-PA. This is due to the polar groups of PA, which absorb significant amounts of water, and PE and PE-g-MAH have little ability to absorb water [40]. However, the presence of moisture can significantly reduce the strength and dimensional stability of the products [41]. In the case of blends, water absorption decreased with the addition of PE-g-MAH compared to polyamide. For blends containing 1 and 3 wt.%, the reduction in water absorptivity was similar to the blend without PE-g-MAH. This is due to the homogeneity of the structure under the influence of the compatibilizing agent, which reduces the interactions of the strongly bound water molecules in the polyamide with polar amide groups. Hence, the less polar the polyamide particle, the lower the water absorption of the R-PE/R-PA blend [41], and for this reason the dimensional stability of products made with this blend is much higher.

3.4. Viscosity Evaluation

The flow curve (complex viscosity η^* as a function of angular frequency ω) of the studied materials at 230 °C is presented in Figure 2.

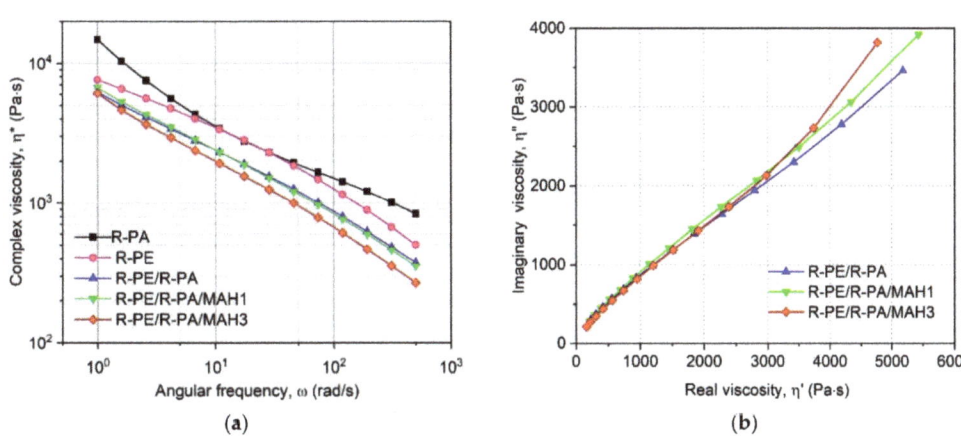

Figure 2. (a) Complex viscosity (η^*) as a function of angular frequency (ω) and (b) imaginary part of viscosity (η'') as a function of real part of viscosity (η') of studied materials at 230 °C.

As shown in Figure 2a, all tested samples presented shear-thinning behavior in the applied range of angular viscosity, which is rather typical for these materials [29]. Even though the course of the η^* vs. ω curves obtained for both polymers and their blends is similar, some differences can be seen. The most interesting behavior is presented by the R-PA sample: the slope of the curve becomes steeper in the lower frequency range. It can be presumed that this material does not approach the Newtonian plateau. Very similar results were obtained by Bai et al. for polyamide 12 at 230 °C [42]. This behavior can be explained by the fact that at higher shear rates, the polymeric chains move relative to each other more easily and their entanglement decreases. The complex viscosity of the unfilled

R-PE also decreases with the angular frequency, but the slope of the curve becomes less steep in the lower ω range, which results from the different molecular structures of the two polymers.

The R-PE/R-PA blend can be characterized by similar rheological properties as its components: the curve presented in Figure 2a is parallel to the one of R-PA but shifted toward lower viscosity values. This result can indicate the low miscibility of both polymers; the presence of polyamide macromolecules does not limit the possibility of movement of polyethylene chains and vice versa.

The addition of 1 wt.% MAH did not significantly change the run of the η^* vs. ω curve, except for a small increase in the lower angular frequency range. This may indicate a slight enhancement of interactions between phases [43,44]. In the case of the R-PE/R-PA/MAH-3 sample, decreased viscosity can be observed, which is probably caused by the lower content of highly viscous polyamide in the modified blend and a higher percentage of molten maleic anhydride, which flows easily and acts as a plasticizer. Similar results were obtained by Bai and Dou for polypropylene/polylactide blends modified by maleic anhydride-grafted polypropylene [42]. It can be concluded that the presence of the modifier should not cause difficulties related to the flow of the polymers, but rather will enhance the processability of the blends.

Cole–Cole plots representing imaginary viscosity η'' as a function of real viscosity η' are shown in Figure 2b. Such plots are commonly used to evaluate the miscibility of polymeric blends. Good compatibility of the studied components is indicated by the semicircular shape of the plot [45]. As can be seen in Figure 1b, the plot of the R-PE/R-PA blend is linear rather than semicircular, which confirms that the two components create a non-homogeneous mixture. Adding the compatibilizing agent does not result in a more circular shape of the plot. Due to the complex composition and morphology of the recycled materials, the Cole–Cole plot may not be the best method to evaluate the miscibility of the blends.

3.5. Fourier Transform Infrared (FT-IR) Analysis

Fourier transform infrared (FT-IR) spectroscopy was used to identify peak shifts between the two polymers, to detect whether the type of interaction between the materials was strong or weak [46]. Figure 3 shows the FT-IR spectra of pure materials R-PE and R-PA, the R-PE/R-PA blend, and the R-PE/R-PA blend with the addition of PE-g-MAH compatibilizer.

As shown in Figure 3, pure R-PE shows characteristic strong peaks at 2914 and 2847 cm^{-1} associated with the asymmetric and symmetric aliphatic $–CH_2$ stretching vibration of carbon–hydrogen bonds [47,48] and bending vibrations at 1463 and 718 cm^{-1}. The absorption at 1463 cm^{-1} corresponds to the stretching vibration attributed to the CH_2 bond ($–CH_2$ scissoring) and the band at 720 cm^{-1} ($–C–C$ rocking) refers to the stretching vibration of the carbon–carbon bond [49]. The polyamide spectrum presents bands of the ester group, C=O asymmetrical stretching of the imide group at 1738 cm^{-1}, amide I (C=O stretching) at 1637 cm^{-1} amide II (C–N stretching and N–H bending of hydrogen-bonded N–H groups at 1538 cm^{-1} [50]. The band at 3296 cm^{-1} is associated with hydrogen-bonded N–H stretching (amide A). The two strong bands at 2934 and 2865 cm^{-1} correspond to the asymmetric with respect to the symmetric C–H stretching vibrations [47]. R-PE/R-PA blends show N–H stretching at 2915 cm^{-1}, C=O stretching of the imide group at 1740 cm^{-1}, C=O stretching of the amide group at 1633 cm^{-1}, and N–H bonding and C–N stretching of amide II at 1533 cm^{-1} [46]. It can be seen that all R-PE/R-PA blends with PE-g-MAH compatibilizer show similar peaks to the R-PE/R-PA blend, except for the region around 1000 cm^{-1}, where the PE-g-MAH1 blend shows a stronger signal. No additional peaks of ester groups derived from maleic anhydride were observed. According to Tjong et al. [51], this is due to degradation of the MAH group grafted into –COOH during the melting process.

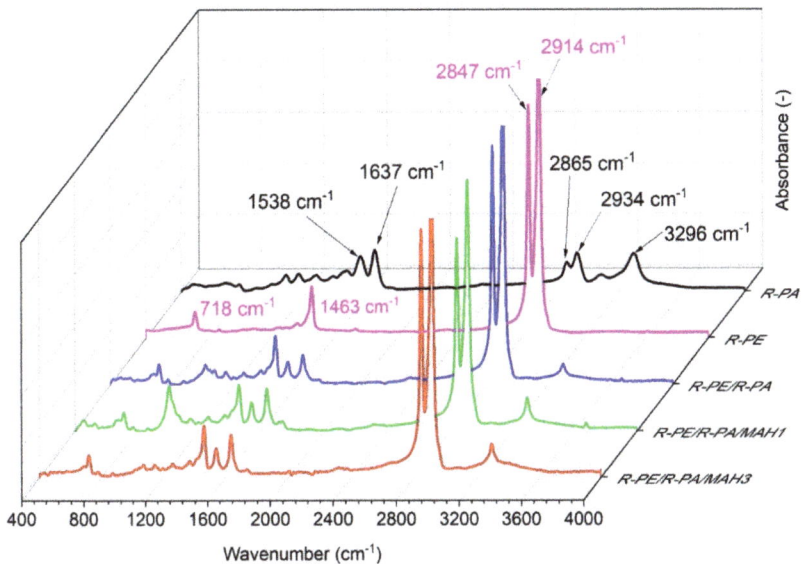

Figure 3. FT-IR spectra of R-PE, R-PA, R-PE/R-PA binary blend, and R-PE/R-PA blends grafted by PE-g-MAH.

3.6. Thermal Differential Scanning Calorimetry (DSC) Results

Heating and cooling DSC thermograms of the investigated samples are presented in Figure 4. The melting and crystallization temperatures and the melting enthalpy ΔH_m for R-PE, R-PE/R-PA, and R-PE/R-PA/MAH blends are summarized in Table 4.

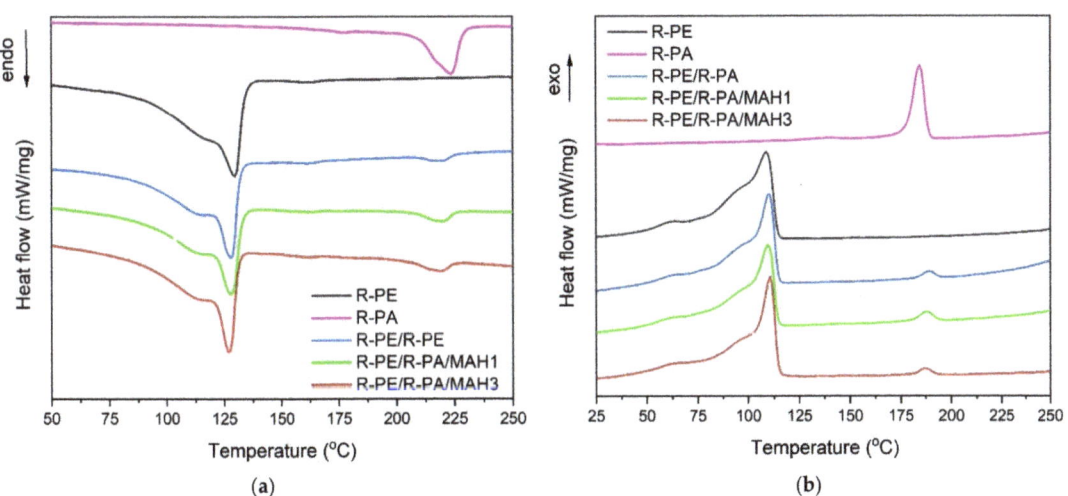

Figure 4. DSC (**a**) melting and (**b**) crystallization curves of R-PE, R-PA, R-PE/R-PA binary blend, and R-PE/R-PA blends grafted by PE-g-MAH.

Table 4. DSC data of R-PE, R-PA, R-PE/R-PA, and R-PE/R-PA/MAH blends (second heating at 10 °C/min); X_c (%) is the crystallinity degree of polyethylene phase in the blend.

Sample	T_{m1} [°C]	ΔH_{m1} [J/g]	T_{m2} [°C]	T_{cr1} [°C]	T_{cr2} [°C]	X_c [%]
R-PA	223.8	−29.8	-	-	184.6	15.7 ± 0.5
R-PE	125.0	−135.4	-	109.4	-	46.2 ± 1.1
R-PE/R-PA	126.0	−99.8	218.9	111.3	189.6	42.6 ± 0.7
R-PE/R-PA/MAH1	126.7	−110.9	219.3	111.5	189.4	47.3 ± 0.6
R-PE/R-PA/MAH3	127.0	−102.1	220.1	112.6	190.1	44.5 ± 1.1

As can be seen in Figure 4, all blends reveal two melting peaks T_{m1} and T_{m2}, characteristic of immiscible polymer mixtures with two phases [52]. The first one, T_{m1} at about 125 °C, and the second, T_{m2} at about 224 °C, correspond to the melting of polyethylene (R-PE) with respect to polyamide 6 (R-PA). The presence of these two values indicates the heterogeneous structure of the blends [19]. From these data, a slightly higher melting temperature (T_{m1}) of R-PE/R-PE at 3 wt.% PE-g-MAH loading was observed, which indicates the interfacial interaction between the blend in the presence of the compatibilizer. Moreover, two separate crystallization peaks (T_{cr1} and T_{cr2}) during the DSC cooling run (see Figure 4) were observed. The temperature of 184.6 °C corresponds to polyamide crystallization and about 109.0 °C to polyethylene crystallization. The presence of two crystallization peaks is also confirmation of the heterogeneous structure of the R-PE/R-PA/MAH blends. The polyethylene crystallization temperature (~ 109 °C) the slightly increased, up to about 113 °C with the PE-g-MAH concentration, in particular for 3 wt.% of the R-PE/R-PA/MAH3 blend, indicating that the PE-g-MAH may act as a nucleating agent that affects the polyethylene matrix. As shown in Table 4, a distinct increase in the degree of crystallization (X_c), as well as the heat of melting for R-PE/R-PA blends were noted. The X_c value, for the R-PE/R-PA/MAH1 blend increased from 43% for the R-PE/R-PA binary blend, up to about 47% for the R-PE/PA blend modified with 1 wt.%. of PE-g-MAH, indicating the chemical interactions between the PE-g-MAH and polyamide, which implied the molecular chain mobility of the R-PA that is needed for crystallization, resulting in an increase in polyamide crystallinity. A similar trend was observed by Zhang et al. [53], who investigated the effect of multiwall carbon nanotubes (MWCNTs) on the crystallization of polyamide in a blend with polypropylene. They found that the addition of MWCNTs to a R-PA/R-PP blend resulted in increased crystallinity of the PA phase. Similar results were reported by Yousfi et al. [54] in the case of PA/PP/PP-g-MAH blends.

3.7. Thermogravimetric Analysis (TGA) Results

The characteristic TGA curve and its first derivative differential of thermogravimetry (DTG) curve are shown in Figure 5. Thermogravimetric measurements are usually applied to detect the composition of blends, copolymers, and composites and to characterize the degradation process.

In our case, the temperature $T_{5\%}$ and $T_{50\%}$ of R-PE/R-PA/PE-g-MAH samples, the maximum temperature of decomposition of the blends T_{max}, and the residues at 800 °C are taken into account (see Table 5).

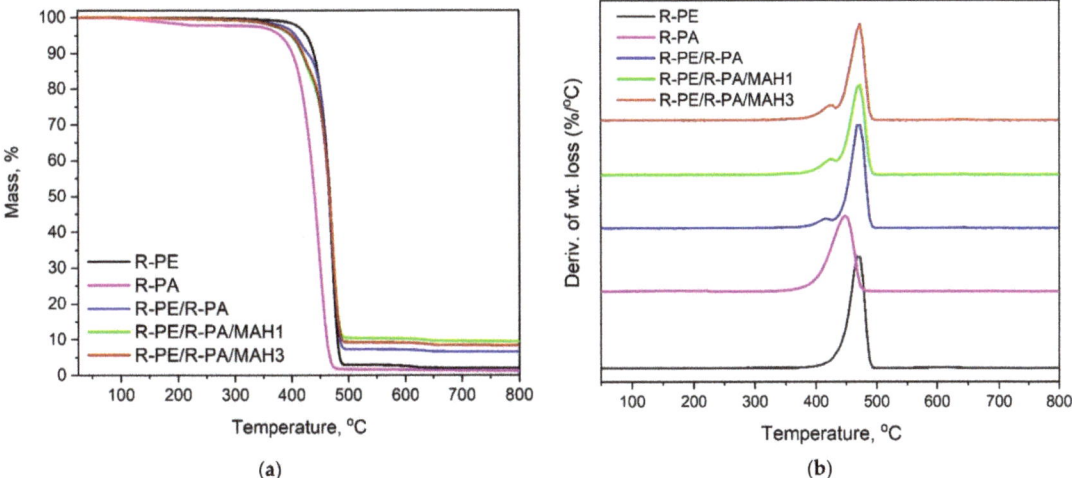

Figure 5. (a) TGA and (b) DTG curves of R-PE, R-PA, R-PE/R-PA binary blend, and R-PE/R-PA blends grafted by PE-g-MAH (heating rate: 10 °C/min; nitrogen atmosphere).

Table 5. Temperature $T_{5\%}$ and $T_{50\%}$ at mass loss of 5 wt.% and 50 wt.%, respectively, maximum peak temperature decomposition, and residual mass at 800 °C for R-PA, R-PE, and their blends.

Sample	$T_{5\%}$ [°C]	$T_{50\%}$ [°C]	Decomposition Temperature (Peak) [°C]	DTG [%/°C]	Residue at 800 °C [wt.%]
R-PA	382.0	440.5	448.7	−20.7	1.2
R-PE	439.0	466.0	470.3	−31.4	2.0
R-PE/R-PA	408.4	467.4	467.6	−28.5	6.5
R-PE/R-PA/MAH1	400.0	466.5	473.0	−24.9	9.4
R-PE/R-PA/MAH3	402.0	466.8	475.0	−28.2	8.3

From Table 5, we can find that the polyethene (R-PE) and polyamide (R-PA) undergo a single-step degradation process with onset at 470 and 450 °C, respectively. In the case of R-PE/R-PA blends, we can see a two-step decomposition process with a decomposition temperature starting at about 430 °C, corresponding to the degradation of the PA phase, and the second region of the main decomposition stage at a temperature of 470 °C. Additionally, it can be seen in Table 5 that the char residue at 800 °C for R-PE/R-PA blends grafted with PE-g-MAH is about 9 mass %, which means it is higher compared to R-PE/R-PA binary blend and polyethylene matrix (by about 2%), indicating that the compatibilized R-PE/R-PA/MAH blends have better char-forming ability than polymer matrix.

From Figure 5a we can see that the decomposition temperature ($T_{5\%}$) increases when PE-g-MAH is added to the PE/PA blend. When 3 wt.% of PE-g-MAH is added, the decomposition temperature (as measured at the point of 5% mass loss) increases from 401 to about 408 °C. This means that the R-PE/R-PA/MAH blend has the highest thermal stability compared with polyethylene and polyamide. This may be due to the good dispersion of the polyamide in the polyethylene matrix with the participation of the compatibilizer. According to Zhai, the reason for this may be that the thermal properties of PE-g-MAH are much better than those of polyethylene [55]. The result is consistent with what was found by Yousi et al. [54] in case of talc-filled PP/PA nanocomposites. The authors noted that the addition of talc fillers induced a significant decrease in the size of the PA6 domains, which is related to the higher thermal stability of the PP/PA blends [54]. A similar effect was observed by Araújo et al. in a mixture of polyamide with high-density polyethylene and the compatibilizer [56].

3.8. Tensile Testing and Hardness Results

Figure 6 presents the tensile stress–strain curves for R-PA, R-PE, and R-PE/R-PA blends based on the procedure described in the previous section. Regranulated R-PE/R-PA blends grafted with 1 and 3 wt.% PE-g-MAH as compatibilizer were investigated. We can see that the R-PE curves are ductile and those of R-PE with R-PA blend based on wastes are rigid and brittle, like polyamide waste. However, significant differences can be seen in the curves for the PE-g-MAH modified blends. The strain–stress curves of the R-PE/R-PE blends show a significant increase in elongation at break compared to polyethylene/polyamide blend.

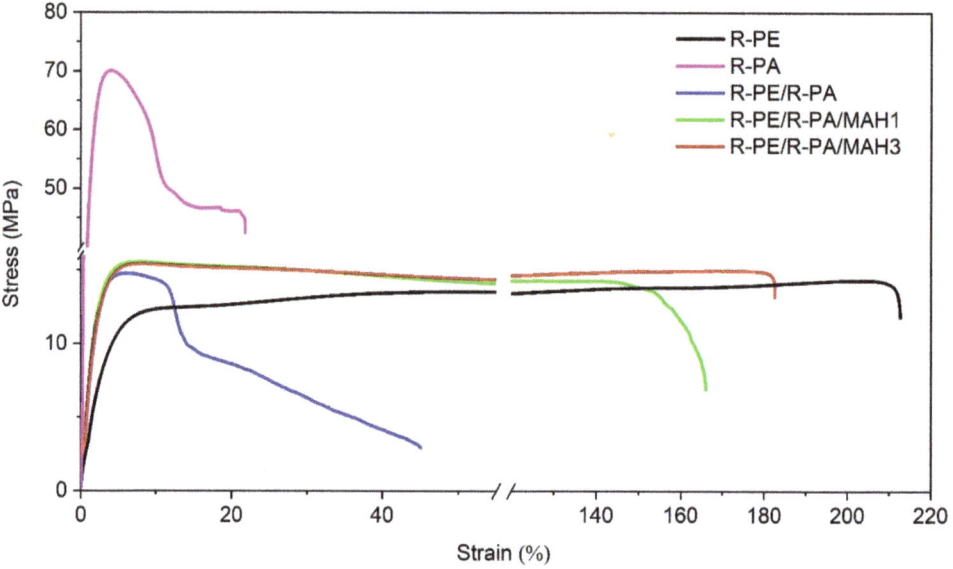

Figure 6. Stress–strain curves of R-PE, R-PA, and R-PE/R-PA blend with varying compatibilizer content.

The effects of different PE-g-MAH content on the Young's modulus (E), ultimate tensile strength (σ_M), tensile stress at break (σ_B), elongation at break (ε_B), and Shore'a hardness of all samples are summarized in Table 6.

Table 6. Results of mechanical testing.

Sample Code	Young's Modulus, E [MPa]	Ultimate Tensile Strength, σ_M [MPa]	Tensile Stress at Break, σ_B [MPa]	Elongation at Break, ε_B [%]	Hardness Shore'a [°Sh]
R-PA	1130 ± 15	70.4 ± 0.3	42.6 ± 1.1	25 ± 12	80.3 ± 0.5
R-PE	254 ± 17	11.9 ± 0.2	10.2 ± 1.0	210 ± 11	54.2 ± 0.3
R-PE/R-PA	371 ± 12	19.7 ± 0.1	2.9 ± 0.1	48 ± 18	56.1 ± 0.2
R-PE/R-PA/MAH1	456 ± 16	20.5 ± 0.4	18.5 ± 1.7	165 ± 15	57.2 ± 0.1
R-PE/R-PA/MAH3	474 ± 11	21.7 ± 0.2	20.1 ± 0.6	187 ± 23	61.3 ± 0.3

The results show that ultimate tensile strength for the R-PE/R-PA blends increased with the addition of PE-g-MAH as a compatibilizer. Tensile stress at break for R-PE/R-PA blends increased for samples with MAH compatibilizer, indicating that a small amount of 1 wt.% PE-g-MAH leads to better interfacial adhesion between the blend components.

As shown in Figure 6, the tensile strength at break of R-PE decreased when recycled polyamide was added, and then increased with the addition of PE-g-MAH. In the blend that contained 3 wt.% MAH, Young's modulus and tensile strength increased by 190 and 19%, respectively, compared to recycled polyethylene. The result shows that the hardness of samples increased for the blends with R-PE/R-PA and PE-g-MAH.

The increased strength and hardness of R-PE/R-PA blends with the addition of a compatibilizer in relation to the two-component mixture may indicate the compatibilizing effect of PE-g-MAH. It can be concluded that during the reactive extrusion process, block copolymers can form at the phase boundary, causing the formation of strong interfacial bonds between PE-g-MAH groups and end groups of polyamide 6 [57].

Similar results were presented by Hamid et al. [46], who noted that incorporating HDPE-g-MAH into blends improved hardness as a result of increased interfacial adhesion in the blends. As seen in Table 6, for the R-PE/R-PA blend with PE-g-MAH, we observed an increase in both tensile strength and elongation at break. The elongation at break of the R-PE/R-PA/MAH3 compatibilized blends increased from 48 to 186%, which corresponds to an increase of approximately 300%. The increase in tensile stress at break of the R-PE/R-PA/PE-g-MAH blend can be attributed to the increased degree of crystallinity, due to the increased miscibility of thermodynamic non-miscibility of polyethylene and polyamide mixture wastes. There is good agreement with the calorimetric results.

The results show a significant increase of Young's modulus for the R-PE blend with PA, which corresponds to an increase in stiffness to about 100% compared to the recycled polyethylene. The higher tensile modulus values of the R-PE/R-PA regranulate shows that the addition of the PE-g-MAH by extrusion to the polyethylene/polyamide binary blends improved partial miscibility and compatibility of these blends. A similar behavior in reactive compatibilization polyamide 6/olefin block copolymer blends has been reported by Lin, et al. [38].

4. Conclusions

The investigations carried showed a significant influence of PE g-MAH on morphology and mechanical properties of the R-PE/R-PA blends. In this study, the R-PE/R-PA blends that contained PE-g-MAH was prepared by reactive extrusion in a twin-screw extruder. It was found that the addition of PE-g-MAH allows for an advantage combination of immiscible polymers come from waste. Thus, it was proven that this connection is possible, and from the obtained blends, it is possible to produce regranulates for industrial applications.

A morphological study showed that the PE-g-MAH significantly improved the structure of the R-PE/RPA/MAH blend compared to R-PE/R-PA binary blend, resulting in reduction dimension of R-PA domains size of the disperse phases and better interfacial adhesion. SEM presented that adding a small (1 wt.%) amount of PE-g-MAH compatibilizer to the R-PE/R-PA waste blend increases the interfacial interactions of this composition, thus ensuring its structural coherence.

R-PE/R-PA blends containing 1 wt.% compatibilizer showed reduced water absorption compared to the polyamide, due to the homogeneity of the R-PE/R-PA blend grafted by PE-g-MAH structure under the influence of the compatibilizing agent, which reduces the interactions of the strongly bound water molecules in the polyamide with polar amide groups. Viscosity analysis showed that in the case of blends made of R-PE/R-PA waste with the addition of a compatibilizer, there is a slight decrease in the viscosity of the blend in relation to R-PE or R-PE/R-PA binary blend, due to the lower viscosity of PE-g-MAH compared with polyethylene and polyamide. However, the decreased MFR shows that the influence of the compatibilizing agent on the rheological properties of the blend depends on the applied shearing conditions. It can also be concluded that the application of PE-g-MAH should not cause problems during processing of the blends. The results of the mechanical properties show a significant increase in Young's modulus for the R-PE blend with R-PA grafted by PE-g-MAH, which corresponds to an increase of about 100%

compared to recycled polyethylene. The higher tensile properties in R-PE/R-PA/MAH were related to an increase in crystalline structure. This demonstrates its improved stiffness and strength when considering using this material for films with increased strength and resistance to destruction, due to the increased degree of crystallinity of R-PE/R-PA with compatibilizer compared to polyethylene matrix.

The proposed method of waste management with defined characteristics and structure to produce regranulates based on unmixed polymers of polyethylene and polyamide with the addition of a compatibilizer may be attractive in terms of application in the packaging industry for the production of films. In particular, the use of recycled polyethylene and polyamide can be a useful technique for processing regranulates based on immiscible polymers, creating a sustainable solution for an environmental problem. This study proves that R-PE/R-PA regranulates are waste plastics that can be combined despite their incompatibility to create materials for further applications, which is in line with the circular economy principle.

Author Contributions: Conceptualization, D.C.-K. and J.N.-G.; methodology, D.C.-K. and J.N.-G.; formal analysis, D.C.-K., investigation, D.C.-K., M.T. and O.M.; resources, O.M., K.G., M.T. and D.C.-K.; data curation, D.C.-K.; writing—original draft preparation, K.G., O.M. and D.C.-K.; writing—review and editing, K.G., O.M., J.N.-G. and D.C.-K.; supervision, D.C.-K. and K.G. and.; project administration, D.C.-K. and J.N.-G.; funding acquisition, D.C.-K. and K.G. All authors have read and agreed to the published version of the manuscript.

Funding: This research was funded by the Ministry of Science and Higher Education of the Republic of Poland, grant no. 0613/SBAD/4710 and grant no.1/S/KPBMIM/20.

Institutional Review Board Statement: Not applicable.

Informed Consent Statement: Not applicable.

Data Availability Statement: The data presented in this study are available on request from the corresponding author.

Acknowledgments: The authors would like to thank Mikolaj Poplawski from the Poznan University of Technology, Poland for SEM microscopic images.

Conflicts of Interest: The authors declare no conflict of interest.

References

1. PlasticsEurope. Available online: https://www.plasticseurope.org/pl/focus-areas/circular-economy (accessed on 25 April 2021).
2. Laguna, O.; Collar, P.E.; Taranco, J. Reuse of plastics recovered from solid wastes. Thermal and morphological studies from HDPE/LDPE blends. *J. Appl. Polym. Sci.* **1989**, *38*, 667–685. [CrossRef]
3. Czarnecka-Komorowska, D.; Wiszumirska, K. Sustainability design of plastic packaging for the Circular Economy. *Polimery* **2020**, *65*, 8–17. [CrossRef]
4. Warguła, Ł.; Wieczorek, B.; Kukla, M.; Krawiec, P.; Szewczyk, J. The Problem of Removing Seaweed from the Beaches: Review of Methods and Machines. *Water* **2021**, *13*, 736. [CrossRef]
5. Texts Adopted—A European Strategy for Plastics in a Circular Economy. Available online: https://www.europarl.europa.eu/doceo/document/TA-8-2018-0352_PL.html (accessed on 25 April 2021).
6. Corona, B.; Shen, L.; Reike, D.; Carreón, J.R.; Worrell, E. Towards Sustainable Development through the Circular Economy—A Review and Critical Assessment on Current Circularity Metrics. *Resour. Conserv. Recycl.* **2019**, *151*, 104498. [CrossRef]
7. Aderogba, K.A. Polymer Wastes and Management in Cities and Towns of Africa and Sustainable Environment: Nigeria and European Experiences. *Soc. Sci.* **2014**, *3*, 79. [CrossRef]
8. Kwiatek, G. Manufacturing Process Analysis of Polyamide/Polyolefins Blends. BSc Thesis, Faculty of Mechanical Engineering, Poznan University of Technology, Poznań, Poland, 2019. (In Polish).
9. Dorigato, A. Recycling of polymer blends. *Adv. Ind. Eng. Polym. Res.* **2021**, *4*, 53–69. [CrossRef]
10. Garofalo, E.; Claro, M.; Scarfato, P.; Di Maio, L.; Incarnato, L. Upgrading of Recycled Plastics Obtained from Flexible Packaging Waste by Adding Nanosilicates. *AIP Conf. Proc.* **2015**, *1695*, 020053. [CrossRef]
11. Czarnecka-Komorowska, D.; Wiszumirska, K.; Garbacz, T. Films LDPE/LLDPE made from Post—Consumer Plastics: Processing, structure, mechanical properties. *Adv. Sci. Technol. Res. J.* **2018**, *12*, 134–142. [CrossRef]
12. Czarnecka-Komorowska, D.; Sterzynski, T.; Andrzejewski, J. Evaluation of Structure and Thermomechanical Properties of Polyoxymethylene Modified with Polyhedral Oligomeric Silsesquioxanes (POSS). *Przem. Chem.* **2013**, *92*, 2129–2132.

13. Barczewski, M.; Matykiewicz, D.; Mysiukiewicz, O.; Maciejewski, P. Evaluation of polypropylene hybrid composites containing glass fiber and basalt powder. *J. Polym. Eng.* **2018**, *38*, 281–289. [CrossRef]
14. Czarnecka-Komorowska, D.; Wiszumirska, K.; Garbacz, T. Manufacturing and Properties of Recycled Polyethylene Films with an Inorganic Filler by the Extrusion Blow Moulding Method. In *Advances in Manufacturing II—Mechanical Engineering*; Gapinski, B., Szostak, M., Ivanov, V., Eds.; Springer: Berlin/Heidelberg, Germany, 2019; Volume 4, pp. 625–638. ISBN 978-3-030-16943-5.
15. Dobrzyńska-Mizera, M.; Dutkiewicz, M.; Sterzyński, T.; Di Lorenzo, M.L. Isotactic polypropylene modified with sorbitol-based derivative and siloxane-silsesquioxane resin. *Eur. Polym. J.* **2016**, *85*, 62–71. [CrossRef]
16. Czarnecka-Komorowska, D.; Grześkowiak, K.; Popielarski, P.; Barczewski, M.; Gawdzińska, K.; Popławski, M. Polyethylene Wax Modified by Organoclay Bentonite Used in the Lost-Wax Casting Process: Processing–Structure–Property Relationships. *Materials* **2020**, *13*, 2255. [CrossRef]
17. Merijs Meri, R.; Zicans, J.; Maksimovs, R.; Ivanova, T.; Kalnins, M.; Berzina, R.; Japins, G. Elasticity and Long-Term Behavior of Recycled Polyethylene Terephthalate (RPET)/Montmorillonite (MMT) Composites. *Compos. Struct.* **2014**, *111*, 453–458. [CrossRef]
18. Goodship, V.; Hulme, A.; Smith, G.F. Improving Processing and Material Characteristics of Contaminated End-of-Life Polypropylene by Compatibilisation. *Polym. Recycl.* **2000**, *5*, 205–211.
19. Jeziorska, R. Recycling of multilayer films by reactive extrusion. *Polimery* **2003**, *48*, 130–133. [CrossRef]
20. Utracki, L.A.; Mukhopadhyay, P.; Gupta, R.K. Polymer Blends: Introduction. In *Polymer Blends Handbook*; Utracki, L.A., Wilkie, C.A., Eds.; Springer: Dordrecht, The Netherlands, 2014; pp. 3–170. ISBN 978-94-007-6064-6.
21. Utracki, L.A. Compatibilization of Polymer Blends. *Can. J. Chem. Eng.* **2002**, *80*, 1008–1016. [CrossRef]
22. Janik, J.; Krala, G.; Pilawka, R.; Gorący, K. Effect of Compatibilizer and Processing Condition on Properties of PC/PP Blends. *Plast. Process.* **2012**, *18*, 578–581.
23. Mengual, A.; Juarez, D.; Balart, R.; Ferrandiz, S. PE-g-MA, PP-g-MA and SEBS-g-MA compatibilizers used in material blends. *Procedia Manuf.* **2017**, *13*, 321–326. [CrossRef]
24. Jeziórska, R.; Szadkowska, A. Wpływ PE-g-MAH na właściwości mieszanin poliamidu i odpadów włókienniczych zawierających poli(tereftalan etylenu). Zeszyty Naukowe Politechniki Poznańskiej. *Budowa Masz. Zarządzanie Prod.* **2007**, *Nr 4*, 75–80. (In Polish)
25. Jeziórska, R. Wpływ metylomaleinianu rycynolo-2-oksazoliny na strukturę oraz właściwości mieszanin poliamidu 6 i polietylenu małej gęstości. *Polimery* **2004**, *49*, 623–633. (In Polish)
26. Psarski, M.; Pracella, M.; Galeski, A. Crystal phase and crystallinity of polyamide 6/functionalized polyolefin blends. *Polymer* **2000**, *41*, 4923–4932. [CrossRef]
27. Wahab, M.K.A.; Ismail, H.; Othman, N. Compatibilization Effects of PE-g-MA on Mechanical, Thermal and Swelling Properties of High Density Polyethylene/Natural Rubber/Thermoplastic Tapioca Starch Blends. *Polym. Technol. Eng.* **2012**, *51*, 298–303. [CrossRef]
28. Majid, R.A.; Ismail, H.; Taib, R.M. Effects of PE-g-MA on Tensile Properties, Morphology and Water Absorption of LDPE/Thermoplastic Sago Starch Blends. *Polym. Technol. Eng.* **2009**, *48*, 919–924. [CrossRef]
29. Huitric, J.; Ville, J.; Médéric, P.; Moan, M.; Aubry, T. Rheological, morphological and structural properties of PE/PA/nanoclay ternary blends: Effect of clay weight fraction. *J. Rheol.* **2009**, *53*, 1101. [CrossRef]
30. SIGMA-ALDRICH: Product Specification—Polyethylene-Graft-Maleic Anhydride. Available online: https://api.sigmaaldrich.com/deepweb/assets/sigmaaldrich/quality/spec/279/292/456632-BULK_____ALDRICH__.pdf (accessed on 15 June 2021).
31. Michalska-Pożoga, I.; Rydzkowski, T.; Mazur, P.; Sadowska, O.; Thakur, V.K. A Study on the Thermodynamic Changes in the Mixture of Polypropylene (PP) with Varying Contents of Technological and Post-User Recyclates for Sustainable Nanocomposites. *Vacuum* **2017**, *146*, 641–648. [CrossRef]
32. ISO 1183-1:2004. Plastics—Methods for Determining the Density of Non-Cellular Plastics—Part 1: Immersion Method, Liquid Pyknometer Method and Titration Method. Available online: https://www.iso.org/standard/27790.html (accessed on 26 April 2021).
33. ISO 62:2008. Plastics—Determination of Water Absorption. Available online: https://www.iso.org/standard/41672.html (accessed on 26 April 2021).
34. ISO 1133-1:2011. Plastics—Determination of the Melt Mass-Flow Rate (MFR) and Melt Volume-Flow Rate (MVR) of Thermoplastics—Part 1: Standard Method. Available online: https://www.iso.org/cms/render/live/en/sites/isoorg/contents/data/standard/42/44273.html (accessed on 15 July 2021).
35. Wunderlich, B. *Macromolecular Physics*; Academic Press: New York, NY, USA, 1976; Volume 2.
36. ISO 527-2:2012. Plastics—Determination of Tensile Properties—Part 2: Test Conditions for Moulding and Extrusion Plastics. Available online: https://www.iso.org/cms/render/live/en/sites/isoorg/contents/data/standard/05/60/56046.html (accessed on 15 July 2021).
37. ISO 868:2003. Plastics and Ebonite—Determination of Indentation Hardness by Means of a Durometer (Shore Hardness). Available online: https://www.iso.org/cms/render/live/en/sites/isoorg/contents/data/standard/03/48/34804.html (accessed on 15 July 2021).
38. Lin, X.; Qian, Q.; Xiao, L.; Chen, Q.; Huang, Q.; Zhang, H. Influence of Reactive Compatibilizer on the Morphology, Rheological, and Mechanical Properties of Recycled Poly(Ethylene Terephthalate)/Polyamide 6 Blends. *J. Macromol. Sci. Part B* **2014**, *53*, 1543–1552. [CrossRef]

39. Scaffaro, R.; Botta, L.; Mistretta, M.C.; La Mantia, F.P. Processing—Morphology—Property relationships of polyamide 6/polyethylene blend–clay nanocomposites. *Express Polym. Lett.* **2013**, *7*, 873–884. [CrossRef]
40. Chow, W.S.; Abu Bakar, A.; Ishak, Z.A.M. Water absorption and hygrothermal aging study on organomontmorillonite reinforced polyamide 6/polypropylene nanocomposites. *J. Appl. Polym. Sci.* **2005**, *98*, 780–790. [CrossRef]
41. Monson, L.; Braunwarth, M.; Extrand, C.W. Moisture absorption by various polyamides and their associated dimensional changes. *J. Appl. Polym. Sci.* **2007**, *107*, 355–363. [CrossRef]
42. Bai, Z.; Dou, Q. Rheology, Morphology, Crystallization Behaviors, Mechanical and Thermal Properties of Poly(lactic acid)/Polypropylene/Maleic Anhydride-Grafted Polypropylene Blends. *J. Polym. Environ.* **2018**, *26*, 959–969. [CrossRef]
43. Basseri, G.; Mazidi, M.M.; Hosseini, F.; Aghjeh, M.K.R. Relationship among microstructure, linear viscoelastic behavior and mechanical properties of SBS triblock copolymer-compatibilized PP/SAN blend. *Polym. Bull.* **2014**, *71*, 465–486. [CrossRef]
44. Chow, W.; Abu Bakar, A.; Ishak, Z.M.; Karger-Kocsis, J.; Ishiaku, U. Effect of maleic anhydride-grafted ethylene–propylene rubber on the mechanical, rheological and morphological properties of organoclay reinforced polyamide 6/polypropylene nanocomposites. *Eur. Polym. J.* **2005**, *41*, 687–696. [CrossRef]
45. Chen, Y.; Zou, H.; Cao, Y.; Liang, M. Melt miscibility of HDPE/UHMWPE, LDPE/UHMWPE, and LLDPE/UHMWPE blends detected by dynamic rheometer. *Polym. Sci. Ser. A* **2014**, *56*, 630–639. [CrossRef]
46. Hamid, F.; Akhbar, S.; Halim, K.K. Mechanical and Thermal Properties of Polyamide 6/HDPE-g- MAH/High Density Polyethylene. *Procedia Eng.* **2013**, *68*, 418–424. [CrossRef]
47. González-Henríquez, C.M.; Terraza, C.; Sarabia, M.A.; Vera, A.M.; Aliaga, Á.E. Structure correlation of silylated dicarboxylic acid monomer and its respective oligomeric polyamide-imide using experimental and theoretical vibrational spectra. *Spectrosc. Lett.* **2017**, *50*, 30–38. [CrossRef]
48. Grzeskowiak, K.; Czarnecka-Komorowska, D.; Sytek, K.; Wojciechowski, M. Influence of Waxes Remelting Used in Investment Casting on Their Thermal Properties and Linear Shrinkage. *Metalurgija* **2015**, *54*, 350–352.
49. Charles, J.; Ramkumaar, G.R. Qualitative Analysis of High Density Polyethylene Using FTIR Spectroscopy. *Asian J. Chem.* **2009**, *21*, 8.
50. Krylova, V.; Dukštienė, N. The structure of PA-Se-S-Cd composite materials probed with FTIR spectroscopy. *Appl. Surf. Sci.* **2019**, *470*, 462–471. [CrossRef]
51. Tjong, S.C.; Bao, S. Fracture toughness of high density polyethylene/SEBS-g-MA/montmorillonite nanocomposites. *Compos. Sci. Technol.* **2007**, *67*, 314–323. [CrossRef]
52. Koning, C. Strategies for compatibilization of polymer blends. *Prog. Polym. Sci.* **1998**, *23*, 707–757. [CrossRef]
53. Zhang, L.; Wan, C.; Zhang, Y. Investigation on the multiwalled carbon nanotubes reinforced polyamide 6/polypropylene composites. *Polym. Eng. Sci.* **2009**, *49*, 1909–1917. [CrossRef]
54. Yousfi, M.; Livi, S.; Dumas, A.; Crépin-Leblond, J.; Greenhill-Hooper, M.; Duchet-Rumeau, J. Compatibilization of polypropylene/polyamide 6 blends using new synthetic nanosized talc fillers: Morphology, thermal, and mechanical properties. *J. Appl. Polym. Sci.* **2014**, *131*, 40453. [CrossRef]
55. Zhai, H.; Xu, W.; Guo, H.; Zhou, Z.; Shen, S.; Song, Q. Preparation and characterization of PE and PE-g-MAH/montmorillonite nanocomposites. *Eur. Polym. J.* **2004**, *40*, 2539–2545. [CrossRef]
56. Araújo, J.R.; Vallim, M.R.; Spinacé, M.A.S.; De Paoli, M.-A. Use of postconsumer polyethylene in blends with polyamide 6: Effects of the extrusion method and the compatibilizer. *J. Appl. Polym. Sci.* **2008**, *110*, 1310–1317. [CrossRef]
57. Darie, R.N.; Vasile, C.; Kozlowski, M. The effect of compatibilization in reactive processing low density polyethylene/polyamide 6/EPDM blends. *Polimery* **2006**, *51*, 656–661. [CrossRef]

Article

Chemical Recycling of Vacuum-Infused Thermoplastic Acrylate-Based Composites Reinforced by Basalt Fabrics

Inès Meyer zu Reckendorf [1,2], Amel Sahki [1,2], Didier Perrin [1,*], Clément Lacoste [1], Anne Bergeret [1], Avigaël Ohayon [2] and Karynn Morand [2]

[1] Polymers Composites and Hybrids (PCH), IMT Mines Alès, 30100 Alès, France; ines.meyer-zu-reckendorf@mines-ales.fr (I.M.z.R.); amel.sahki@mines-ales.fr (A.S.); clement.lacoste@mines-ales.fr (C.L.); anne.bergeret@mines-ales.fr (A.B.)

[2] SEGULA Technologies, 69200 Vénissieux, France; avigael.ohayon@segula.fr (A.O.); karynn.morand@segula.fr (K.M.)

* Correspondence: didier.perrin@mines-ales.fr

Abstract: The objective of this work was to compare the material recovered from different chemical recycling methodologies for thermoplastic acrylate-based composites reinforced by basalt fabrics and manufactured by vacuum infusion. Recycling was done via chemical dissolution with a preselected adapted solvent. The main goal of the study was to recover undamaged basalt fabrics in order to reuse them as reinforcements for "second-generation" composites. Two protocols were compared. The first one is based on an ultrasound technique, the second one on mechanical stirring. Dissolution kinetics as well as residual resin percentages were evaluated. Several parameters such as dissolution duration, dissolution temperature, and solvent/composite ratio were also studied. Recycled fabrics were characterized through SEM observations. Mechanical and thermomechanical properties of second-generation composites were determined and compared to those of virgin composites (called "first-generation" composites). The results show that the dissolution protocol using a mechanical stirring is more adapted to recover undamaged fabrics with no residual resin on their surface. Moreover, corresponding second-generation composites display equivalent mechanical properties than first generation ones.

Keywords: recycling; thermoplastic composites; basalt fibers; ultrasound; mechanical properties

1. Introduction

1.1. Thermoplastic Composite Materials in the Automotive Industry

In 2019, the global market of polymer-reinforced composites materials reached 17.7 megatons in volume with a value of $86 billion. In virtually every region of the world and every application sector, the composites market is growing in both volume and value. China (28%) and North America (26%) remain the largest markets in terms of volume, ahead of Europe (21%) and the rest of Asia (19%) [1].

The main industries using composites, in volume, are transportation (28%), ahead of construction (20%), electronics and electrical (16%), and pipes and tanks (15%) [1]. The automotive industry is one of the largest users of composites in the transportation sector owing to demand for weight reduction in vehicles; this is mainly driven by the demand for better fuel efficiency and reduced CO_2 emissions in order to comply with EU legislation. From 1 January 2020, this legislation set an EU fleet-wide target of 95 gCO_2/km for the average emissions of new passenger cars and an EU fleet-wide target of 147 gCO_2/km for the average emissions of new light commercial vehicles registered in the EU [2]. Moreover, a reduction in weight also implies lower environmental impacts (from <130 gCO_2/km in 2015 to <95 gCO_2/km by 2021). Composites can offer some lightening compared to other traditionally dominant structural materials (steel, aluminum, etc.) ranging from 15–25% for glass fiber reinforced polymers (GFRP) to nearly 25–40% for carbon fiber reinforced

polymers (CFRP). The benefits of lightweight solutions can be translated into potential savings of 8 million tons of CO_2 per year in the EU wide vehicle fleet [3].

Consequently, since the 2010s, car makers have been using composite materials for passenger cells (CFRP), car roofs (both CFRP and GFRP), and fluid filter module (GFRP) of their cars [3,4]. Currently choosing CFRP as a chassis material for vehicles is a good solution because of its high strength and low weight. As the production cost of this type of material tends to decrease, different types of CFRP are present in a vehicle chassis, such as the header above the windshield, door sills, transmission tunnel, front-to back and left-to-right roof reinforcement tubes and bows, the B-pillar between front and rear doors which provides structural support for the vehicle roof panel, the C-pillar (the vertical structure behind the rear door), and rear parcel shelf [4]. GFRP have been gradually used for impact-absorbent automotive parts (such as instrument panel and inner door modules).

Glass fibers are by far the most used reinforcement (88%) all composite markets combined [1], but other mineral fibers have been emerging over these last years. Among them, basalt fibers are one of the most promising.

Basalt fibers are made of solidified lava. The extrusion of this mineral allows the production of basalt fibers. This type of fiber has been widely used in the production of reinforced concrete for floors, construction, and even roads. Currently, this fiber is attracting increasing interest because of its lower cost than glass fibers for equivalent mechanical properties and its better alkaline resistance than E-glass fibers. In a review on basalt fibers, Fiore et al. [5] reported higher mechanical properties for basalt fibers compared to glass fibers with an elastic modulus of 89 GPa, and a tensile strength of 2.8 GPa (76 GPa and 1.8–2.5 GPa, respectively, for E-glass fibers). As the density of basalt fibers (2800 kg/m^3) is higher than those of glass fibers (2560 kg/m^3), equivalent specific mechanical properties are obtained for basalt fibers compared to glass fibers (around 30 GPa per kg/m^3).

In the field of composite materials, thermoset resins are the most used (61%) ahead of thermoplastic resins (38%). Thermoplastic composites have become a real industrial alternative to traditional thermoset composites due to several benefits—among them, the fact that they are recyclable and take part to circular economy [6]. At present, the more commonly used thermoplastic matrices in the automotive industry are polypropylene (PP), polyamide 6 (PA6), polyamide 6.6 (PA66), polyoxymethylene (polyacetal) (POM), high density polyethylene (HDPE), poly(methyl methacrylate) (PMMA), polycarbonates (PC) and polyvinyl chloride (PVC), acrylonitrile butadiene styrene (ABS), and polyetheretherketone (PEEK) [7,8].

Elium® resin is a new thermoplastic polymer produced by Arkema Co. (Paris Villepinte, France) since 2014. It is obtained from a low viscosity reactive mixture of polymer solutions of methacrylic monomers (methyl methacrylate (MMA), alkyl acrylic) and acrylic copolymer chains (viscosity of 100 cPs) and possibly other comonomers. Its polymerization is activated by a dibenzoyl peroxide as a thermal initiator and can be carried out at room temperature in the presence of an iron salt catalyst [9]. Elium® resin is a resin adapted to the vacuum infusion process at room temperature. Regarding its mechanical properties, the tensile modulus is around 3.3 GPa and tensile strength is around 76 MPa with an elongation of 6% [6].

Elium®/basalt fibers composites appear to be a promising composite alternative, combining good specific mechanical properties for transportation applications and recycling potential.

1.2. Fiber Reinforced Composites Recycling Techniques

Different types of material recoveries could be employed to value the components of composites.

Firstly, mechanical recycling processes consist of crushing the materials after an initial cutting step into small pieces, a step common to all recycling techniques. This mechanical process does not separate the fiber from the resin. Mechanical shredding has been applied more to GFRC, particularly SMC (Sheet Molding Compound) and BMC (Bulk Molding Compound), but studies on CFRC also exist [10].

In addition, several thermal recycling processes allow the recovery of fibers by pyrolysis, combustion, co-incineration, fluidized bed, and molten salt pyrolysis. The principle is to expose the material to high temperatures under oxidant or inert atmospheric conditions to decompose the matrix.

The third technique of material recovery is chemical recycling, which is often associated to solvolysis, defined by the capability of the solvent to break the macromolecular chains as chemical dissolution of the matrix corresponds to the disentanglement of the chains thanks to solvation. There is therefore no breaking of macromolecular chains unlike solvolysis. This technique offers a large number of possibilities with a wide range of solvents, temperatures, pressures, and catalysts.

Chemical recycling performed on acrylate (Elium®) based composites reinforced by basalt fabrics will be evaluated in this work.

1.3. Chemical Recycling of Acrylate-Based Composites

The solvents considered able to dissolve PMMA are also able to dissolve Elium® resin because of the similar chemical structure of both acrylic polymers. The properties of the different solvents can be controlled by the operating conditions (temperature, pressure, and reaction time) imposed on the process.

Evchuk et al. [11] studied the possibility of existing correlation between the PMMA dissolution rate and the structure or properties of the solvents, including polarity. Solvents such as trichloromethane, trichloroethylene, 1,4-dioxane, cyclohexanone, acetophenone, ethyl acetate, pentyl acetate, and dimethyl formamide were studied by these authors at a temperature range of 30–70 °C. However, no correlation was highlighted. Trichloroethylene was pointed out by these authors to be the best solvent for PMMA, while trichloromethane was the poorest. For Evchuk et al. [11], polar solvents, such as ethyl acetate, tetrahydrofuran, and cyclohexanone, are adapted to dissolve PMMA.

In the works by Tschentscher et al. [12] and Gebhardt et al. [13], these latter solvents were also used to recycle CFRP based on Elium® 150 resin involving specimens of 200 mm × 180 mm. These latter were placed in a closed container with a selected solvent (acetone, acetophenone, ethyl acetate, or xylene). These composites were tested for two "composite:solvent" ratios (1:10 and 1:20) and for three dissolution times (24 h, 48 h, 72 h) according to the solvent nature. The authors stated that a solvent ratio of 1:20 is recommended for a time-saving dissolution process (24 h). Ethyl acetate represents an environmentally friendly alternative to the established room temperature-based recycling process using acetone. The amount of Elium® recovered as well as the amount of Elium® remaining on the fibers were close for both solvents. Nevertheless, Elium® glass transition was nearer to virgin Elium® one in the case of ethyl acetate than acetone. As concerns acetophenone, it was considered by the authors to be a good solvent at higher temperatures.

The review of Oliveux et al. [10] reported several technologies for composites recycling in order to recover fibers. Among them, Adherent Technologies, Inc. (ATI) (Albuquerque, NM, USA) in USA [14,15] use a wet chemical breakdown to recover carbon fibers from industrial waste composites as well as from end-of-life composite materials. From comparison with different recycling techniques, the authors conclude that a process using a low temperature and low pressure is the most interesting in terms of quality of the recovered fibers and of price. In France, Innoveox [16] proposes a technology based on supercritical hydrolysis with the possibility to control the solvent properties and the reaction rates by conducting pressure manipulations. Supercritical water (temperature > 374 °C and pressure > 221 bar) has been mainly applied to CFRC in order to recover carbon fibers of good quality without paying much attention to the products of the resin degradation. SACMO (Holnon, France) [17] has also been interested in the composite solvolysis process and filed in 2014 a patent proposing a device for solvolysis treatment of a solid composite material to extract fibers from the treated material. The reactor can have a volume from 25 to 400 L. Pressure and heating can be applied. In addition, Panasonic Electric Works Co. (Tokyo, Japan) [18] has shown a willingness to exploit their hydrolysis process to recycle 200 tons of GFRC (based on unsaturated polyester resin) manufacturing waste per year. Wastes are treated at 230 °C with subcritical water at 2.8 MPa with additives (NaOH, KOH) for 4 h.

1.4. The Use of Ultrasounds as Processing Aids for Solvolysis

For more efficient technologies and in an environmental friendlier approach, mechanical stirring and heating are unfavorable techniques that need a non-negligible amount of energy to be processed. It is the reason why ultrasound was preferentially used to replace a commonly used stirring mechanism during the recycling protocol as a greener alternative.

Das et al. [19] proposed a sono-chemical method to recover carbon fiber from CFRP. Epoxy/carbon fibers composites are treated with a mixture of dilute nitric acid and hydrogen peroxide in the presence of ultrasound. Specimens of 30 mm × 25 mm × 2 mm are immersed for a pretreatment in the mixture until the samples are fully swollen. The solid-to-fluid ratio is maintained at 1:60. Then the sonication (470 kHz, 65 °C) begins in a water bath. The composite layer separation begins after 4 h. The total duration of the treatment is 8 h. The recovered fibers have little or no epoxy resin on the surface and their tensile strength is comparable to virgin fibers. According to Das et al., the use of ultrasound in aqueous solutions leads to cavitation. Microbubbles implode on the surface of the composite and facilitate the degradation of the resin first on the surface and then in the core of the composite, also by solvent diffusion.

Desai et al. [20] have already studied ultrasounds in order to depolymerize polymers such as polypropylene in p-xylene and decalin solvents. The degradation of PMMA using ultrasounds has also been studied. The authors have investigated the effect of alkyl groups' (such as methyl, ethyl or butyl groups) substituent on the ultrasonic degradation of PMMA, PEMA, and PBMA at 30 °C in toluene. The degradation rate constantly increases with an increase in the number of carbon atoms in the alkyl group [21].

In terms of degradation mechanism, homolytic and/or heterolytic cleavage of a covalent bond may occur. However, the breakage of a C–C bond in the macromolecule is the most common mechanism. Yan et al. [22] established a degradation mechanism of aqueous solution of carboxylic curdlan by ultrasound treatment. The ultrasound treatment (15 min) implies a shear force that is responsible for the disaggregation of polymer clusters. Non-covalent intra- and intermolecular bonds are broken. As the time of the ultrasound treatment increases (30 min), chains are further split, and shorter random coil chains are created. A degradation process thus occurs.

This literature review revealed that, to the best of our knowledge, no investigations have been carried on chemical recycling of Elium®/basalt fabrics composites as well on an academic and industrial point of view. Therefore, this paper provide new results on this challenging new composite material and the best way to recover either the Elium® resin or basalt fabrics to produce new composites that will be called "second generation" composites. To lower the carbon footprint of the process, an environmental-friendly solvent and room temperature dissolution procedure will be chosen. In addition, this paper will analyze the influence of ultrasound on the kinetics dissolution of Elium® resin to provide a greener approach, as compared to mechanical stirring and heating.

2. Materials and Methods

2.1. Materials

This study focuses on basalt woven sized fibers in plain weave. This fabric supplied by Basaltex (Wevelgem, Belgium) has a basis weight of 220 g/m^2 with a density of 2.65 g/cm^3 and a fiber diameter of 13 µm. The resin Elium® 150 was provided by Arkema (Lacq, France). The resin has a low viscosity of 100 mPa.s and a density of 1.01 g/cm^3. According to the producer, its composition is a mixture of a methyl methacrylate copolymer (70–90 wt%), citral or 3,7-dimethyl-2,6-octadienal ($\leq$5 wt%), hydro-treated light paraffinic distillates (petroleum) (≤2 wt%), and other additives. Perkadox® CH-50X dibenzoyl peroxide, provided by AkzoNobel, was added to the resin to initiate the radical polymerisation (2 wt%). After 30 min at room temperature, an exothermic reaction occurred so that no post-curing is required.

2.2. Composites Processing

"First generation" composites of 41 cm × 41 cm were manufactured by resin infusion at room temperature with 10 plies of fabric. Elium® resin and initiator (2 wt%) were mixed, and then infused under vacuum outlet until the surface of the fibres was covered with resin. The infused plate was left to polymerize at room temperature for 24 h without any further post-curing. The composites obtained have an average thickness of about 1.68 mm and a mass fraction of 71% and volume fraction of 48% of fibres. The density was measured through Micrometrics gas pycnometer with helium gas. It was 2.01 ± 0.02 g/cm^3.

"Second generation" composites were also manufactured by resin infusion at room temperature with 10 plies of fabric. For each sample, all the 10 plies belong to the same recovered sample from the first-generation composites. The composites from recycled fabrics have been infused in the dimensions necessary for the tensile tests, i.e., 125 mm × 25 mm under the same processing parameters used for the first-generation counterparts.

2.3. Dissolution Parameters

Parameter selection. Different parameters were studied: the choice of solvent, the composite/solvent ratio, the dissolution time, the dissolution temperature, and the presence of agitation or not. In this article are presented the results of the dissolution in acetone, comparing the influence of different composite/solvent ratio, time of dissolution and the use of mechanical agitation, ultrasound, or simply dipped in the solvent and are described in Table 1. For this purpose, the dissolution devices were adapted according to the protocol in order to obtain optimal results.

Table 1. Conditions of dissolution.

Parameters	Conditions
Samples dimension	6 cm × 10 cm or 13.5 cm × 3.5 cm
Container	1 or 2 L reactor
Solvent	Acetone 99.9% purity
"Composite:solvent" ratio	1:4, 1:10, 1:40
Dissolution time	7, 16, 24, 48, 72 h
Temperature	25 °C
Mechanical agitation	Pale notched, 60 rpm
Ultrasounds	3 min applied at the beginning or end of dissolution

Sample preparation. For dissolution experiments, composites were cut using a circular raw. The dimensions were limited to a few square centimeters (from 6 cm × 10 cm and 12.5 cm × 2.5 cm) for a good fixation into the reactor.

Dissolution devices. Figure 1 shows the experimental device used for dissolution experiments using mechanical stirring (60 rpm). A mesh size of 1–2 mm sieve with a folded into a cylinder shape is placed at the bottom of the reactor, around the dipping paddle, in order to avoid contact of the paddle with the composite samples. For dissolution experiments using ultrasounds, two composite samples (13.5 cm × 3.5 cm) were dissolved in an 800 mL beaker with 800 mL acetone, as shown in Figure 2. The power of ultrasounds was fixed at 550 W, and the frequency was around 50/60 Hz (Elmasonic S, Lille, France).

Figure 1. Experimental device used for dissolution experiments using mechanical stirring.

Figure 2. Experimental device used for dissolution experiments using ultrasounds.

Different experimental conditions summarized in Table 1 were applied during this study in order to compare: (i) the effect of the composite/solvent ratio, (ii) the effect of the use of mechanical agitation versus the use of ultrasounds, and (iii) the time of dissolution. The dissolutions were carried out in acetone SLR, Fisher Chemical. For each dissolution experiment, five samples were used, and each experiment was investigated in triplicate

Parameter determined after dissolution. After each dissolution time, the basalt fibers were extracted and dried at 60 °C in an oven in order to evaporate any remaining acetone. The solvent also containing the dissolved resin were placed in a rotary evaporator in order to extract the resin in solid form and then dried at 60 °C in an oven.

The percentage dissolution is defined as the rate of resin recovered from the initial composite before dissolution and noted %D. It was calculated after each test, as per Equation (1):

$$\%D = \frac{m_{recovered\ resin}}{m_{initial\ resin}} \times 100 \qquad (1)$$

A dissolution series corresponds to five samples dissolved during five different dissolution times. Each series is repeated three times. The samples in a series are taken from the same initial composite plate.

2.4. Characterization Techniques

2.4.1. Determination of Fiber and Residual Resin Contents

Loss on ignition tests were performed on composites to determine either the reinforcement content of first-generation composites or the residual resin content on recycled fibres after resin dissolution. The tests were carried out according to the NF EN ISO 1172 standard.

About 2 g of samples were placed in a ceramic crucible and calcinated at 600 °C for 30 min. Each sample was tested in triplicate.

The weight percentage of fiber and resin, and volume fibre contents of the composites were calculated according to Equations (2)–(4), respectively:

$$W_f = \frac{m_f}{m_c} \times 100 = \frac{\text{mass composite after test}}{\text{mass composite before test}} \times 100 \quad (2)$$

$$W_f = \frac{m_f}{m_c} \times 100 = \frac{\text{mass loss after test}}{\text{mass composite before test}} \times 100 \quad (3)$$

$$V_f = \frac{W_f/\rho_f}{W_f/\rho_f + W_r/\rho_r} \times 100 \quad (4)$$

where W_f (%) is the weight percentage of fibre, W_r (%) is the weight percentage of resin, and V_f (%) the volume percentage of fibre in the composite. M_r (g) and m_f (g) are the masse of resin and fibres, respectively. ρ_r and ρ_f are the density of the resin and fibres of basalt (g/cm^3).

The percentage of residual resin is defined as the rate of resin present on the fibers resulting from the dissolution, noted %RR, and measured by the loss ignition test. It was calculated with Equation (5):

$$\%RR = \frac{\text{mass of fibers before test} - \text{mass of fibers after test}}{\text{mass of initial composite}} \times 100 \quad (5)$$

2.4.2. Physicochemical Characterization of Recovered Resin after Dissolution

Fourier Transform InfraRed Spectrometry

FTIR (Fourier Transform InfraRed) analysis was performed using a Vertex 70 FT MIR spectrometer from Bruker with an ATR (Attenuated Total Reflection) unit equipped with a diamond crystal. The resolution was at 4 cm^{-1}, 16 scans for background acquisition, and 32 scans for the sample spectrum. Most of the samples were directly analyzed on the crystal. Some were cut to obtain better acquisition, as ATR is sensitive to surface aspects. Analyzed surfaces were cleaned with ethanol and left to dry. Spectra were acquired from 4000 to 400 cm^{-1} and analyzed, thanks to the OPUS software provided with the spectrometer.

Gel Permeation Chromatography

(GPC) was performed in order to determine the macromolecular chain weight of Elium® resin after dissolution. The tests were performed on the Varian 390-LC device with refractive index (RI) and viscometry detectors. Two columns were used: PLgel 5 microns and Mixed-D 300 mm × 7.5 mm. The eluent was THF, and the flow rate was 1 mL/min. The temperature of the columns was 30 °C. The samples of virgin and recycled Elium® were ground and then dissolved in THF at a height of 10 mg for 1 mL of solvent and 0.05% toluene. After half a day and under ultrasound, the samples were dissolved. There was one test for each sample.

2.4.3. Fabrics and Composites Observations through Scanning Electron Microscopy and Porosity Measurements

Scanning electron microscopy (SEM) was investigated to observe the fabric surface after dissolution in order to confirm or not the presence of residual resin.

SEM was also carried out on composites of the first and second generation to examine the resin impregnation of the fabrics.

MEB FEI QUANTA 200 FEG was used in both cases. A special treatment where the samples were inserted into polished sections of the polymer which the surface was polished, and a thin carbon foil was applied using a Balzers CED030 metal coater.

Pycnometry analysis were carried out to measure the rate of porosity within the first- and second-generation composites (2 or 3 rectangles of composites for each measure, Micrometrics AccuPyc 1330, Verneuil en Halatte, France).

2.4.4. Static and Dynamic Mechanical Characterization of Composite Materials

Mechanical Tensile Test

Composites were cut with a circular saw with dimensions of 125 mm × 25 mm for first- and second-generation composites. Aluminium stubs measuring 25 mm × 25 mm × 2 mm were glued onto composite specimens previously sanded. Five tests were performed for each first and second-generation composite.

To measure the mechanical capacities in terms of rigidity, i.e., the elastic behaviour of the material, and the breaking strength, tensile tests were carried out according to the ISO 527-4 standard on MTS Criterion 50 (Créteil, France) equipped with a 100 KN load cell. All the tests were performed at a speed of 10 mm/min until the test piece broke. The variation in elongation was measured by a laser extensometer. The higher the elongation, the more ductile the material.

Dynamic Mechanical Analysis

Dynamic Mechanical Analysis was used to determine the thermomechanical properties and damping of a material according to the standard ASTM D5026. Analyses were performed on first- and second-generation composite samples (dimensions 10 mm × 50 mm × 1.68 mm) in triplicate from room temperature up to 180 °C at a constant frequency of 1 Hz and a dynamic displacement of 10^{-6} µm (DMA 50N O1-db, Limonest, France).

Linearity tests and a frequency sweep between 0.1 and 100 Hz at room temperature were carried out to verify the viscoelastic properties of the material and select the right displacement.

3. Results and Discussion

3.1. Dissolution Kinetics

The aim of this section is to compare the effectiveness of different dissolution protocols in terms of dissolution percentage (%D) and residual resin rate (%RR) on basalt fabrics. The effect of an ultrasonic stirring and the influence of the composite/acetone ratio will be discussed. Ultrasonic stirring is then compared to the application of mechanical stirring.

3.1.1. Influence of an Ultrasonic Stirring Application

An ultrasonic stirring system was applied at different key steps of the dissolution: at the beginning, and at the end of the dissolution, for a "composite:acetone" ratio of 1:4. The efficiency of the dissolution is reported in Figure 3. In each protocol considered below, 3 min of ultrasound were applied. Dissolution profiles are compared with a sample only soaked into acetone (Figure 4).

When no stirring or heating is applied, the induction time is approximately 72 h and the associated %D is 81.3% ± 3.7% (Figure 4). Furthermore, resin clogging can be observed at dissolution times of more than 24 h, due to plasticization of the polymer with the solvent [23].

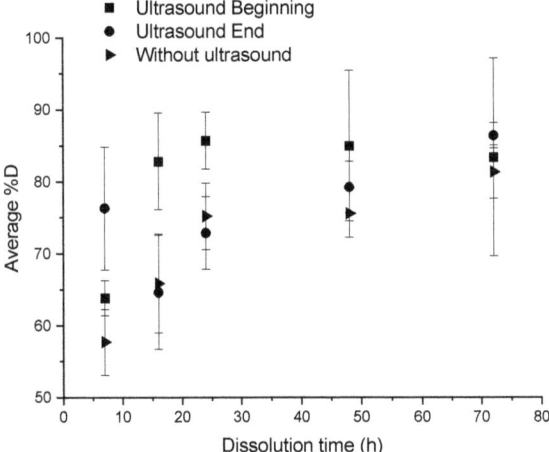

Figure 3. Influence of different ultrasonic conditions on the average %D as a function of dissolution time for a "composite:acetone" ratio of 1:4.

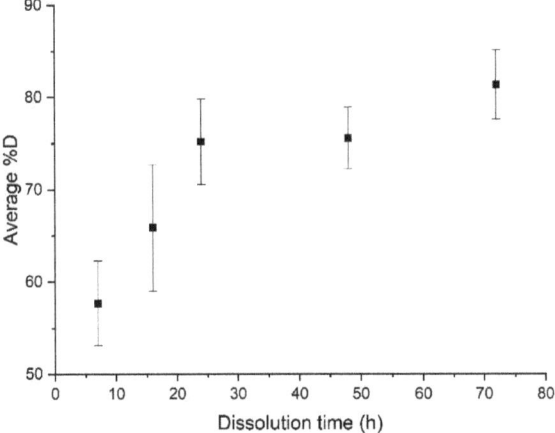

Figure 4. Evolution of the average %D as a function of dissolution time for a simple dissolution of a "composite:acetone" ratio of 1:4.

According to Figure 3, for times less than 24 h, a significant difference is observed according to the fact that ultrasounds are applied at the beginning or at the end of dissolution. Applying ultrasounds at the end of dissolution is more efficient than ultrasound at the beginning of dissolution for a dissolution time of 7 h (78% versus 62% of dissolution, respectively). This trend is reversed from 16 h onwards. Furthermore, the average %D in the case of ultrasound at the beginning of dissolution are higher than those in the case without ultrasound whatever the time of dissolution, compared to the case with ultrasound at the end of dissolution. This proves the effectiveness of ultrasound in general on resin dissolution kinetics. The standard deviation corresponding to each average %D value depends on the capability of the resin to clog on the container. This phenomenon tends to decrease while the solvent volume increases or the dissolution time increases or a mechanical stirring system is applied.

Thus, ultrasound at the beginning of dissolution has a real efficiency for a dissolution time of 16 h, whereas ultrasound at the end of dissolution shows an efficiency on the dissolution of the resin for a dissolution time of 7 h.

The interest of ultrasound at the beginning of the dissolution is that it allows to limit the damage of the basalt fabrics (Figure 5a). Only a few fibers on the fabric edges were disentangled. This is also the case for simple soaking of the samples. On the contrary, Figure 5b shows tufted fibers after dissolution with ultrasound applied at the end of the dissolution. The size of the samples must be considered. It should be noted that the composite surface influences the fabric damage. Thus, if larger samples are considered, it is possible to recover some undamaged fabrics by removing the damaged fabric edges. Figure 5c shows less-damaged fibers; nevertheless, the textile layers stick together because of higher %RR. Thus, these recovered fibers are less interesting in terms of reprocess ability after recycling.

(a)

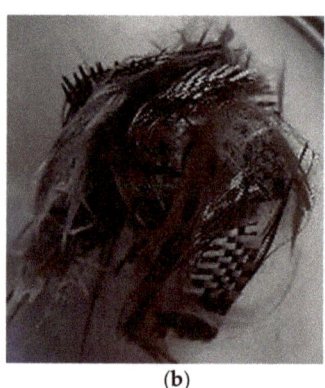
(b)

(c)

Figure 5. Basalt fibers recovered (**a**) after recycling 16 h with ultrasound at the beginning of the dissolution, (**b**) after recycling 7 h with ultrasounds at the end of the dissolution and (**c**) after recycling without ultrasounds. A 1:4 ratio is considered for each photo.

3.1.2. Influence of the "Composite:Acetone" Ratio

The influence of the "composite:solvent" ratio was studied in the case of the application of ultrasound at the end of dissolution for 3 min. Thus, the evolution of the %D as a function of the dissolution time is plotted in each case and allows to understand the dissolution kinetics of the resin, as shown in Figure 6. Figure 7 allows to observe the relative visual aspect of dissolved resin after solvent evaporation. The results show that the average %D increase with the dissolution time whatever the "composite:solvent" ratio even by considering the high standard deviation. Moreover, the higher the solvent volume, the higher the average %D value. As explained by Miller-Chou and Koenig [23], the dissolution of a polymer into a solvent are results of two main transport processes: solvent diffusion and chain disentanglement. In contact with a thermodynamically compatible solvent (PMMA with acetone in our case), the solvent diffuses into the polymer. Firstly, a gel-like swollen layer is formed (Figure 7a). Then, after a certain induction time, the polymer is dissolved (Figure 7c).

A comparison of the three ratios: 1:4, 1:10, and 1:40 makes possible to determine the most interesting one (Figure 6), in the case of ultrasound at the end of dissolution. After 7 h of dissolution, there was no significant difference between the average dissolution rate for the ratios 1:4 and 1:40, which allows to reach a satisfactory %D over 60%. From a duration of 16 h onwards, the profiles of dissolution become more distinct. When the 1:40 ratio already achieves 100% dissolution, the 1:10 and 1:4 ratios only allow to reach 80% and circa 60% of dissolution, respectively. At times greater than 48 h, the average %D value converges towards over 80%, even close to 100% for 1:40 and 1:10 ratios. Some %D values are higher than 100% and this is attributed to a slightly accumulation of resin on the glassware after several dissolutions.

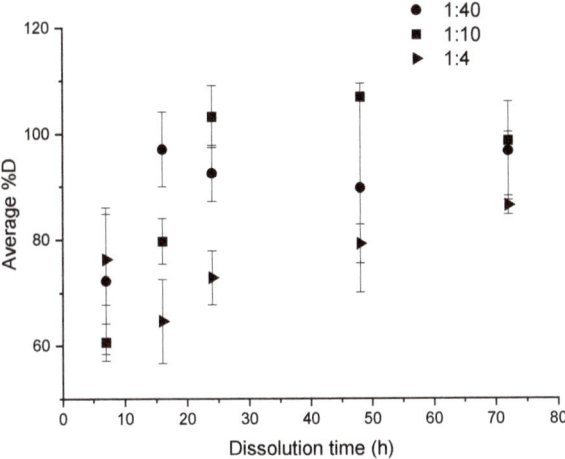

Figure 6. Influence of "composite:acetone" ratio on the average %D as a function of dissolution time for ultrasound at the end of dissolution.

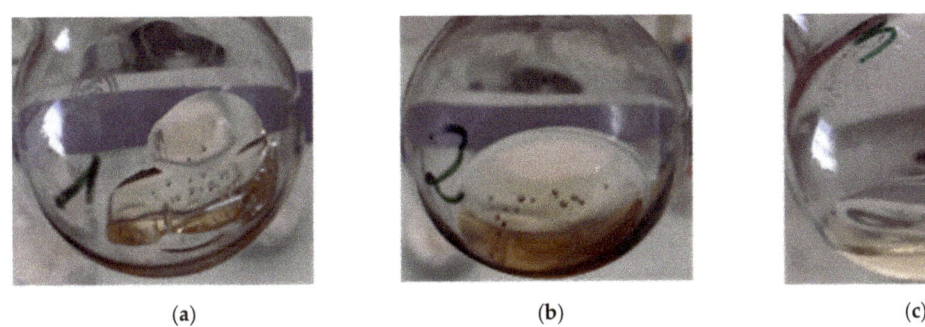

Figure 7. Dissolution of Elium® resin during 24 h with acetone at different "resin:solvent" ratios: (a) 1:1, (b) 1:2 and (c) 1:4.

A ratio of 1:4 and to a lesser extent a ratio of 1:10 are interesting dissolution conditions for economic and environmental concerns for an upscaling production because of the reduction of solvent used. However, the kinetic of dissolution is then clearly lower, especially in the case of the 1:4 ratio. For this latter, 48 h are required to achieve an average %D over 80%. Thus, the 1:10 ratio appears to be a good compromise in terms of dissolution kinetic and volume used with 80% of dissolution reached after 16 h.

3.1.3. Influence of the Mechanical Stirring

The average dissolution profile in Figure 8a shows an average %D close to 80% after 3 h of dissolution and a complete dissolution for a dissolution time of 7 h. Slightly higher values than 100% are, as previously, attributed to the fact that small amounts of resin are accumulated on the glassware during the dissolution process, despite careful washing. Then %D values decrease for dissolution times above 7 h. The optimal dissolution time required for total dissolution of the resin in this protocol is about 7 h. After recycling, no damage of the fabrics (Figure 8b) is revealed.

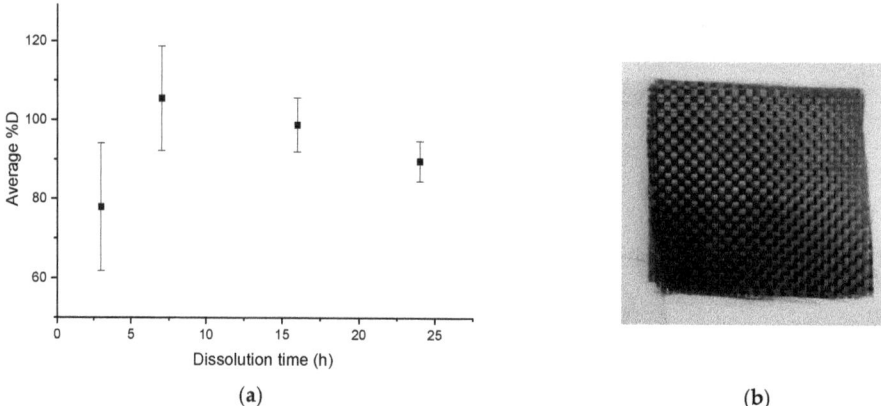

Figure 8. (a) Evolution of the average %D as a function of dissolution time for mechanical stirring for a ratio of 1:40, (b) Fabrics obtained after the complete dissolution.

3.1.4. Summary of Dissolution Experiments

A simple dissolution in acetone of Elium®/basalt fabrics requires a too long dissolution time (72 h) in the case of ratio of 1:4. To reduce the dissolution time, it would be necessary to increase largely the volume of solvent, which is not advantageous from an environmental point of view. The device using ultrasound at the beginning of the dissolution at a ratio of 1:4 makes it possible to combine dissolution efficiency (82.8% ± 6.7%), reasonable dissolution time (16 h), and low damaged fabrics. Later in this paper, the protocol using ultrasounds is adapted to reduce the dissolution time: the dissolution lasts 7 h at a ratio of 1:40 of acetone using 3 min of ultrasounds at the beginning. The device comprising mechanical agitation alone is even more efficient than the previous device: a %D of 105.5% ± 13.3% after 7 h of dissolution with a ratio of 1:40 in acetone is reached with no fabric damage. This device requires a larger quantity of solvent but the agitation with the paddle allows a good diffusion of the solvent between the fabric layers. It would also be interesting to compare the energy consumption of a protocol using 3 min of ultrasonic bath and one using 7 h of continuous mechanical agitation. In addition, the acetone/dissolved resin mixture can be processed by evaporation and therefore the acetone can be easily recycled. Thus, a complete recycling cycle of the system can be set up to limit solvent consumption.

3.2. Characterization of the Recovered Basalt Fibers

Two methods are used to characterize the residual resin on the surface of the recycled fibers: loss on ignition in quantitative terms and SEM observations in qualitative terms.

3.2.1. SEM Observations of Recovered Basalt Fabrics

SEM imaging is used to confirm the presence of resin on the surface of the fibers after recycling. To understand the influence of the dissolution process on the %RR, a comparison is made between virgin basalt fibers and basalt fibers after resin dissolution in presence of acetone (1:4 ratio) for 24 h (Figure 9). A residual resin layer can be easily observed on the recycled fibers, in comparison with virgin fibers.

Figure 9. SEM image of (**a**) Virgin basalt fibers, (**b**) Recycled basalt fibers dissolved in acetone for 24 h at a solvent ratio of 1:4.

Figure 10 shows SEM observations of basalt fabrics after dissolution in different conditions. The red circles indicate the localization of residual resin corresponding to the darker area. Figure 10a shows that a very small amount of residual resin remains in the center of a mesh and between the fibers. Acetone seems to have better dissolved the resin on the surface of the fabric layers than between the fibers of the same fabric. This is confirmed by a manual pliability test and the ability to separate fiber layers just after the dissolution. Layers are harder to bend when there is a film of residual resin. The solvent diffusion is in fact harder in the bulk of the composite sample, so that a higher quantity of resin remains in the bulk. However, mechanical stirring tends to limit that phenomenon. On the rest of the recycled fabric, no trace of residual resin was reported.

In the case of recycling with ultrasound at the beginning of dissolution (Figure 10b), clusters of resin are present in the center of the fabric mesh and in slightly greater quantities than in the case of recycling with ultrasounds at the end of dissolution (Figure 10c). In both cases, acetone dissolves the resin better at the crossings between the fibers of the same fabric.

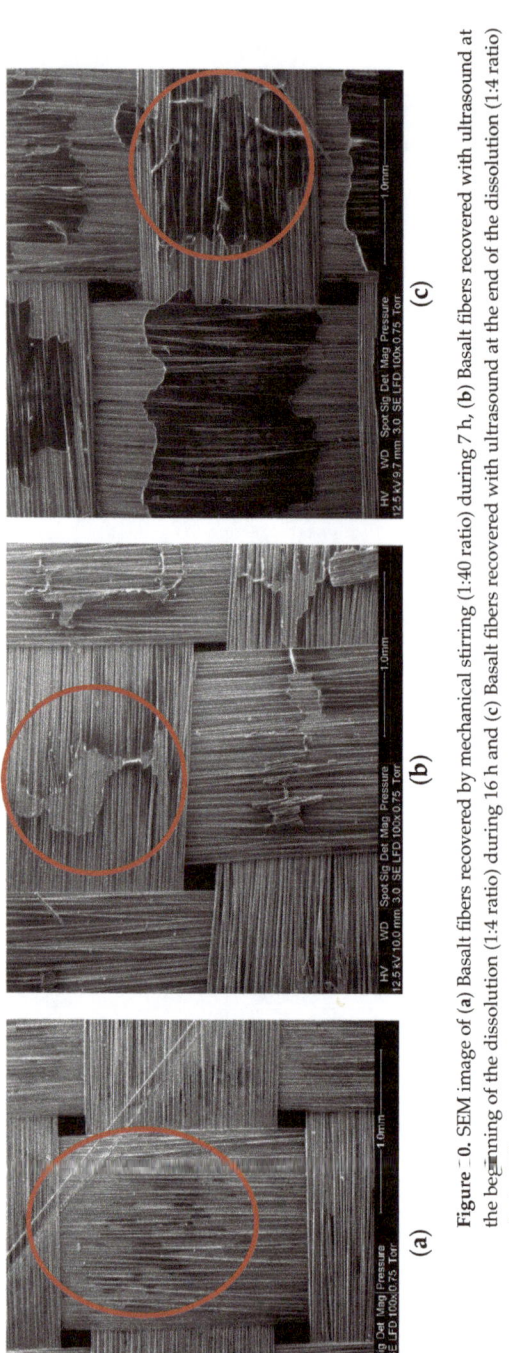

Figure 10. SEM image of (**a**) Basalt fibers recovered by mechanical stirring (1:40 ratio) during 7 h, (**b**) Basalt fibers recovered with ultrasound at the beginning of the dissolution (1:4 ratio) during 16 h and (**c**) Basalt fibers recovered with ultrasound at the end of the dissolution (1:4 ratio) during 7 h.

3.2.2. Residual Resin Rate on Basalt Fabrics

This loss on ignition test is performed on basalt fabrics after resin dissolution in different conditions but for a common ratio of 1:4.

According to Figure 11, in the three conditions, the average %RR is stable around 2%. However, it remains higher in the case of recycling with ultrasound at the beginning of dissolution, especially at 7 h where the %RR is very high (11%) and where the standard deviation is also very high. From 16 h onwards, there is no great difference in the evolution of the residual resin rate according to the dissolution conditions.

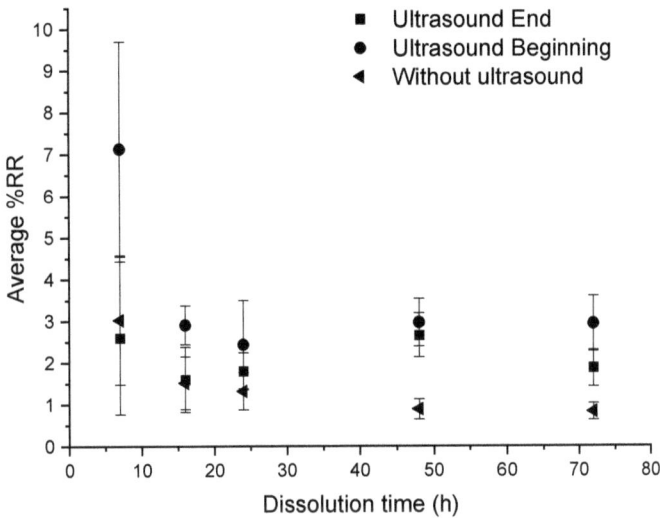

Figure 11. Influence of different ultrasonic conditions on the average %RR as a function of dissolution time for composite: acetone ratio of 1:4.

From these results, no real effectiveness of ultrasound on the residual resin rate can be discerned. Moreover, in the case of recycling without ultrasound, the residual resin content is always lower than in the case of recycling with ultrasound.

In general, the residual resin amount onto fibers decreases with the dissolution time, which is consistent with the average dissolution profiles (Figure 3). Ultrasound does not significantly reduce the residual resin content on fibers. However, the use of ultrasound at the beginning of the dissolution process seems to increase the average %D, regardless of the dissolution time compared to a device without ultrasound. Ultrasound seems to improve the recovery of the resin in the medium without necessarily decreasing the rate of residual resin on the surface of the fibers.

In the case of recycling with mechanical stirring, a very small amount of resin is on the fibers, even at 3 h of recycling (Figure 12), which is confirmed by SEM analysis.

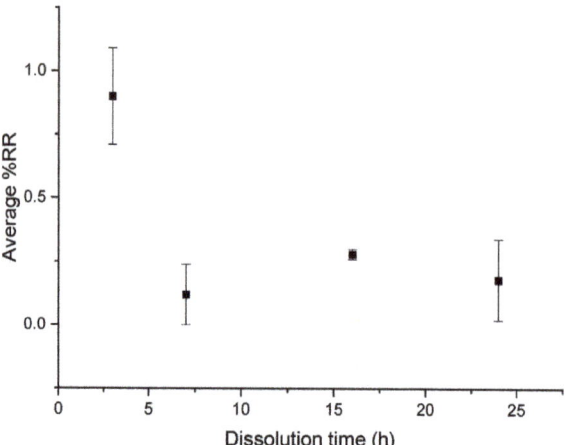

Figure 12. Evolution of the average %RR as a function of time for recycling with mechanical stirring and acetone at a ratio of 1:40.

3.3. Characterizations of Recovered Resin

3.3.1. ATR FTIR

Infrared analysis allows to understand the influence of acetone dissolution on the chemical composition of the resin. A simple dissolution was performed for 24 h at a "composite:acetone" ratio of 1:4 (Figure 13). The reference peak is the peak at 1720 cm^{-1}, which is associated with the C=O bonds from the methacrylate functional group.

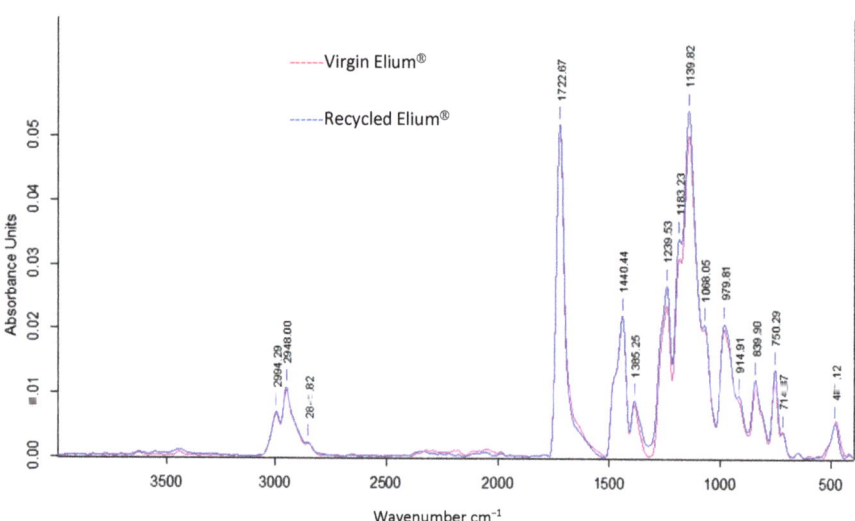

Figure 13. ATR FTIR absorbance spectrum of virgin Elium® and Elium® after 24 h of simple dissolution in acetone 1:4.

An identification of the peaks allows to find common chemical functions between recycled and virgin Elium®. The peak at 1140 cm^{-1} is associated with the C–O–C bond, at 1450 cm^{-1} with the C–H bond (bending), and especially the four distinct peaks of CH3 and CH2 bonds (C–H stretching) between 2800 and 3000 cm^{-1} into a consolidated envelope curve.

It can be observed that the three peaks at 1361 cm^{-1}, 911 cm^{-1}, and 715 cm^{-1} have disappeared in the spectrum of the recycled Elium®. The first peak is attributed to C–H Csp3 bonds and the second to C–H Csp2 bonds. Otherwise, the other peaks in the two spectra merge. In addition, small chains, monomer residues, and some additives such as citral can probably exfoliate and migrate out of the polymer. In conclusion, the chemical composition of Elium® recovered after acetone dissolution is close to that of virgin Elium®.

3.3.2. SEC

The dissolution of resin samples in acetone coupled with mechanical stirring or ultrasounds may have an influence on the chemical structure of the polymer. Size exclusion chromatography allows to understand whether a polymer degradation can occur. Thus, the evolution of Mn, Mw, and also the polydispersity (Table 2) and molecular weight distribution (Figure 14) helps to get an overview of degradation mechanisms during dissolution.

Table 2. Average molecular weights and polydispersity of recycled and virgin resin after dissolution.

	Mn (g/mol)	Mw (g/mol)	PD
Ultrasounds at the beginning acetone 1:4 16 h	67,790	189,140	2.8
Mechanical stirring acetone 1:40 7 h	61,890	217,440	3.5
Virgin Elium®	85,600	213,200	2.5

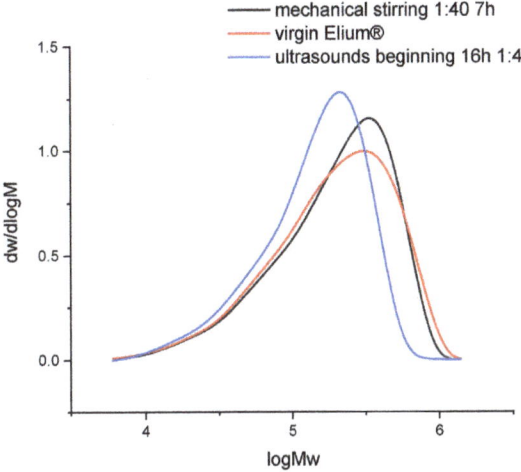

Figure 14. Molecular weights distribution of virgin Elium® and recycled Elium® resin after dissolution.

In both recycling cases, PD increases. The molecular weight distribution curves do not show the formation of smaller molecules but rather larger ones in the case of recycling with ultrasounds. Thus, in that case, recombination or intermolecular reactions may take place. After recycling, the molecular weight distribution remains unimodal. In the case of recycling with mechanical stirring, the size of molecules is similar to that of virgin Elium®.

In the case of mechanical stirring alone, Mw remains constant and Mn decreases (−28%). Thus, this stirring system does not imply any degradation of the polymer chains but allows a faster disentanglement of the chains during solvation.

In the case of recycling with ultrasound at the beginning of dissolution, Mw (−21%) and Mn (−11%) decrease. Thus, the polymer could degrade and involve the creation of smaller molecules. According to [23], only long chains can be affected by ultrasound. Thus, the heterogeneity in terms of molecular weights decreases. Indeed, the polydispersity PD remains close to that of virgin Elium®. However, as ultrasound is only applied for

3 min at the beginning of dissolution, it is not sufficient to really create a degradation. Recombination of molecules can also be at the origin of the formation of larger molecules.

3.4. Mechanical Properties of Second-Generation Composites

3.4.1. Static Tensile Tests

Mechanical results of uniaxial tensile tests performed on Elium®/basalt composites are reported in Table 3. The pore volume of the composites was also measured and are presented in Table 3.

Table 3. Modulus, tensile strength, and porosity of first and second generation Elium®/basalt composites recycled using mechanical stirring and ultrasounds at the beginning of the dissolution.

	Modulus (GPa)	Stress (MPa)	Porosity (%)	Fiber Weight Fraction (%)
1st generation	19.9 ± 2.7	508.0 ± 34.8	8.6 ± 5.9	71.4 ± 1.0
2nd generation: mechanical stirring	24.4 ± 4.7	586.8 ± 22.1	4.2 ± 2.7	71.4 ± 0.4
2nd generation: ultrasounds at beginning	20.6 ± 2.4	474.3 ± 24.5	2.9 ± 2.0	67.1 ± 0.8

Second generation composites after a dissolution process using mechanical stirring have a 22% higher modulus (24.4 ± 4.7 GPa against 19.9 ± 2.7 GPa in the initial state) and a 15% higher stress (586.8 ± 22.1 MPa against 508 ± 34.8 MPa). After a dissolution process using ultrasound, second generation composites have tensile properties comparable to first generation composites with a modulus (−3%) of 20.6 ± 2.4 GPa against 19.9 ± 2.7 GPa and a stress of 474.3 ± 24.5 MPa against 508.0 ± 34.8 MPa (−6%). The lower fabric degradation after dissolution process using mechanical stirring (Figure 6) compared to those using ultrasound (Figure 5) may explain the better performance in tensile tests.

However, for dissolution conditions, the porosity is lower than for first generation composites, −51% for mechanical stirring despite an equal fiber weight ratio ≅ 71.4%. Nevertheless, composites from ultrasounds have less than −66% of porosity with a fiber weight fraction much lower (67.1 ± 0.8%). In addition, despite a higher porosity rate (4.2% by mechanical stirring vs. 2.9% by ultrasounds), the composite recycled by mechanical stirring has superior tensile performance (+15% in modulus and +19% in stress), compared to the composite recycled by ultrasounds.

In addition, the SEM images (Figure 15a–c) show no difference in interfacial adhesion between the two types of recycling—good impregnation and interfacial adhesion, which contribute to the good mechanical properties of second-generation composites.

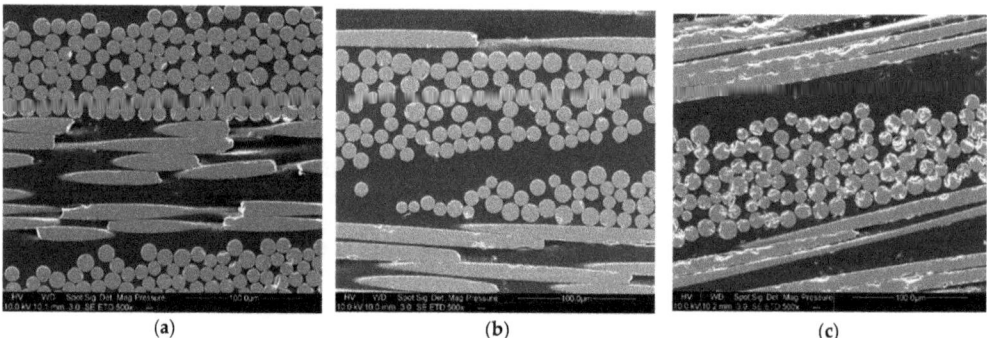

Figure 15. SEM images of (**a**) first generation composite (**b**) second generation composite after dissolution using mechanical stirring and (**c**) second generation composite after dissolution using ultrasound in the beginning.

3.4.2. Dynamic Mechanical Analysis

The tests were carried out on first- and second-generation composites (Figure 16). For the latter, the dissolution was performed using ultrasound for 3 min in the beginning for 16 h (ratio 1:4) and mechanical stirring for 7 h (ratio 1:40). Table 4 is a summary of the data measured from the DMA tests, where E'160 and E'40 are the rubber and vitreous conservation modules, respectively.

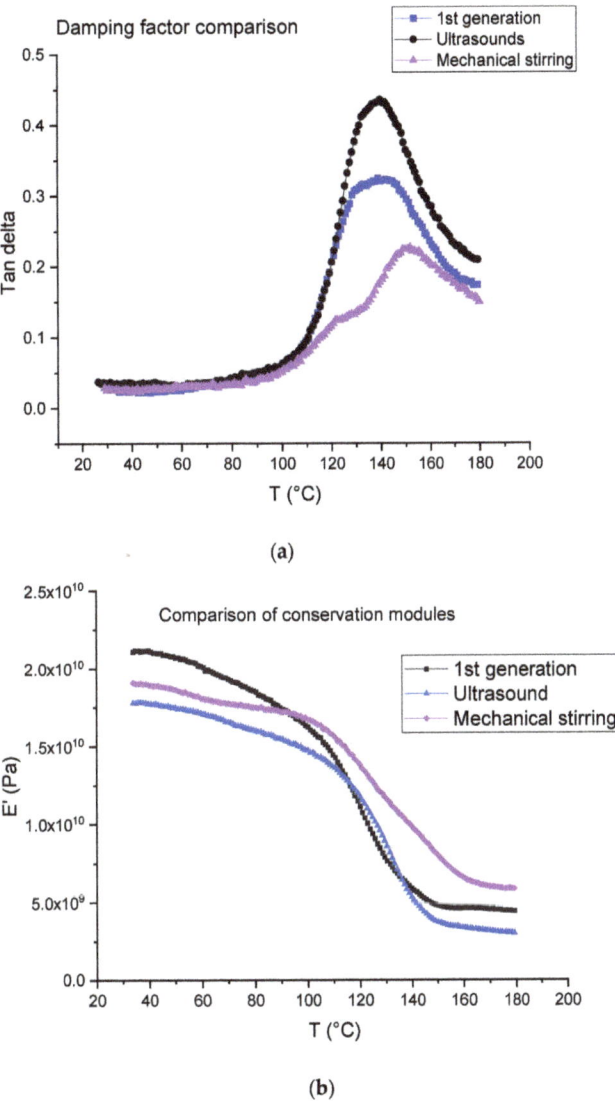

Figure 16. (a) Tangent delta; (b) conservation modulus of first-generation composites and second-generation composites obtained after dissolution using mechanical stirring and ultrasounds.

Table 4. DMA results of first generation composites and those recycled by mechanical stirring and ultrasound.

	1st Generation	2nd Generation Mechanical Stirring	2nd Generation: Ultrasounds
E'40 (GPa)	19.87 ± 1.87	18.99 ± 0.27	16.25 ± 1.18
E'160 (GPa)	4.25 ± 0.69	6.30 ± 0.35	3.03 ± 0.20
tan d1	/	0.13 ± 0.01	/
T1 (°C)	/	122.30 ± 1.45	/
tan dα	0.32 ± 0.01	0.21 ± 0.02	0.40 ± 0.03
Tα (°C)	139.27 ± 2.12	151.11 ± 1.16	138.81 ± 4.97

The first generation composite shows a main relaxation peak related to glass transition temperature located at about 139 °C with an intensity of 0.32 At the same time, the modulus drops from 19.87 ± 1.87 GPa to 4.25 ± 0.69 GPa.

In the case of a dissolution using mechanical stirring, second generation composites provide evidence of a shoulder on the main relaxation peak located at 122 °C, i.e., at a lower temperature (151 °C) than the main relaxation. The intensity of both peaks is 0.13 (shoulder) and 0.21 (main peak). In the case of a dissolution using ultrasounds, no shoulder is observed and the main relaxation is located at about 139 °C as for the first-generation composites. Nevertheless, a higher intensity is determined (0.40). As concerns storage modulus, it drops from 18.99 ± 0.27 GPa to 6.30 ± 0.3 GPa when mechanical stirring is used and from 16.25 ± 1.18 GPa to 3.03 ± 0.20 when ultrasounds are used. First, it can be concluded that the storage modulus at 40 °C is not affected by recycling and corresponds to the same order than the first-generation composites.

Differences may be explained by the presence of residual resin on the fabrics after resin dissolution. This resin may form a thin homogeneous layer when mechanical stirring is used, as it seems to be a more heterogeneous cluster when ultrasounds are used (Figure 10), even if the residual resin rate is higher in the case of ultrasonic recycling in comparison with the mechanical method. Further investigation will be carried out to confirm this assumption.

4. Conclusions

The objective of this work was to compare the material recoveries resulting from different chemical recycling methodologies for thermoplastic acrylate-based composites reinforced by basalt fabrics and manufactured by vacuum infusion. The recycling was processed via chemical dissolution with a preselected adapted solvent i.e., acetone. The main goal of the study was to recover undamaged basalt fabrics to reuse them as reinforcements for second generation composites.

A protocol based on an ultrasound technique was compared to another one using mechanical stirring. The results show that the dissolution protocol using a mechanical stirring is more adapted to recover undamaged fabrics with no residual resin on their surface.

Several parameters, such as dissolution duration, dissolution temperature, and solvent/composite ratio were also studied. The results show that in the case of a 1:40 ratio, the optimal dissolution time required for total dissolution of the resin is about 7 h. Moreover, the higher the solvent volume, the higher the dissolution rate.

FTIR and SEC analysis show that there is no degradation of the resin after dissolution by acetone, but recombination phenomena do not modify the chemical composition of the Elium® resin. The stirring system allows disentanglement of long polymer chains, and this has a low impact on the molecular mass distribution.

Finally, second generation composites were elaborated with these recycled fabrics and mechanical and thermomechanical properties were determined and compared to those of first-generation composites. Corresponding second generation composites displayed equivalent mechanical properties than first generation ones.

Heating was avoided in this study for lower energy consumption and lower safety risks, but it should be interesting to evaluate the effect of temperature. As the protocol with mechanical agitation requires a larger amount of solvent (50 mL of acetone for 1 g of composite), whereas the protocol with ultrasound can be limited to 5 mL of solvent for 1 g of composite, an up-scale study should be performed. Greener solvent should be considered in further investigations.

Author Contributions: Conceptualization, I.M.z.R., A.S., A.B., C.L., D.P., K.M. and A.O.; methodology, I.M.z.R., A.S., A.B., C.L., D.P., K.M. and A.O.; software, I.M.z.R. and A.S.; validation, A.B., C.L., D.P., K.M. and A.O.; formal analysis, I.M.z.R. and A.S.; investigation, I.M.z.R.; resources, I.M.z.R. and A.S.; data curation, I.M.z.R. and A.S.; writing—original draft preparation, I.M.z.R. and A.S.; writing—review and editing, A.B., C.L., D.P., K.M. and A.O.; visualization, I.M.z.R. and A.S.; supervision, A.B., C.L., D.P., K.M. and A.O.; project administration, D.P.; funding acquisition, K.M. All authors have read and agreed to the published version of the manuscript.

Funding: This research received no external funding.

Institutional Review Board Statement: Not applicable.

Informed Consent Statement: Not applicable.

Data Availability Statement: Data cannot be shared, as it is a part of an ongoing study.

Acknowledgments: The authors would like to acknowledge Segula Technologies and National Association of Research and Technology (ANRT) for the financial support. They would also like to thank Arkema for the supply of Elium® resin. The authors also thank Caroline Chaix of the Charles Gerhardt Institute of Montpellier of the University of Montpellier for carrying out the SEC tests. The authors thank Guillaume Ienny at CM2A of IMT mines Alès for their expertise on infusion.

Conflicts of Interest: The authors declare no conflict of interest.

References

1. JEC Composites. Available online: https://www.jeccomposites.com/press/le-jec-observer-une-vision-globale-du-marche-des-composites/ (accessed on 11 October 2021).
2. EUR-Lex. Available online: https://eur-lex.europa.eu/legal-content/EN/TXT/?uri=CELEX%3A02019R0631-20211202 (accessed on 27 December 2021).
3. Komornicki, J.; Bax, L.; Vasiliadis, H.; Ong, K. Polymer composites for automotive sustainability. *Innov. Manager SusChem Secr.* **2015**.
4. Marmol, G.; Ferreira, D.P.; Fangueiro, R. 24—Automotive and construction applications of fiber reinforced composites. In *Woodhead Publishing Series in Composites Science and Engineering, Fiber Reinforced Composites*; Joseph, K., Oksman, K., George, G., Wilson, R., Appukuttan, S., Eds.; Woodhead Publishing: Sawston, UK, 2021; pp. 785–819.
5. Fiore, V.; Scalici, T.; Di Bella, G.; Valenza, A. A review on basalt fiber and its composites. *Compos. Part B* **2015**, *74*, 74–94. [CrossRef]
6. Galindo, B.; Ruiz, R.; Ulldemolins, C.; Losada, C. Development of long fiber thermoplastic composites for industrial applications. *Reinf. Plast.* **2021**. [CrossRef]
7. Balaji, K.V.; Shirvanimoghaddam, K.; Rajan, G.S.; Ellis, A.V.; Naebe, M. Surface treatment of Basalt fiber for use in automotive composites. *Mater. Today Chem.* **2020**, *17*, 100334.
8. Mallick, P.K. 5—Thermoplastics and thermoplastic–matrix composites for lightweight automotive structures. In *Woodhead Publishing Series in Composites Science and Engineering, Materials, Design and Manufacturing for Lightweight Vehicles*; Mallick, P.K., Ed.; Elsevier: Amsterdam, The Netherlands, 2010; pp. 174–207.
9. Gaumond, B. Compréhension des Interfaces/Interphases Formées dans les Composites PPS/Fibres de Carbone et PPS/Fibres de Basalte Réalisés à Partir de Mèches Comélées et Retordues. Matériaux. Ph.D. Thesis, Université de Lyon, Lyon, France, 2020.
10. Oliveux, G.; Dandy, L.O.; Leeke, G.A. Current status of recycling of fibre reinforced polymers: Review of technologies, reuse and resulting properties. *Prog. Mater. Sci.* **2015**, *72*, 61–99. [CrossRef]
11. Evchuk, I.Y.; Musii, R.I.; Makitra, R.G.; Pristanskii, R.E. Solubility of Polymethyl Methacrylate in Organic Solvents. *Russ. J. Appl. Chem.* **2005**, *78*, 1576–1580. [CrossRef]
12. Tschentscher, C.; Gebhardt, M.; Chakraborty, S.; Meiners, D. Recycling of Elium CFRPs for high temperature dissolution: A study with different solvents. In Proceedings of the Symposium Materialtechnik, Clausthal-Zellerfeld, Germany, 25–26 February 2021.
13. Gebhardt, M.; Chakraborty, S.; Manolakis, I.; Meiners, D. Recycling of CFRP composites based on a thermoplastic matrix (Elium 150) which can be infused and cured at room temperature—Recovery of the matrix by dissolution processes and testing of fibre damage by means of SEM and single fibre tensile tests. In Proceedings of the Forschungsfeldkolloquium 2020 Forschungsfeld Rohstoffsicherung und Ressourceneffizienz, Clausthal-Zellerfeld, Germany, June 2020.

14. Adherent-Technologies. Available online: http://www.adherent-tech.com/recycling_technologies (accessed on 12 January 2022).
15. Gosau, J.-M.; Wesley, T.F.; Allred, R. Integrated Composite Recycling Process. In Proceedings of the 38th SAMPE Technical Conference, Dallas, TX, USA, 7–9 November 2006.
16. Morin, C.; Loppinet-Serani, A.; Cansell, F.; Aymonier, C. Near- and supercritical solvolysis of carbon fibre reinforced polymers (CFRPs) for recycling carbon fibres as a valuable resource: State of the art. *J. Supercrit. Fluids* **2012**, *66*, 232–240. [CrossRef]
17. Baudu, C. Device for Treating a Solid Composite Material by Solvolysis with a View to Extracting Fibres from the Treated Material, Issued 16 October 2014. Available online: https://patentscope.wipo.int/search/en/detail.jsf?docId=WO2014166962 (accessed on 20 June 2021).
18. Nakagawa, T.; Itoh, T.; Hidaka, M.; Urabe, T.; Yoshimura, T. Enhanced and Horizontal Recycling Technology of FRP Using Sub-Critical Water. *J. Netw. Polym.* **2008**, *29*, 158–165.
19. Das, M.; Varughese, S. A Novel Sonochemical Approach for Enhanced Recovery of Carbon Fiber from CFRP Waste Using Mild Acid–Peroxide Mixture. *ACS Sustain. Chem. Eng.* **2016**, *4*, 2080–2087. [CrossRef]
20. Desai, V.; Shenoy, M.A.; Gogate, P.R. Degradation of polypropylene using ultrasound-induced acoustic cavitation. *Chem. Eng. J.* **2008**, *140*, 483–487. [CrossRef]
21. Daraboina, N.; Madras, G. Kinetics of the ultrasonic degradation of poly (alkyl methacrylates). *Ultrason. Sonochemistry* **2009**, *16*, 273–279. [CrossRef] [PubMed]
22. Yan, J.K.; Pei, J.J.; Ma, H.L.; Wang, Z.B. Effects of ultrasound on molecular properties, structure, chain conformation and degradation kinetics of carboxylic curdlan. *Carbohydr. Polym.* **2015**, *121*, 64–70. [CrossRef] [PubMed]
23. Miller-Chou, B.A.; Koenig, J.L. A review of polymer dissolution. *Prog. Polym. Sci.* **2003**, *28*, 1223–1270. [CrossRef]

Article

Effects of Recycled Fe_2O_3 Nanofiller on the Structural, Thermal, Mechanical, Dielectric, and Magnetic Properties of PTFE Matrix

Ahmad Mamoun Khamis [1], Zulkifly Abbas [1,2,*], Raba'ah Syahidah Azis [1,3], Ebenezer Ekow Mensah [4] and Ibrahim Abubakar Alhaji [1]

[1] Department of Physics, Faculty of Science, Universiti Putra Malaysia, Serdang 43400, Selangor, Malaysia; akhameis@yahoo.com (A.M.K.); rabaah@upm.edu.my (R.S.A.); alhaji259@gmail.com (I.A.A.)
[2] Introp, University Putra Malaysia, Serdang 43400, Selangor, Malaysia
[3] Institute of Advanced Material, University Putra Malaysia, Serdang 43400, Selangor, Malaysia
[4] Faculty of Science Education, University of Education, Winneba, P.O. Box 40 Mampong, Ashanti, Ghana; ebenekowmensah@aamusted.edu.gh
* Correspondence: za@upm.edu.my

Citation: Khamis, A.M.; Abbas, Z.; Azis, R.S.; Mensah, E.E.; Alhaji, I.A. Effects of Recycled Fe_2O_3 Nanofiller on the Structural, Thermal, Mechanical, Dielectric, and Magnetic Properties of PTFE Matrix. *Polymers* 2021, 13, 2332. https://doi.org/10.3390/polym13142332

Academic Editor: Didier Perrin

Received: 15 June 2021
Accepted: 29 June 2021
Published: 16 July 2021

Publisher's Note: MDPI stays neutral with regard to jurisdictional claims in published maps and institutional affiliations.

Copyright: © 2021 by the authors. Licensee MDPI, Basel, Switzerland. This article is an open access article distributed under the terms and conditions of the Creative Commons Attribution (CC BY) license (https://creativecommons.org/licenses/by/4.0/).

Abstract: The purpose of this study was to improve the dielectric, magnetic, and thermal properties of polytetrafluoroethylene (PTFE) composites using recycled Fe_2O_3 (rFe_2O_3) nanofiller. Hematite (Fe_2O_3) was recycled from mill scale waste and the particle size was reduced to 11.3 nm after 6 h of high-energy ball milling. Different compositions (5–25 wt %) of rFe_2O_3 nanoparticles were incorporated as a filler in the PTFE matrix through a hydraulic pressing and sintering method in order to fabricate rFe_2O_3–PTFE nanocomposites. The microstructure properties of rFe_2O_3 nanoparticles and the nanocomposites were characterized through X-ray diffraction (XRD), field emission scanning electron microscopy (FESEM), and high-resolution transmission electron microscopy (HRTEM). The thermal expansion coefficients (CTEs) of the PTFE matrix and nanocomposites were determined using a dilatometer apparatus. The complex permittivity and permeability were measured using rectangular waveguide connected to vector network analyzer (VNA) in the frequency range 8.2–12.4 GHz. The CTE of PTFE matrix decreased from $65.28 \times 10^{-6}/°C$ to $39.84 \times 10^{-6}/°C$ when the filler loading increased to 25 wt %. The real (ε') and imaginary (ε'') parts of permittivity increased with the rFe_2O_3 loading and reached maximum values of 3.1 and 0.23 at 8 GHz when the filler loading was increased from 5 to 25 wt %. A maximum complex permeability of $(1.1 - j0.07)$ was also achieved by 25 wt % nanocomposite at 10 GHz.

Keywords: PTFE; nanoparticles; microwave; recycled Fe_2O_3; complex permittivity; complex permeability

1. Introduction

Mill scale scrap is considered as a solid waste generated in the steel and iron industries. In general, nearly (8–15) kg of mill scale waste is produced per 1000 kg of steel, and the annual production of mill scale waste is over 13.5 million tons according to preliminary statistics. Moreover, most of this waste is stockpiled in factories due to its limited utilization. The iron protoxide in this waste can be dissolved in acid rain and subsequently transferred into soil and ground waters. Mill scale waste has a dual concern with respect to environmental pollution and land occupancy. Therefore, it is crucial to develop a cleaner method to solve the abundant mill scale resource issue [1].

The study of synthetic techniques for ferrite nanoparticles production has recently attracted considerable attention in view of their excellent magnetic and electrical properties [2]. Up to date, various techniques have been used for the synthesis of Fe_2O_3 nanoparticles (Reference code: 00-024-0072) such as electrospinning technique [3], reduction-oxidation process [4], and hydrolysis process [5]. Indeed, the latter techniques have many disadvantages such as: friability of inorganic nanofibers after calcination [6], non-cost-effective for high contaminant concentrations, and slow reaction speed, respectively. However, mill scale waste materials have never been used to obtain Fe_2O_3 (00-024-0072) yet.

The microwave absorbing materials (MAMs) are in high demand to absorb the unwanted electromagnetic waves via dielectric and magnetic losses [7]. Microwave materials are used in the frequency range (300 MHz–300 GHz) [8]. MAMs are extensively used for many military and commercial applications such as electromagnetic shielding [9]. Therefore, the development of polymer composites (PCs) with tunable dielectric and magnetic properties have triggered researchers in recent years. The PCs reinforced with nanofillers have been widely studied to provide improved radiation protection properties compared to their metal counterparts. In addition, the electronics, packaging, household, energy, sports, communication, and leisure industries are fully involved in the use of PCs for various applications [10]. The advanced polymer composites can be fabricated based on the requirement by selecting quantity, shape, size, and type, of fillers as the reinforcements in different polymer matrix [11].

Polytetrafluoroethylene (PTFE) is a thermoplastic polymer, and PTFE is the best solvent chemical with the best resistance among thermoplastics. The properties of PTFE such as high service temperature, low moisture absorption, chemical inertness, etc., are crucial for many microwave applications [12]. However, PTFE also has certain disadvantages such as high coefficient of thermal expansion (CTE) [13,14], and low relative complex permittivity (ε^*) [15]. In addition, several studies have used Fe_2O_3 as the filler in PTFE matrix. Madusanka et al. synthesized thick film gas sensor using PTFE and α-Fe_2O_3 nanoparticles [16] while Kang et al. fabricated porous Fe_2O_3-PTFE nanofiber membranes for photocatalysis applications [17]. Moreover, PTFE-Al-Fe_2O_3 composites were prepared to be used as reactive materials [18]. The electro-Fenton system was synthesized by combining multiwall carbon nanotubes and Fe@Fe_2O_3 nanowires with PTFE [19]. It is also worth noting that Fe_2O_3 nanoparticles have been widely used as fillers to enhance the dielectric and magnetic properties of various matrixes such as polycaprolactone (PCL) [2], polyaniline (PANI) [20], silica [21], and poly(methyl methacrylate) (PMMA) [22]. However, rFe_2O_3 nanoparticles have never been used as fillers in PTFE matrix for microwave absorption applications yet.

In this study, both CTE and ε^* were improved by incorporating rFe_2O_3 into a PTFE matrix. The rFe_2O_3 nanofiller with enhanced relative complex permittivity can be embedded with a non-conducting polymer matrix such as PTFE to prepare a novel absorber which could have promising attenuation properties for electromagnetic interference (EMI) suppression. This absorber can be further enhanced by the magnetic nature of the Fe_2O_3 nanoparticles and with their high interfacial density due to their good compactness [23].

Knowledge of the complex permittivity and complex permeability of materials has sparked a great interest in industrial and scientific applications [24]. The interaction between a dielectric sample and high frequency electromagnetic signal can be expressed by the relative complex permittivity equation: $\varepsilon^* = \varepsilon' - j\varepsilon''$, where ε' and ε'', respectively, represent the real and imaginary parts. The ratio of $tan\delta = \varepsilon''/\varepsilon'$ represents the loss tangent of a sample and higher values indicate higher attenuation properties [25]. In Addition, the interaction effect between a specimen and the magnetic field component of EM waves can be characterized by the relative complex permeability (μ^*) which can be expressed by the equation: $\mu^* = \mu' - j\mu''$ where μ' and μ'' represent the real and imaginary parts, respectively [26]. The real part determines the amount of energy that the material has stored from an external magnetic field, while the imaginary part measures the attenuation of the magnetic field by the material. The magnetic loss tangent ($tan\delta_\mu$) is calculated using the equation $tan\delta_\mu = \frac{\mu''}{\mu'}$ [27], which represents the power lost versus power stored in a sample [28].

In this research, a low-cost, less complicated and non-chemical synthesis of recycled ferrite was done from mill scale waste. Then, the particle size of rFe_2O_3 decreased to nano-size via high energy ball milling (HEBM) in order to enhance the dielectric properties. The current study also involves the processes of synthesizing five nanocomposites of rFe$_2$O$_3$—PTFE by varying the rFe$_2$O$_3$ percentage in the nanocomposites. Afterwards, the resultant effects of rFe$_2$O$_3$ loadings on the phase composition, thermal expansion tensile strength, density, complex permittivity, and complex permeability of PTFE matrix were investigated.

2. Materials and Methods

2.1. Materials

The materials used for the preparation of the rFe$_2$O$_3$ nanoparticles and rFe$_2$O$_3$–PTFE nanocomposites were: Mill scale flakes (Perwaja Sdn. Bhd., Chukai, Terengganu, Malaysia) and PTFE molding powder (Fujian Sannong New Materials Co., LTD, Sanming, China) with average particle size 50–110 µm.

2.2. Synthesis of rFe$_2$O$_3$ Nanopartilces from Mill Scale

Mill scale waste was initially cleaned by removing the impurities and then crushed manually into powder using mortar. This was followed by the application of the magnetic separation technique (MST) and Curie temperature separation technique (CTST) to further purify the powdered mill scale. The materials used for these processing techniques were an electromagnet which produced 1 Tesla magnetic field intensity, a thin cylindrical glass tube open at both ends and deionized water of density 1 g/cm^3. The CTST was used to separate the wustite (FeO) and magnetite (Fe$_3$O$_4$) contained in the mill scale slurry collected in the cold water separation process [23]. In this separation step, the clamped glass tube closed at the bottom end with a glass stopper, was filled with hot deionized water (100 °C) followed by a small quantity of the slurry. The particles that remained attracted to the magnetic field were inferred to be magnetite (Fe$_3$O$_4$) because of its higher Curie temperature of 585 °C. The rFe$_3$O$_4$ powder obtained was milled via the mechanical alloying technique using the high energy ball mill (SPEX Sample Prep 8000D, Metuchen, NJ, USA) operated by a 1425 rpm 50 Hz motor at a clamp speed of 875 cycles/minute. The powder to balls ratio of 1:10 was used. A total of 7 gm of the Fe$_3$O$_4$ powder was poured into each of the two vials containing the 70 g of steel balls and milled for 6 h to produce nanoparticles. During the ball milling Fe$_3$O$_4$ was converted to Fe$_2$O$_3$ due to high temperature and the content of oxygen in the vials. The preparation process of rFe$_2$O$_3$ nanoparticles is shown in Figure 1.

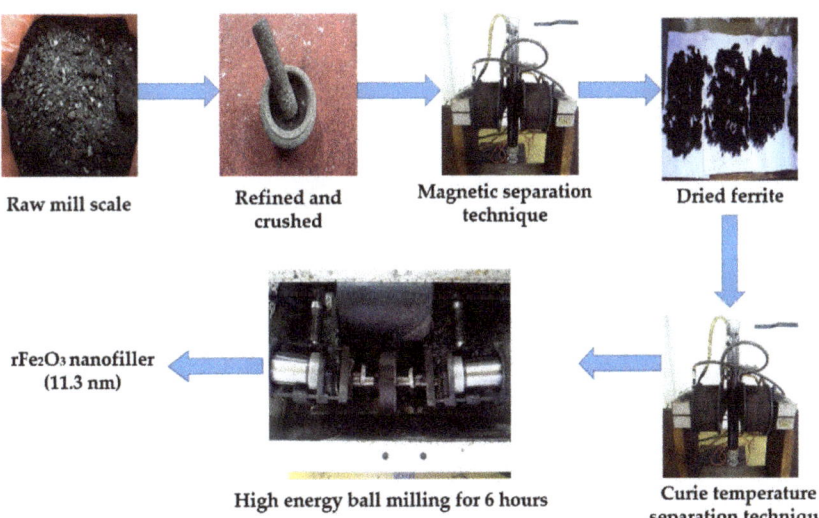

Figure 1. Preparation process of rFe$_2$O$_3$ nanoparticles.

2.3. Fabrication of rFe$_2$O$_3$–PTFE Nanocomposites

The rFe$_2$O$_3$–PTFE nanocomposites shown in Figure 2 were fabricated by mixing PTFE powder with different mass percentages (5–25%) of the rFe$_2$O$_3$ nano-powder as listed in Table 1. The composites were prepared through a hydraulic pressing and sintering method.

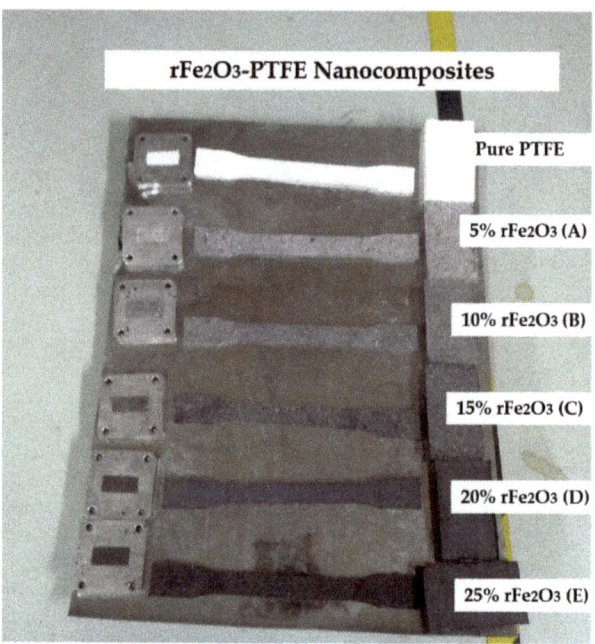

Figure 2. rFe$_2$O$_3$–PTFE nanocomposites.

Table 1. Composition of rFe$_2$O$_3$–PTFE nanocomposites.

Composite Code	Recycled Fe$_2$O$_3$ Nano-Powder		PTFE		Total Mass (g)
	Percentage %	Mass (g)	Percentage %	Mass (g)	
Control	0	0	100	60	60
A	5	3	95	57	60
B	10	6	90	54	60
C	15	9	85	51	60
D	20	12	80	48	60
E	25	15	75	45	60

Three different shapes of nanocomposites were fabricated for each mass percentage in order to suite different characterization apparatuses. The raw materials were weighed using analytical micro balance (A&D Company, Ltd., GR-200, Tokyo, Japan) which has an accuracy of ±0.00007 g. The mixing was performed via Wing mixer/blender for 10 min. Then, the mixed powders were placed into three different molds and pressed using a hydraulic pressing machine for 5 min under 10 MPa load. Finally, the compact rFe$_2$O$_3$–PTFE nanocomposites were sintered as follows: the compact nanocomposites were heated from room temperature to 380 °C in a furnace at 2.97 °C/min and then maintained for 1 h to allow the PTFE particles to coalesce completely. The cooling process was 1 °C/min from 380 °C to room temperature. The rFe$_2$O$_3$–PTFE nanocomposites preparation is shown in Figure 3. After cooling, the rFe$_2$O$_3$–PTFE nanocomposites were ready for characterizations.

Figure 3. rFe$_2$O$_3$–PTFE nanocomposites preparation.

2.4. Characterization

rFe$_2$O$_3$ nanoparticles, PTFE powder, rFe$_2$O$_3$–PTFE nanocomposites were characterized using the following techniques:

2.4.1. X-ray Diffraction (XRD)

The phase composition structure of the samples was analyzed using X-ray diffraction (XRD). The data were collected using a fully automated Philips X'pert system (Model PW3040/60 MPD, Amsterdam, The Netherlands) with Cu–Kα radiation operating at a current of 40 mA, a wavelength of 1.5405 Å, and a voltage of 40 kV. The diffraction patterns were recorded with a scanning speed of 2°/min in the 2θ range 10 to 70°. The collected data were subjected to the Rietveld analysis on PANalytic X'Pert Highscore Plus v.3.0 software (PANalytical B.V., Almelo, The Netherlands). The diffraction peaks were compared with the Inorganic Crystal Structure Database (ICSD). The rFe$_2$O$_3$–PTFE samples were cut from rFe$_2$O$_3$–PTFE nanocomposites prepared in advanced in a solid form, while the rFe$_2$O$_3$ was in the form of a fine powder.

2.4.2. High-Resolution Transmission Electron Microscopy (HRTEM)

The size and shape of rFe$_2$O$_3$ nanoparticles were investigated using HRTEM (JEM-2100F, JEOL, Tokyo, Japan). Drops of rFe$_2$O$_3$ nanoparticles were dispersed and placed on TEM grids for drying. The dried specimen was then placed in a high-vacuum chamber of the microscope for viewing and analysis of the nanoparticles.

2.4.3. Field Emission Scanning Electron Microscopy (FESEM)

The dispersion of rFe$_2$O$_3$ nanoparticles in PTFE matrix was evaluated using FESEM (JEOL Ltd. JSM-7600, Tokyo, Japan). Carbon tapes were used to cover the aluminums stubs before placing the specimens on them. Then, the samples were coated with titanium using an auto-fine coater (JEOL Ltd. JEC-3000FC, Tokyo, Japan) to increase their conductivity and eliminate electromagnetic charges. Finally, the stubs were placed on the FESEM chamber for analysis.

2.4.4. Electronic Densitometer

The effects of rFe$_2$O$_3$ nanoparticles on the density of PTFE matrix was determined using electronic densitometer (Alfa Mirage Co., Ltd. Model MD-300S0, Osaka, Japan) which employs Archimedes' principle in order to calculate the density. Distilled water was used as the immersion fluid. The density was determined using the following equation [29]:

$$\rho = \frac{W_{air} \times \rho_{dis.water}}{W_{air} - W_{dis.water}} \quad (1)$$

where $W_{dis.water}$ is the composite weight in distilled water, W_{air} is the composite weight in air, and $\rho_{dis.water}$ is the density of the distilled water (1 gm/cm^3).

2.4.5. Dilatometer Linseis L75 Platinum

Dilatometer is one of the main techniques used for coefficients of thermal expansion (CTEs) measurements. In general, to determine the CTE, two physical quantities (temperature and displacement) must be measured on a specimen that is undergoing a thermal cycle. In this study, dilatometer (Linseis L75 Platinum, Selb, Germany) was used to determine the CTEs of composites. The sample was enclosed in the furnace and then heated up to 200 °C. When the sample expands, it pushes the central rod along the axis of the tube, and the relative movement can be characterized [30]. In addition, from an atomic perspective, the CTE represents the increment in the average distance among atoms with increasing temperature and, generally, weaker bonds have a higher CTE value [31].

2.4.6. Extensometer (Shimadzu AGS-X 100kN)

rFe$_2$O$_3$–PTFE nanocomposites were fabricated as per the ASTM-D638 standard in order to find out the ultimate tensile strength (UTS) [32]. The influence of rFe$_2$O$_3$ nanoparticles on the UTS of PTFE matrix was examined using the Shimadzu (AGS-X 100kN, Kyoto, Japan) with industry-leading (Shimadzu, TRAPEZIUM X, Kyoto, Japan) data processing software. The pneumatic grips were displaced with a rate of 5 mm/min at room temperature. The data were obtained from the software and data acquisition system monitor continuously until the specimen reached its UTS and broke.

2.4.7. Rectangular Waveguide (RWG)

The measurements of relative complex permittivity and permeability were carried out in the frequency range of 8.2–12.4 GHz using rectangular waveguide (RWG) connected to a Keysight (E5063A) vector network analyzer (Keysight Technologies, Santa Rosa, CA, USA). As shown in Figure 4, the sample holder containing the sample was carefully attached to the flanges of RWG in order to totally avoid the air gap. The complex permittivity and permeability of the nanocomposites were deduced depending on the transmission and reflection of the waves throughout the composites. The measurement model of RWG to measure the permeability and permittivity was the poly-reflection–transmission μ and ε model. This model used an optimization technique which gives constant values for both ϵ' and μ' throughout the whole frequency range. The transmission–reflection method is widely used in the EMI characterization because of the field-focusing ability which ensures an accurate characterization at the microwave frequency. The VNA was calibrated by implementing a standard full two-port calibration technique for 201 frequency points in the frequency range 8.2 GHz to 12.4 at room temperature. The characterization procedures were also described in our previous study [33].

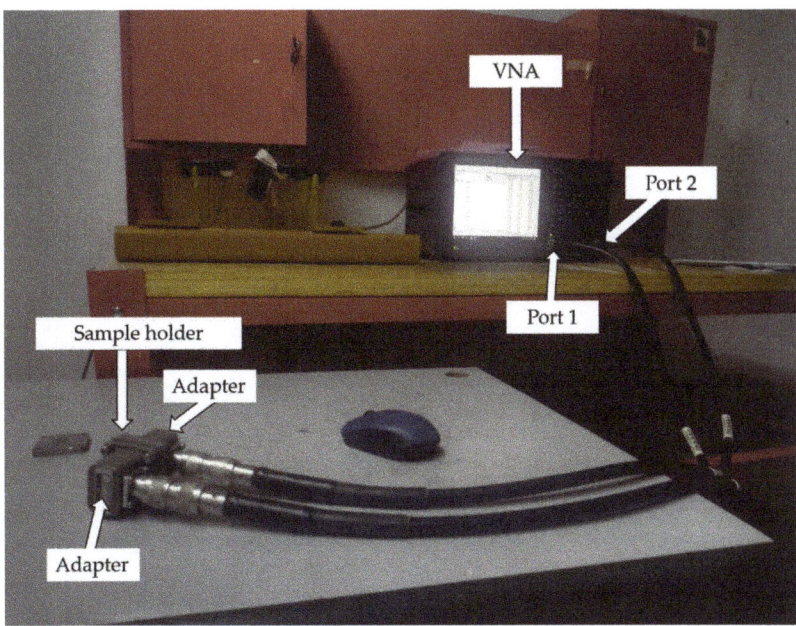

Figure 4. Measurement of complex permeability and permittivity using rectangular waveguide technique.

3. Results and Discussion

3.1. Microstructural Characterization

The X-ray diffraction (XRD) technique was used to confirm the phase structurers of PTFE matrix, rFe$_2$O$_3$ nanopowder, and rFe$_2$O$_3$–PTFE nanocomposites. The XRD results are shown in Figure 5. By comparing the obtained diffractograms with the standard patterns from the Inorganic Crystal Structure Database (ICSD), the Bragg peaks of obtained recycled powder were all identified as rhombohedral crystal structure of single phase hematite (Fe$_2$O$_3$, Reference code: 00-024-0072) with R—3 space group. The diffractogram of rFe$_2$O$_3$ nanopowder showed the following peaks in the pattern at 2θ of 24.1°, 33.1°, 35.6°, 40.8°, 43.5°, 49.4°, 54°, 57.5°, 62.4°, and 63.9° corresponding to the miller indices (hkl): (012), (104), (110), (113), (202), (024), (116), (122), (214), and (300), respectively. On the other hand, the diffractogram of PTFE matrix showed six peaks in the pattern at 2θ of 18.04°, 31.54°, 36.60°, 37°, 41.22°, and 49.11°, which agreed with the study carried out by Yamaguchi et al. [34]. The mentioned peaks corresponded to the miller indices (hkl): (100), (110), (200), (107), (108), and (210), respectively. The diffractograms of rFe$_2$O$_3$–PTFE nanocomposites show dominant crystalline phase characteristics of PTFE at low concentration of rFe$_2$O$_3$ content in the nanocomposites. However, the peaks corresponding to the rFe$_2$O$_3$ nanopowder increased in intensity as the content increased. All peaks shown in rFe$_2$O$_3$–PTFE profiles belongs to the materials used for the preparation of the composites. The patterns did not show new peaks suggesting the nanocomposites were pure and there was no conversion of PTFE or rFe$_2$O$_3$ nanopowder into other materials. The absence of any new peaks of the nanocomposites also suggests that the rFe$_2$O$_3$ nanoparticles did not chemically interact with the PTFE matrix and that the mixture was physical in nature.

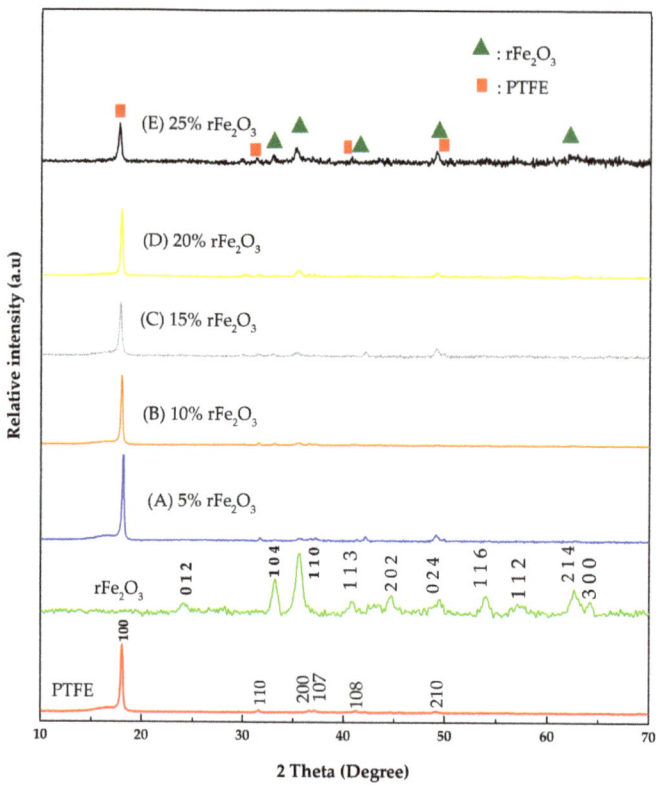

Figure 5. XRD diffractograms of the rFe$_2$O$_3$ nanoparticles, PTFE, and rFe$_2$O$_3$–PTFE nanocomposites.

The microstructure characterization of the rFe$_2$O$_3$ nanoparticles was performed using high resolution transmission electron microscopy (HRTEM) to investigate the size and shape of rFe$_2$O$_3$ particles. The HRTEM results confirmed that the nanoparticles were highly crystalline as shown in Figure 6, which agreed with the study conducted by Wen et al. [35]. It can also be seen in Figure 6 that the rFe$_2$O$_3$ nanoparticles are aggregated and spherical, which may be attributed to the long-range magnetic dipole–dipole interaction between the particles. This observation may also be caused by the drying process during TEM sample preparation [36]. The rFe$_2$O$_3$ nanoparticles sizes were within the range 7.4–15.24 nm after 6 h of ball milling with an average particle size of 11.3 nm.

The cross-sectional surface images of 5% rFe$_2$O$_3$ (A), 15% rFe$_2$O$_3$ (C), and 25% rFe$_2$O$_3$ (E) nanocomposite are shown in Figure 7. It can be seen in 5% nanocomposite (A) that there are two main types of patterns observed in PTFE structure. The first pattern is small outgrowths which are uniformly dispersed on the surface of PTFE and was called "warts". The second pattern is "ribbon" like shapes which entirely cover the PTFE surface and was called "dendrites". These observations are in accordance with the study conducted by Glaris et al. [37]. The images confirmed that with an increase in rFe$_2$O$_3$ loadings in the composites, the adhesion to PTFE matrix increased gradually. There were no agglomerates observed in rFe$_2$O$_3$–PTFE nanocomposites. The uniform dispersal of rFe$_2$O$_3$ nanoparticles in the PTFE matrix gives an indication that the nanoparticles were completely implanted in the nanocomposites and offered interfacial bonding which can enhance the complex permittivity and permeability.

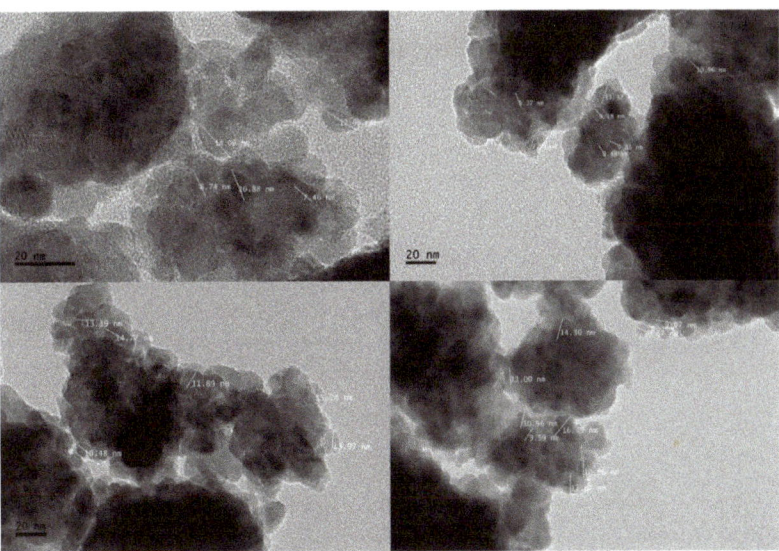

Figure 6. HRTEM images of rFe$_2$O$_3$ nanoparticles.

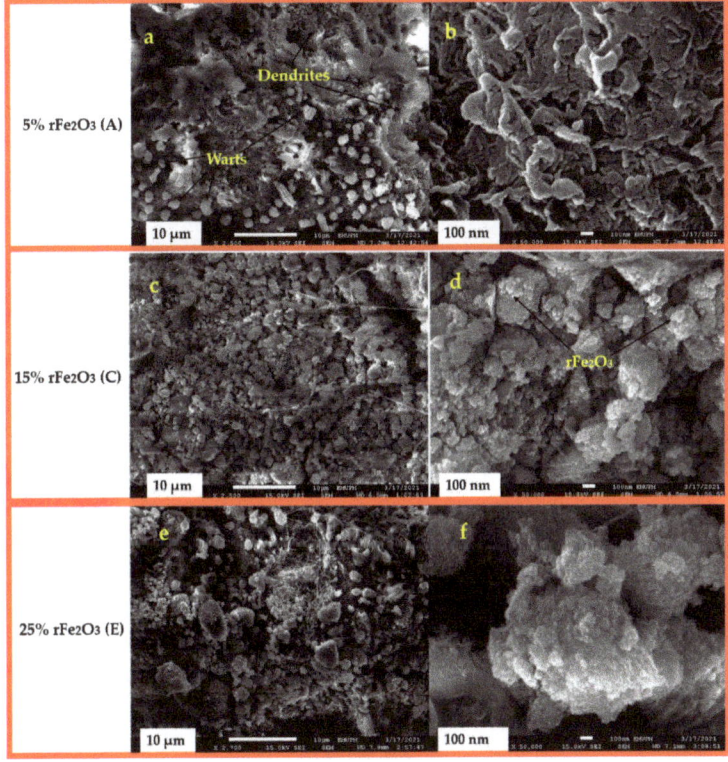

Figure 7. FESEM of rFe$_2$O$_3$–PTFE nanocomposites: 5% rFe$_2$O$_3$ (**A**) nanocomposite (**a**,**b**), 15% rFe$_2$O$_3$ (**C**) nanocomposite (**c**,**d**), and 25% rFe$_2$O$_3$ (**E**) nanocomposite (**e**,**f**).

3.2. Coefficients of Linear Thermal Expansion (CTEs)

The variation in CTEs with temperature for rFe_2O_3–PTFE nanocomposites in the temperature range 30–200 °C is presented in Figure 8. It was observed that the values of CTEs were low for the nanocomposites with higher content of rFe_2O_3 nanofiller. The decrease in CTEs of rFe_2O_3–PTFE nanocomposites is due to the difference between the CTEs of PTFE matrix and rFe_2O_3 nanofiller (Fe_2O_3 has a lower CTE comparing to PTFE). Figure 8 also shows that the CTEs of all nanocomposites increased as temperature increased throughout the measurement range. This behavior can be attributed to the increase in molecular vibrations with temperature, and a linear relationship exists between the CTE and the thermal capacitance per unit volume. The incorporation of low-CTE rFe_2O_3 in the PTFE matrix increased the thermal stability of the nanocomposites over a wide temperature range. Composites with low CTE are highly desirable for precision structures [38] such as microwave applications, where temperature fluctuations at the operating environment could cause substantial change in dimensions leading to destruction in the application.

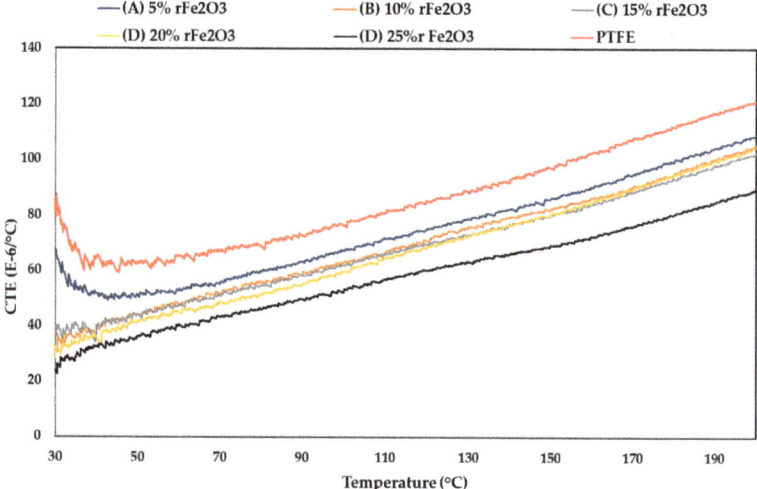

Figure 8. Thermal expansion coefficients of rFe_2O_3–PTFE nanocomposites.

3.3. Tensile Strength and Density of rFe_2O_3–PTFE Nanocomposites

The variation in tensile strength with filler content for different rFe_2O_3–PTFE nanocomposites is presented in Figure 9. The tensile strength of polymer composites depends on the properties of nanoparticle–matrix interaction which plays an important role in the level of dissipated energy by different damaging mechanisms which take place at the nanoscale [39]. As shown in Figure 9, the tensile strength of nanocomposites decreased with increasing rFe_2O_3 nanofiller content. This behavior was expectable for polymer composites with inorganic filler and it was reported by a study conducted by Jiang et al. [40]. This decrease could be attributed to the rFe_2O_3 nanofiller which were not able to support the stress transferred from PTFE matrix which weakened the nanocomposite [14].

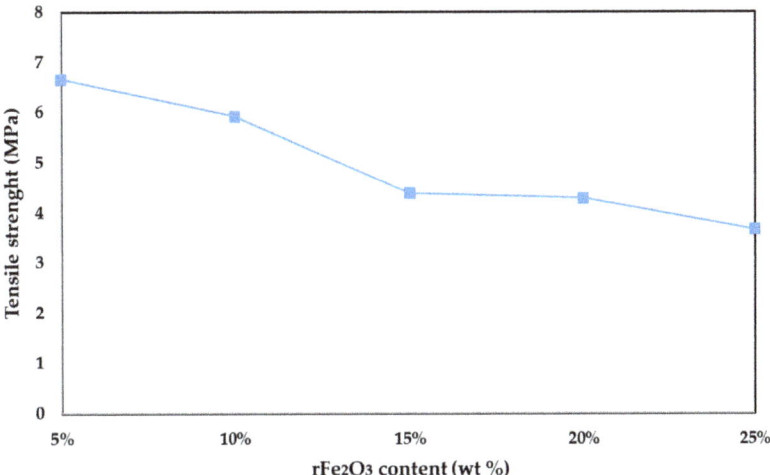

Figure 9. Tensile strength vs. various rFe_2O_3 nanofiller content for rFe_2O_3–PTFE nanocomposites.

The variation of density values with different rFe_2O_3 wt % content in the PTFE matrix is shown in Figure 10. The 5% (A), 10% (B), 15% (C), 20% (D), and 25% (E) rFe_2O_3 nanocomposites had the respective density values of 2.2, 2.32, 2.4, 2.49, 2.53 gm/cm^3. Therefore, increasing the rFe_2O_3 nanofiller corresponded to increased density values of rFe_2O_3–PTFE nanocomposites. The increase in density with the increase in Fe_2O_3 nanofiller is supported by a study carried out by Ghasemi-Kahrizsangi et al. [41]. This behavior can be attributed to the higher true density of Fe_2O_3 (5.24 gm/cm^3) [42], in comparison to the PTFE (2.2 gm/cm^3) [43]. The increase in density can cause an observed increase in dielectric constant [44]. Moreover, Leyland and Maharaj found that the relationship between density and dielectric constant is directly proportional for various materials [45]. The denser rFe_2O_3–PTFE nanocomposites results higher number of molecules per unit volume. A larger number of molecules per unit volume means that there is more interaction with the electric fields and therefore an increase in the complex permittivity.

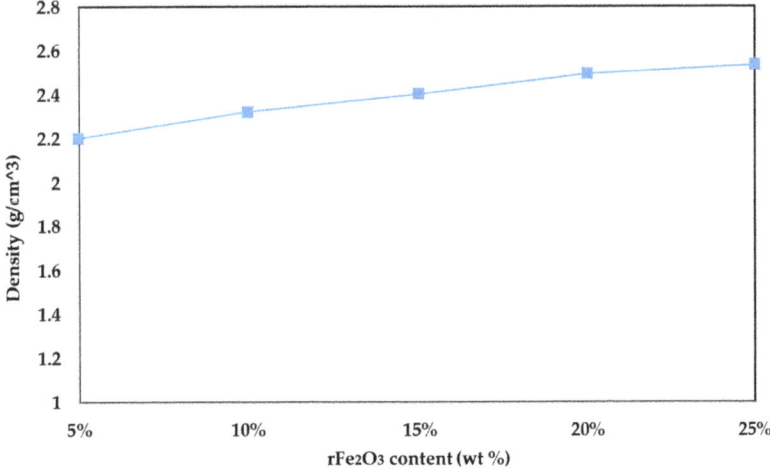

Figure 10. Density vs. various rFe_2O_3 nanofiller content for rFe_2O_3–PTFE nanocomposites.

3.4. Complex Permittivity of rFe$_2$O$_3$–PTFE Nanocomposites

The measured complex permittivity values (real and imaginary parts) of rFe$_2$O$_3$–PTFE nanocomposites in the frequency range of 8.2–12.4 GHz using rectangular waveguide are shown in Figure 11; Figure 12 while loss tangent is shown in Figure 13. The highest ε' value of rFe$_2$O$_3$–PTFE nanocomposites was 3.1 for 25% rFe$_2$O$_3$ nanocomposite along the frequency range 8.2 GHz to 12.4 GHz while the lowest ε' value was 2.01 for the PTFE sample. By increasing the rFe$_2$O$_3$ nanofiller in the nanocomposites, both ε' and ε'' increased throughout the measurement range. It can be noticed that ε' and ε'' increased from 2.29 and 0.10 for 5% rFe$_2$O$_3$ nanocomposite (A) to 2.39 and 0.14, respectively for 10% rFe$_2$O$_3$ nanocomposite (B) at 8.2 GHz. The same behavior of increasing complex permittivity by increasing the rFe$_2$O$_3$ nanofiller was observed in all nanocomposites. The ε' and ε'' rely on the contribution of the atomic, electronic, interface, and orientation polarization in the sample [46]. The interface polarization increased due to the differences in polarization of the rFe$_2$O$_3$ nanofiller and PTFE matrix.

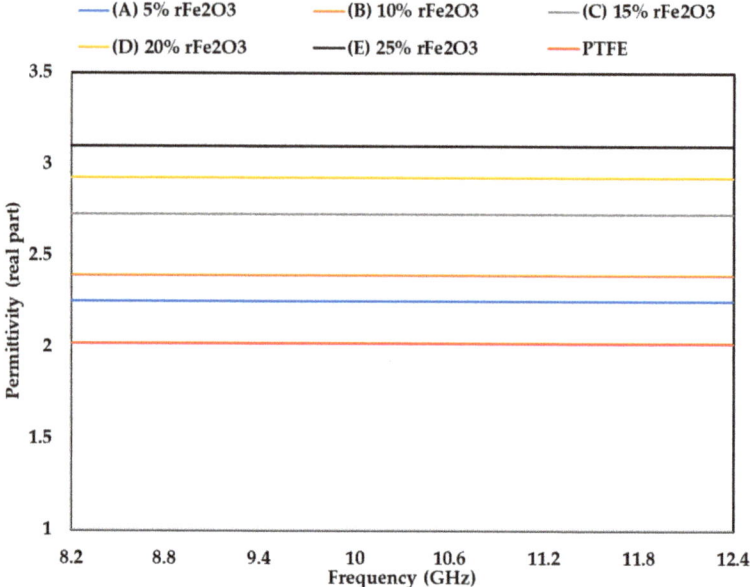

Figure 11. Variation in permittivity (real part) of rFe$_2$O$_3$–PTFE nanocomposites using rectangular waveguide.

The increment in complex permittivity with increasing rFe$_2$O$_3$ nanofiller can be attributed to the polarization process due to the enhanced conductivity and interfacial polarization in the composite and hopping exchange of charges between localized states [23]. One more reason for the influence on the relative complex permittivity of the nanocomposites is the particle size of rFe$_2$O$_3$ nanofiller since the nano size had high specific surface area that enabled good contact between the particles and the matrix [47]. The increment in ε' and ε'' at low frequency can be attributed to the dominant role of dipolar and interfacial polarization. However, the increment at high frequencies can be attributed to the electronic and ionic polarization of the system [48].

It can be also seen in Figure 12 that ε'' of rFe$_2$O$_3$–PTFE nanocomposites decreased by increasing the frequency from 8.2 to 12.4 GHz. This decrement can be attributed to the interfacial dipoles having less time to align in the direction of the external field. The molecules were able to do a complete orientation at low frequency but they were not able to achieve the same orientation at high frequency [49]. However, the ε'' of the PTFE sample did not change with frequency because PTFE is a nonpolar polymer [50], which means its complex permittivity is not dependent on frequency.

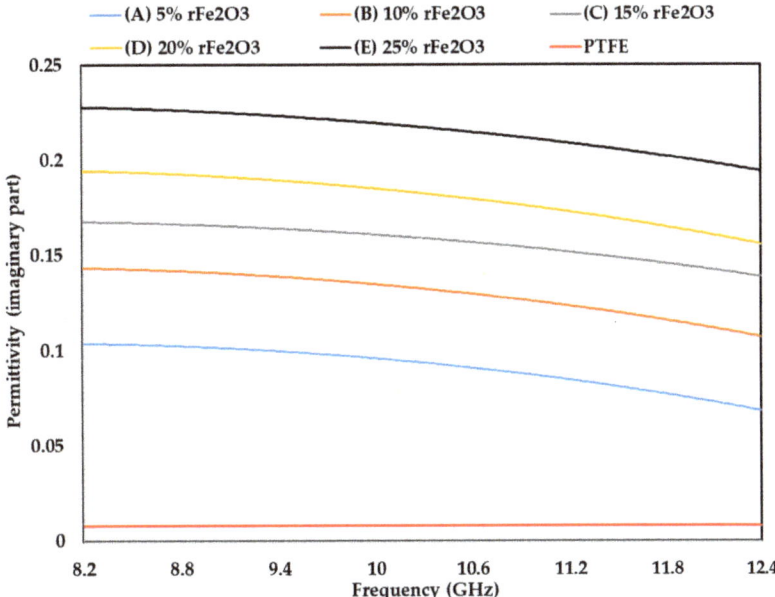

Figure 12. Variation in permittivity (imaginary part) of rFe$_2$O$_3$–PTFE nanocomposites using rectangular waveguide.

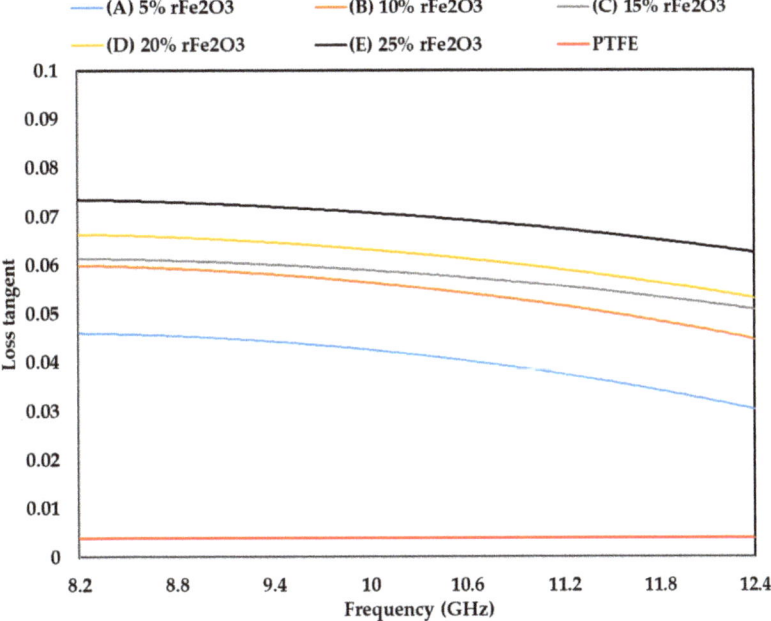

Figure 13. Variation in loss tangent of rFe$_2$O$_3$–PTFE nanocomposites.

3.5. Complex Permeability of rFe$_2$O$_3$–PTFE Nanocomposites

The permeability characteristics of rFe$_2$O$_3$–PTFE nanocomposites can only be attributed to the rFe$_2$O$_3$ nanoparticles because PTFE is a non-magnetic material [51]. The variations in μ', μ'', and $tan\delta_\mu$ of rFe$_2$O$_3$–PTFE nanocomposites are shown in Figures 14–16, respectively. It is clear that the values of μ', μ'', and $tan\delta_\mu$ were higher for rFe$_2$O$_3$–PTFE nanocomposites having higher rFe$_2$O$_3$ content.

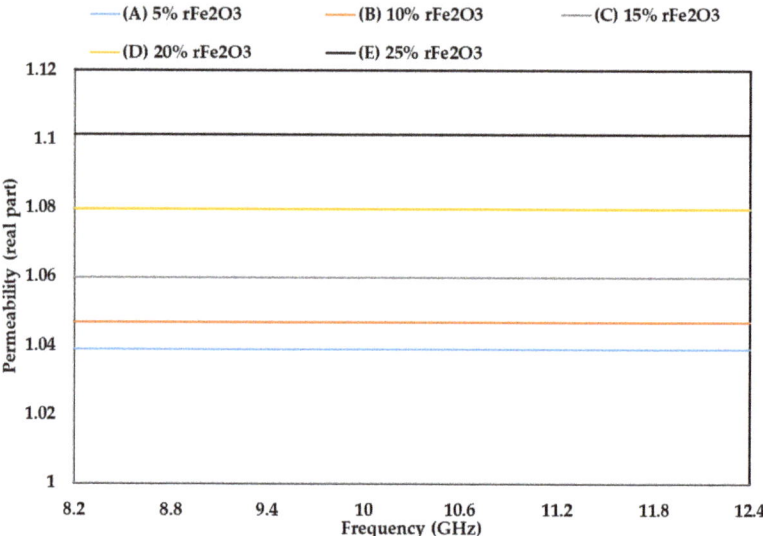

Figure 14. Variation in permeability (real part) of rFe$_2$O$_3$–PTFE nanocomposites using rectangular waveguide.

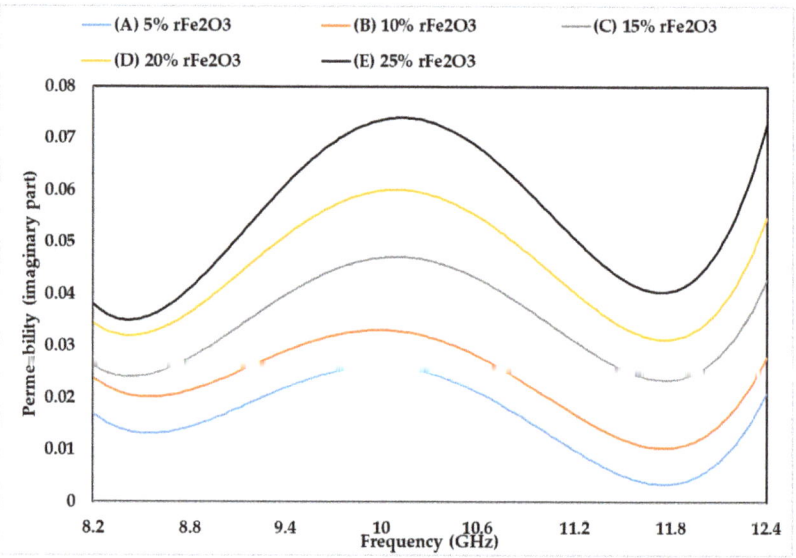

Figure 15. Variation in permeability (imaginary part) of rFe$_2$O$_3$–PTFE nanocomposites using rectangular waveguide.

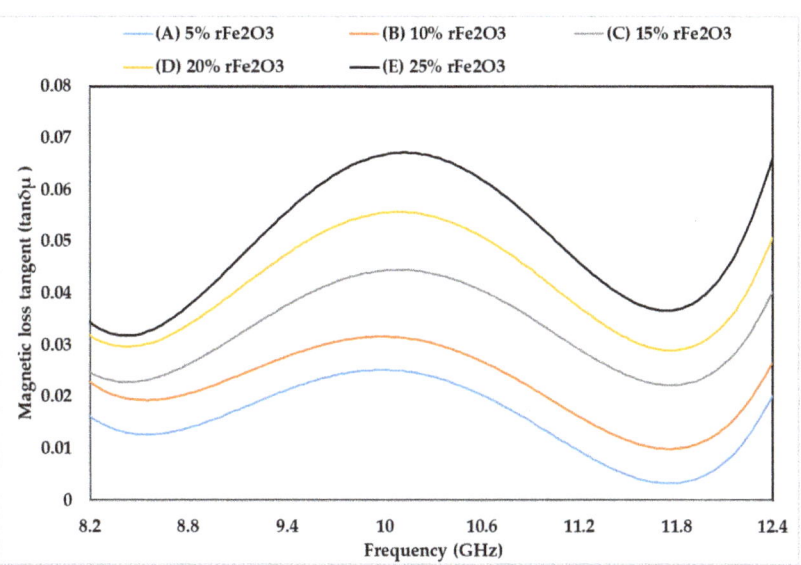

Figure 16. Variation in magnetic loss tangent of rFe$_2$O$_3$–PTFE nanocomposites.

As the content of rFe$_2$O$_3$ nanofiller increased in the rFe$_2$O$_3$–PTFE nanocomposites, the spacing in the nanocomposites reduced due to the high specific surface area of the rFe$_2$O$_3$ nanofiller. This made it more difficult for the magnetic field to pass through the rFe$_2$O$_3$–PTFE nanocomposites, thereby increasing the values of the relative complex permeability.

The real part (μ') of the relative permeability values were 1.039, 1.047, 1.06, 1.8 and 1.1 for 5% (A), 10% (B), 15% (C), 20% (D) and 25% (E) rFe$_2$O$_3$ nanocomposites, respectively, within the frequency range of 8.2–12.4 GHz while the corresponding values of μ'' were 0.017, 0.023, 0.026, 0.034 and 0.038 at 8.2 GHz. A resonant peak can be clearly seen in each μ'' curve between 10 and 10.6 GHz. The increment in μ' and μ'' of the polymer composites with increase in Fe$_2$O$_3$ is supported by studies conducted by Abdalhadi et al. [52] and Ahmad et al. [53]. In addition, the magnetic loss tangent in Figure 16 followed the behavior of μ'' because it was calculated using the equation $tan\delta_\mu = \frac{\mu''}{\mu'}$.

It can be seen that μ' in Figure 14 shows constant values while μ'' in Figure 15 shows non-steady behavior with the frequency. The μ' is associated with the material storage capacity of the magnetic field [54], and it did not vary with frequency because of the RWG measurement model described in Section 2.4.7. On the other hand, the fluctuations in μ'' with the frequency can be attributed to the magnetic loss. In general, the magnetic loss of a sample originates because of one of these four reasons: the hysteresis loss, domain-wall resonance, eddy current effect, and natural resonance [55]. The first reason (hysteresis loss) which originates from the irreversible magnetization is negligible in a weak applied magnetization field [56]. The second reason (domain-wall resonance) normally occurs in the (1–100) MHz range, therefore, this reason can be excluded in this study [57]. The third reason (eddy current effect) can usually be estimated using the equation of $C_0 = \mu'(\mu'')^2 f^{-1}$. When the magnetic loss originates for this reason, the C_0 should be almost constant as the frequency varied. However, it can be seen in Figure 17 that the C_0 values for the rFe$_2$O$_3$–PTFE nanocomposites varied considerably with the frequency; so, this reason can also be excluded. Hence, these resonant peaks are associated with the natural ferromagnetic resonance [58]. The trend of μ'' is also supported by the study conducted by Liang [59].

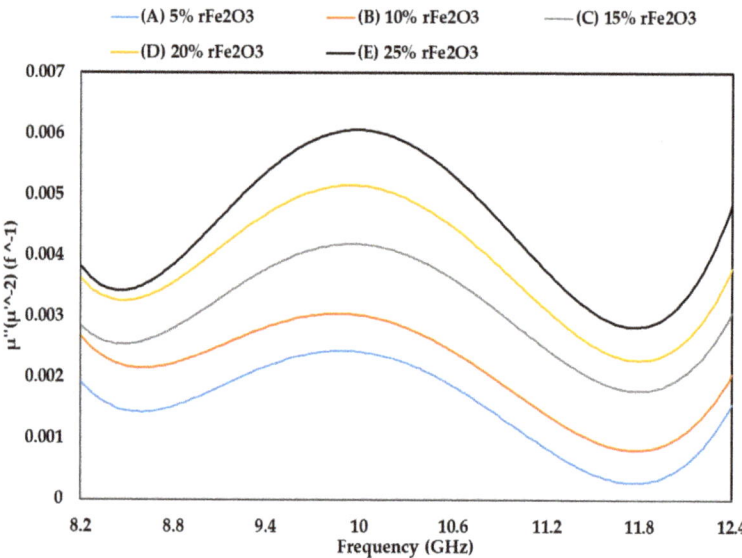

Figure 17. The values of $\mu'(\mu'')^2 f^{-1}$ as a function of frequency for rFe$_2$O$_3$–PTFE nanocomposites.

4. Conclusions

In this research, rFe$_2$O$_3$–PTFE nanocomposites were successfully fabricated using rFe$_2$O$_3$ nanopowder as the filler and PTFE as the matrix. The effect of rFe$_2$O$_3$ loading on the structural, mechanical, thermal, dielectric, and magnetic properties of PTFE matrix was investigated. rFe$_2$O$_3$ nanofiller was dispersed uniformly with complete implantation in the PTFE matrix as presented in the FESEM results. The density increased but the tensile strength decreased with the increase of rF$_2$O$_3$ nanofiller in the nanocomposites. The rFe$_2$O$_3$ nanoparticles improved the thermal properties of the PTFE matrix. The complex permittivity and permeability were also enhanced by embedding rFe$_2$O$_3$ nanofiller in the matrix throughout the measurement range. The important dielectric, magnetic, thermal and mechanical properties of rFe$_2$O$_3$–PTFE nanocomposites linked to the content of the rFe$_2$O$_3$ in the nanocomposites can be employed in possible applications requiring tunable characteristics.

Author Contributions: Conceptualization, A.M.K. and Z.A.; methodology, R.S.A.; software, A.M.K. and E.E.M.; validation, Z.A and R.S.A.; formal analysis, A.M.K.; investigation, A.M.K. and I.A.A.; resources, Z.A. and R.S.A.; data curation, A.M.K. and I.A.A.; writing—original draft preparation, A.M.K.; writing—review and editing, A.M.K.; visualization, A.M.K.; supervision, Z.A.; project administration, Z.A.; funding acquisition, Z.A. All authors have read and agreed to the published version of the manuscript.

Funding: This work is supported by the Universiti Putra Malaysia (UPM) and Ministry of Higher Education (MOHE) Fundamental Research Grant Scheme FRGS/1/2015/ICT05/UPM/02/4.

Institutional Review Board Statement: Not applicable.

Informed Consent Statement: Not applicable.

Data Availability Statement: Not applicable.

Acknowledgments: The authors specially thank Physics Department (UPM), UPM Holdings & RMC for research management facilities, Sarra Gobal Sdn Bhd and MOSTI for AG TF0315D031 project, Sunkoo Machine Tech Ltd. for PTFE powder supply, and Ministry of Higher Education (MOHE) Fundamental Research Grant Scheme (FRGS).

Conflicts of Interest: The authors declare no conflict of interest.

References

1. Liu, B.; Zhang, L.; Zhang, Y.; Han, G.; Zhang, B. Innovative methodology for co-treatment of mill scale scrap and manganese ore via oxidization roasting-magnetic separation for preparation of ferrite materials. *Ceram. Int.* **2021**, *47*, 6139–6153. [CrossRef]
2. Mensah, E.E.; Abbas, Z.; Azis, R.S.; Ibrahim, N.A.; Khamis, A.M. Complex Permittivity and Microwave Absorption Properties of OPEFB Fiber–Polycaprolactone Composites Filled with Recycled Hematite (α-Fe_2O_3) Nanoparticles. *Polymers* **2019**, *11*, 918. [CrossRef] [PubMed]
3. Eid, C.; Brioude, A.; Salles, V.; Plenet, J.-C.; Asmar, R.; Monteil, Y.; Khoury, R.; Khoury, A.; Miele, P. Iron-based 1D nanostructures by electrospinning process. *Nanotechnology* **2010**, *21*, 125701. [CrossRef] [PubMed]
4. El-Leathy, A.; Danish, S.N.; Al-Ansary, H.; Jeter, S.; Al-Suhaibani, Z. Experimental study of compatibility of reduced metal oxides with thermal energy storage lining materials. In Proceedings of the AIP Conference Proceedings, Cape Town, South Africa, 13–16 October 2015; p. 050013. Available online: https://aip.scitation.org/toc/apc/1734/1?size=all (accessed on 31 May 2016).
5. Müller, M.; Villalba, J.C.; Mariani, F.Q.; Dalpasquale, M.; Lemos, M.Z.; Huila, M.F.G.; Anaissi, F.J. Synthesis and characterization of iron oxide pigments through the method of the forced hydrolysis of inorganic salts. *Dye. Pigment.* **2015**, *120*, 271–278. [CrossRef]
6. Shi, X.; Zhou, W.; Ma, D.; Ma, Q.; Bridges, D.; Ma, Y.; Hu, A. Electrospinning of nanofibers and their applications for energy devices. *J. Nanomater.* **2015**, *2015*, 122. [CrossRef]
7. Bai, H.; Yin, P.; Lu, X.; Zhang, L.; Wu, W.; Feng, X.; Wang, J.; Dai, J. Recent advances of magnetism-based microwave absorbing composites: An insight from perspective of typical morphologies. *J. Mater. Sci. Mater. Electron.* **2020**, 1–26. [CrossRef]
8. Raveendran, A.; Sebastian, M.T.; Raman, S. Applications of microwave materials: A review. *J. Electron. Mater.* **2019**, *48*, 2601–2634. [CrossRef]
9. Bhattacharya, P.; Das, C.K. In situ synthesis and characterization of $CuFe_{10}Al_2O_{19}$/MWCNT nanocomposites for supercapacitor and microwave-absorbing applications. *Ind. Eng. Chem. Res.* **2013**, *52*, 9594–9606. [CrossRef]
10. Oladele, I.O.; Omotosho, T.F.; Adediran, A.A. Polymer-Based Composites: An Indispensable Material for Present and Future Applications. *Int. J. Polym. Sci.* **2020**, *2020*, 8834518. [CrossRef]
11. Chandra, R.J.; Shivamurthy, B.; Kulkarni, S.D.; Kumar, M.S. Hybrid polymer composites for EMI shielding application—A review. *Mater. Res. Express* **2019**, *6*, 082008.
12. Wu, K.-T.; Yuan, Y.; Zhang, S.-R.; Yan, X.-Y.; Cui, Y.-R. $ZrTi_2O_6$ filled PTFE composites for microwave substrate applications. *J. Polym. Res.* **2013**, *20*, 223. [CrossRef]
13. Murali, K.; Rajesh, S.; Prakash, O.; Kulkarni, A.; Ratheesh, R. Preparation and properties of silica filled PTFE flexible laminates for microwave circuit applications. *Compos. Part A Appl. Sci. Manuf.* **2009**, *40*, 1179–1185. [CrossRef]
14. Chen, Y.-C.; Lin, H.-C.; Lee, Y.-D. The effects of filler content and size on the properties of $PTFE/SiO_2$ composites. *J. Polym. Res.* **2003**, *10*, 247–258. [CrossRef]
15. Xie, C.; Liang, F.; Ma, M.; Chen, X.; Lu, W.; Jia, Y. Microstructure and dielectric properties of PTFE-based composites filled by micron/submicron-blended CCTO. *Crystals* **2017**, *7*, 126.
16. Madusanka, H.; Samarasekara, P.; Fernando, C. Polytetrafluoroethylene bindered hematite (α—Fe_2O_3) nanostructure for ammonia gas detection at room temperature. *Mater. Res. Express* **2018**, *5*, 125002. [CrossRef]
17. Kang, W.; Li, F.; Zhao, Y.; Qiao, C.; Ju, J.; Cheng, B. Fabrication of porous Fe_2O_3/PTFE nanofiber membranes and their application as a catalyst for dye degradation. *RSC Adv.* **2016**, *6*, 32646–32652. [CrossRef]
18. Huang, J.-Y.; Fang, X.; Li, Y.-C.; Feng, B.; Wang, H.-X.; Du, K. The mechanical and reaction behavior of $PTFE/Al/Fe_2O_3$ under impact and quasi-static compression. *Adv. Mater. Sci. Eng.* **2017**, *2017*, 3540320. [CrossRef]
19. Ai, Z.; Mei, T.; Liu, J.; Li, J.; Jia, F.; Zhang, L.; Qiu, J. Fe@ Fe_2O_3 core-shell nanowires as an iron reagent. 3. Their combination with CNTs as an effective oxygen-fed gas diffusion electrode in a neutral electro-Fenton system. *J. Phys. Chem. C* **2007**, *111*, 14799–14803. [CrossRef]
20. Singh, K.; Ohlan, A.; Kotnala, R.; Bakhshi, S.; Dhawan, S. Dielectric and magnetic properties of conducting ferromagnetic composite of polyaniline with γ-Fe_2O_3 nanoparticles. *Mater. Chem. Phys.* **2008**, *112*, 651–658. [CrossRef]
21. Reda, S. electric and dielectric properties of Fe_2O_3/silica nanocomposites. *Int. J. Nano Sci. Technol.* **2013**, *1*, 17–28.
22. Ul-Haq, Y.; Murtaza, I.; Mazhar, S.; Ullah, R.; Iqbal, M.; Qarni, A.A.; Amin, S. Dielectric, thermal and mechanical properties of hybrid $PMMA/RGO/Fe_2O_3$ nanocomposites fabricated by in-situ polymerization. *Ceram. Int.* **2020**, *46*, 5828–5840. [CrossRef]
23. Mensah, E.E.; Abbas, Z.; Ibrahim, N.A.; Khamis, A.M.; Abdalhadi, D.M. Complex permittivity and power loss characteristics of α-Fe_2O_3/polycaprolactone (PCL) nanocomposites: Effect of recycled α-Fe_2O_3 nanofiller. *Heliyon* **2020**, *6*, e05595. [CrossRef]
24. Paula, A.L.D.; Rezende, M.C.; Barroso, J.J. Experimental measurements and numerical simulation of permittivity and permeability of Teflon in X band. *J. Aerosp. Technol. Manag.* **2011**, *3*, 59–64. [CrossRef]
25. Mensah, E.E.; Abbas, Z.; Azis, R.a.S.; Khamis, A.M. Enhancement of complex permittivity and attenuation properties of recycled hematite (α-Fe_2O_3) using nanoparticles prepared via ball milling technique. *Materials* **2019**, *12*, 1696. [CrossRef]
26. Yusoff, A.; Abdullah, M. Microwave electromagnetic and absorption properties of some LiZn ferrites. *J. Magn. Magn. Mater.* **2004**, *269*, 271–280. [CrossRef]
27. Wang, Y.; Wang, L.; Wu, H. Enhanced microwave absorption properties of α-Fe_2O_3-filled ordered mesoporous carbon nanorods. *Materials* **2013**, *6*, 1520–1529. [CrossRef]

28. Hong, Y.-K.; Lee, J. Ferrites for RF passive devices. *Solid State Phys.* **2013**, *64*, 237–329.
29. Ismail, N.Q.A.; Saat, N.K.; Zaid, M.H.M. Effect of soda lime silica glass doping on ZnO varistor ceramics: Dry milling method. *J. Asian Ceram. Soc.* **2020**, *8*, 909–914. [CrossRef]
30. James, J.; Spittle, J.; Brown, S.; Evans, R. A review of measurement techniques for the thermal expansion coefficient of metals and alloys at elevated temperatures. *Meas. Sci. Technol.* **2001**, *12*, R1. [CrossRef]
31. Wool, R. *Bio-Based Composites from Soybean Oil and Chicken Feathers*; Elsevier: Amsterdam, The Netherlands, 2005.
32. Shabana, R.; Sarojini, J.; Vikram, K.A.; Lakshmi, V. Evaluating the mechanical properties of commonly used 3D printed ABS and PLA polymers with multi layered polymers. *Int. J. Eng. Adv. Technol.* **2019**, *8*, 2351–2356.
33. Khamis, A.; Abbas, Z.; Ahmad, A.; Azis, R.S.; Abdalhadi, D.; Mensah, E.E. Experimental and Computational Study on Epoxy Resin Reinforced with Micro-Sized OPEFB Using Rectangular Waveguide and Finite Element Method. *IET Microw. Antennas Propag.* **2020**, *14*, 752–758. [CrossRef]
34. Yamaguchi, A.; Kido, H.; Ukita, Y.; Kishihara, M.; Utsumi, Y. Anisotropic pyrochemical microetching of poly (tetrafluoroethylene) initiated by synchrotron radiation-induced scission of molecule bonds. *Appl. Phys. Lett.* **2016**, *108*, 051610. [CrossRef]
35. Wen, B.; Li, J.; Lin, Y.; Liu, X.; Fu, J.; Miao, H.; Zhang, Q. A novel preparation method for γ-Fe_2O_3 nanoparticles and their characterization. *Mater. Chem. Phys.* **2011**, *128*, 35–38. [CrossRef]
36. Ang, B.; Yaacob, I.; Nurdin, I. Investigation of Fe_2O_3/SiO_2 nanocomposite by FESEM and TEM. *J. Nanomater.* **2013**, *2013*, 980390. [CrossRef]
37. Glaris, P.; Coulon, J.-F.; Dorget, M.; Poncin-Epaillard, F. Thermal annealing as a new simple method for PTFE texturing. *Polymer* **2013**, *54*, 5858–5864. [CrossRef]
38. Abusafieh, A.A.; Federico, D.R.; Connell, S.J.; Cohen, E.J.; Willis, P.B. Dimensional stability of CFRP composites for space-based reflectors. In Proceedings of the Optomechanical Design and Engineering 2001, San Diego, CA, USA, 29 July–3 August 2001; pp. 9–16.
39. Jahan, M.; Inakpenu, R.O.; Li, K.; Zhao, G. Enhancing the mechanical strength for a microwave absorption composite based on graphene nanoplatelet/epoxy with carbon fibers. *Open J. Compos. Mater.* **2019**, *9*, 230. [CrossRef]
40. Jiang, Z.; Yuan, Y. Effects of particle size distribution of silica on properties of PTFE/SiO2 composites. *Mater. Res. Express* **2018**, *5*, 066306. [CrossRef]
41. Ghasemi-Kahrizsangi, S.; Nemati, A.; Shahraki, A.; Farooghi, M. Densification and properties of Fe_2O_3 nanoparticles added CaO refractories. *Ceram. Int.* **2016**, *42*, 12270–12275. [CrossRef]
42. Jedrzejewska, A.; Sibera, D.; Narkiewicz, U.; Pełech, R.; Jedrzejewski, R. Effect of Synthesis Parameters of Graphene/Fe_2O_3 Nanocomposites on Their Structural and Electrical Conductivity Properties. *Acta Phys. Pol. A* **2017**, *132*, 1424–1429. [CrossRef]
43. Gorshkov, N.; Goffman, V.; Vikulova, M.; Burmistrov, I.; Kovnev, A.; Gorokhovsky, A. Dielectric properties of the polymer–matrix composites based on the system of Co-modified potassium titanate–polytetrafluoroethylene. *J. Compos. Mater.* **2018**, *52*, 135–144. [CrossRef]
44. Lanza, V.; Herrmann, D. The density dependence of the dielectric constant of polyethylene. *J. Polym. Sci.* **1958**, *28*, 622–625. [CrossRef]
45. Leyland, R.; Maharaj, A. Dielectric constant as a means of assessing the properties of road construction materials. *SATC 2010*, *2010*, 487–498.
46. Fahad Ahmad, A.; Aziz, S.H.A.; Abbas, Z.; Mohammad Abdalhadi, D.; Khamis, A.M.; Aliyu, U.S.A. Computational and Experimental Approaches for Determining Scattering Parameters of OPEFB/PLA Composites to Calculate the Absorption and Attenuation Values at Microwave Frequencies. *Polymers* **2020**, *12*, 1919. [CrossRef] [PubMed]
47. Przybyszewska, M.; Zaborski, M. The effect of zinc oxide nanoparticle morphology on activity in crosslinking of carboxylated nitrile elastomer. *Express Polym. Lett* **2009**, *3*, 542–552. [CrossRef]
48. Mandal, S.; Singh, S.; Dey, P.; Roy, J.; Mandal, P.; Nath, T. Frequency and temperature dependence of dielectric and electrical properties of TFe2O4 (T=Ni, Zn, Zn0.5Ni0.5) ferrite nanocrystals. *J. Alloys Compd.* **2016**, *656*, 887–896. [CrossRef]
49. Ahmad, A.F.; Ab Aziz, S.; Abbas, Z.; Obaiys, S.J.; Khamis, A.M.; Hussain, I.R.; Zaid, M.H.M. Preparation of a chemically reduced graphene oxide reinforced epoxy resin polymer as a composite for electromagnetic interference shielding and microwave-absorbing applications. *Polymers* **2018**, *10*, 1180. [CrossRef]
50. Giacometti, J.A.; Wisniewski, C.; Ribeiro, P.; Moura, W.A. Electric measurements with constant current: A practical method for characterizing dielectric films. *Rev. Sci. Instrum.* **2001**, *72*, 4223–4227. [CrossRef]
51. Huashen, W.; Shan, J.; Guodong, W.; Ke, X. Electromagnetic parameters test system based on a refined NRW transmission/reflection algorithm. In Proceedings of the 2007 International Symposium on Microwave, Antenna, Propagation and EMC Technologies for Wireless Communications, Hangzhou, China, 16–17 August 2007; pp. 1276–1280.
52. Abdalhadi, D.M.; Abbas, Z.; Ahmad, A.F.; Matori, K.A.; Esa, F. Controlling the properties of OPEFB/PLA polymer composite by using Fe_2O_3 for microwave applications. *Fibers Polym.* **2018**, *19*, 1513–1521. [CrossRef]
53. Ahmad, A.F.; Abbas, Z.; Obaiys, S.J.; Abdalhadi, D.M. Improvement of dielectric, magnetic and thermal properties of OPEFB fibre–polycaprolactone composite by adding Ni–Zn ferrite. *Polymers* **2017**, *9*, 12. [CrossRef]
54. Barba, A.; Clausell, C.; Jarque, J.; Nuño, L. Magnetic complex permeability (imaginary part) dependence on the microstructure of a Cu-doped Ni–Zn-polycrystalline sintered ferrite. *Ceram. Int.* **2020**, *46*, 14558–14566. [CrossRef]

55. Luo, J.; Shen, P.; Yao, W.; Jiang, C.; Xu, J. Synthesis, characterization, and microwave absorption properties of reduced graphene oxide/strontium ferrite/polyaniline nanocomposites. *Nanoscale Res. Lett.* **2016**, *11*, 141. [CrossRef]
56. Li, W.; Qiu, T.; Wang, L.; Ren, S.; Zhang, J.; He, L.; Li, X. Preparation and electromagnetic properties of core/shell polystyrene@ polypyrrole@ nickel composite microspheres. *ACS Appl. Mater. Interfaces* **2013**, *5*, 883–891. [CrossRef]
57. Liu, P.; Huang, Y.; Yan, J.; Zhao, Y. Magnetic graphene@ PANI@ porous TiO_2 ternary composites for high-performance electromagnetic wave absorption. *J. Mater. Chem. C* **2016**, *4*, 6362–6370. [CrossRef]
58. Li, C.; Ji, S.; Jiang, X.; Waterhouse, G.I.; Zhang, Z.; Yu, L. Microwave absorption by watermelon-like microspheres composed of γ-Fe_2O_3, microporous silica and polypyrrole. *J. Mater. Sci.* **2018**, *53*, 9635–9649. [CrossRef]
59. Liang, K.; Qiao, X.-J.; Sun, Z.-G.; Guo, X.-D.; Wei, L.; Qu, Y. Preparation and microwave absorbing properties of graphene oxides/ferrite composites. *Appl. Phys. A* **2017**, *123*, 445. [CrossRef]

Article

Pyrolysis of Denim Jeans Waste: Pyrolytic Product Modification by the Addition of Sodium Carbonate

Junghee Joo [1], Heeyoung Choi [1], Kun-Yi Andrew Lin [2] and Jechan Lee [1,3,*]

[1] Department of Global Smart City, Sungkyunkwan University, Suwon 16419, Republic of Korea
[2] Department of Environmental Engineering & Innovation and Development Center of Sustainable Agriculture, National Chung Hsing University, Taichung 40227, Taiwan
[3] School of Civil, Architectural Engineering, and Landscape Architecture, Sungkyunkwan University, Suwon 16419, Republic of Korea
* Correspondence: jechanlee@skku.edu

Abstract: Quickly changing fashion trends generate tremendous amounts of textile waste globally. The inhomogeneity and complicated nature of textile waste make its recycling challenging. Hence, it is urgent to develop a feasible method to extract value from textile waste. Pyrolysis is an effective waste-to-energy option to processing waste feedstocks having an inhomogeneous and complicated nature. Herein, pyrolysis of denim jeans waste (DJW; a textile waste surrogate) was performed in a continuous flow pyrolyser. The effects of adding sodium carbonate (Na_2CO_3; feedstock/Na_2CO_3 = 10, weight basis) to the DJW pyrolysis on the yield and composition of pyrolysates were explored. For the DJW pyrolysis, using Na_2CO_3 as an additive increased the yields of gas and solid phase pyrolysates and decreased the yield of liquid phase pyrolysate. The highest yield of the gas phase pyrolysate was 34.1 wt% at 800 °C in the presence of Na_2CO_3. The addition of Na_2CO_3 could increase the contents of combustible gases such as H_2 and CO in the gas phase pyrolysate in comparison with the DJW pyrolysis without Na_2CO_3. The maximum yield of the liquid phase pyrolysate obtained with Na_2CO_3 was 62.5 wt% at 400 °C. The composition of the liquid phase pyrolysate indicated that the Na_2CO_3 additive decreased the contents of organic acids, which potentially improve its fuel property by reducing acid value. The results indicated that Na_2CO_3 can be a potential additive to pyrolysis to enhance energy recovery from DJW.

Keywords: thermochemical conversion process; waste recycling; waste-to-energy; waste treatment; synthetic fiber

1. Introduction

Textiles in municipal solid waste is mainly discarded clothing with other sources such as carpets, footwear, sheets, and towels. According to United States Environmental Protection Agency (U.S. EPA), approximately 17 million tons of textile waste ended up in landfills in 2018 [1]. The U.S. EPA also estimated that nearly 5% of all landfill space is occupied by textile waste. Landfilling textile waste causes environmental issues such as the formation of greenhouse gases upon decomposition and the contamination of groundwater [2]. Some kinds of textiles (e.g., synthetic ones) require longer than 200 years to decompose in landfills [3]. In addition to the generation of tremendous amounts of textile waste, synthetic textile fibers are manufactured using fossil fuel resources (e.g., natural gas and crude oil) as the feedstock. The production, consumption, and postindustrial waste handling of synthetic textile fibers not only generate greenhouse gas emissions, but also release microplastics [4]. Therefore, the strategy for disposal of textile waste has changed from landfill-based solution to recycling-based solution.

Upcycling is a kind of recycling, which converts lower-value substances to higher-value products [5]. Pyrolysis has gained increasing attention as a feasible waste upcycling process because it is an effective process for the treatment of waste materials in a heterogeneous complex nature [6]. Recently, pyrolysis has been widely studied to not only recover energy from waste substances [7] but also transform waste materials into value-added products such as hydrogen gas [8] and commodity chemicals [9]. For instance, a pyrolysis process using a cobalt-based catalyst has been suggested to transform textile waste into combustible products, as a method of waste-to-energy [10,11]. Kim et al. has recently reported that calcium carbonate-based catalysts enhanced nylon monomer recovery from fishing net waste made of nylon [12]. In particular, sodium carbonate (Na_2CO_3) has been used to modify the characteristics of pyrolytic products produced from different carbonaceous feedstocks such as coal [13], pine sawdust [14], and Crofton weed [15,16]. In this respect, pyrolysis in the presence of Na_2CO_3 can be a promising option for the upcycling of textile waste by modifying the pyrolytic products characteristics (e.g., the enhancement of the production of combustible gases and improvement of fuel properties of pyrolytic liquid).

There has been no study on Na_2CO_3-mediated pyrolysis of textile waste so far. This is the first study to explore the impacts of Na_2CO_3 on the pyrolysis of textile waste containing both natural and synthetic fibers such as denim jeans waste (DJW). The DJW was chosen as a textile waste surrogate as the generation of DJW has continuously increased due to the continuous growth of global denim jeans market size [17]. The present study was aimed at supporting the creation of effective energy recovery from textile waste based on the pyrolysis in the presence of Na_2CO_3. It was also expected that this study proposes a potential method to feasibly recover energy from textile waste.

2. Materials and Methods

2.1. Feedstock

DJW was collected at clothing donation centers located in Suwon, Gyeonggi Province, Republic of Korea. After collection, impurities in DJW were thoroughly removed using an air blower followed by cutting to slips with a thickness of 2 cm (Figure S1). Na_2CO_3 (purity: $\geq$99.5%) was purchased, provided from Sigma-Aldrich brand in Merck (Burlington, MA, USA).

Ultimate analysis of DJW was performed using an elemental analyzer (model: 628 series; LECO, St. Joseph, MI, USA). Proximate analysis of DJW (dry basis) was conducted according to ASTM standard methods (D3175 for volatile matter and D3174 for ash). The difference between the initial DJW sample mass and the sum of volatile matter and ash was considered the content of fixed carbon.

2.2. Pyrolysis Experiment and Product Analysis

The pyrolysis experiments were conducted at 400–800 °C in a continuous flow pyrolyser (described in detail in Figure S2). A tubular reactor (outside diameter: 25 mm; inside diameter: 21 mm; length: 0.6 m) made of quartz was heated by a split-hinge tube furnace (Thermo Scientific, Waltham, MA, USA). For a pyrolysis experiment, 1 g of feedstock (i.e., DJW) was loaded in the center of the quartz tube and fixed with quartz wool on both sides of the feedstock. For the pyrolysis experiment conducted in the presence of Na_2CO_3, 10 wt% Na_2CO_3 was added to the feedstock as the Na_2CO_3 loadings below 10 wt% did not have distinct effects on the production of pyrolysates for the DJW pyrolysis.

Nitrogen gas (ultra-high purity) was continuously supplied into the tubular reaction at a flow rate of 100 mL min^{-1}, which was controlled by a mass flow controller (Brooks Instrument, Hatfield, PA, USA). The product stream passes through a condenser composed of an ice trap (-1 °C) and four dry ice/acetone traps (-50 °C each) connected in a series. The condensable fraction of the product stream was collected in the condenser. The collected condensable samples were analyzed using a gas chromatograph equipped with a mass spectrometer (GC/MS; Agilent Technologies, Santa Clara, CA, USA). The non-condensable fraction of the product stream (i.e., the fraction that was not collected in the condenser) further passed through a micro gas chromatograph (micro GC; INFICON, Bad Ragaz, Switzerland) to analyze non-condensable gases evolved from the pyrolysis of DJW. Detailed conditions used for the micro GC and GC/MS analyses are given in Tables S1 and S2.

3. Results and Discussion

3.1. Feedstock Characterization

Table 1 summarizes the results of ultimate and proximate analyses of DJW on dry basis. The DJW feedstock was mostly composed of carbon (45 wt%) and oxygen (32.16 wt%). Considerable amounts of hydrogen (6.35 wt%) and ash (15.09 wt%) were also found. The contents of nitrogen and sulfur were negligible. The proximate analysis of DJW confirmed that the DJW feedstock mostly consisted of volatile matter (84.1 wt%).

Table 1. Results of proximate and ultimate analyses of DJW (dry basis).

Analysis	Composition	Content (wt%)
Ultimate analysis	C	45.00
	H	6.35
	O	32.16
	N	0.20
	S	1.20
	Ash	15.09
	Total	100
Proximate analysis	Volatile matter	84.10
	Fixed carbon	0.81
	Ash	15.09
	Total	100

3.2. Pyrolysis of Denim Jeans Waste

The pyrolysis of DJW resulted in pyrolysates in three different phases, such as pyrolytic gas, pyrolytic oil, and solid residue. In the primary stage of DJW pyrolysis, pyrolytic volatiles were released from the feedstock, and then they were thermally degraded to condensable compounds (i.e., liquid phase pyrolysate) [18]. The condensable compounds further underwent thermal decomposition resulting in lighter molecules, such as non-condensable gases (i.e., gas phase pyrolysate), in the secondary stage of DJW pyrolysis [18]. Solid phase pyrolysate was the residual solid after all pyrolytic volatiles were released from DJW [19].

Figure 1 shows that the yields of three phase pyrolysates produced from DJW without Na_2CO_3 at varied temperatures. A clear trend was observed: an increase in pyrolysis temperature increased the yield of gas phase pyrolysate and decreased the yields of liquid phase pyrolysate and solid residue. For instance, the yield of gas phase pyrolysate increased from 12.2 wt% to 27.4 wt% with increasing the temperature from 400 °C to 800 °C, while the total yield of liquid and solid phase pyrolysates decreased from 87.8 wt% to 72.6 wt%. This clearly indicates that the release of pyrolytic volatiles and thermal degradation of the pyrolytic volatiles were promoted by increasing pyrolysis temperature in the pyrolysis of DJW.

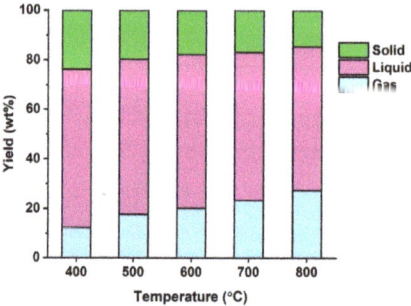

Figure 1. Mass balance of pyrolysate made from DJW without Na_2CO_3 as a function of pyrolysis temperature. Average values of triplicate are reported with and standard deviations of the average values of 3–4%.

The gas phase pyrolysate was composed of combustible gases, such as carbon monoxide (CO), hydrogen (H_2), and C_1–C_4 hydrocarbons, and carbon dioxide (CO_2), as presented in Figure 2. Major components of the gas phase pyrolysate was CO and CO_2 at all temperatures tested. An increase in pyrolysis temperature increased the selectivities toward H_2 and methane (CH_4) in return for the selectivity toward CO_2. For example, when increasing the pyrolysis temperature from 600 °C to 800 °C, the H_2 and CH_4 selectivity increased from 6.6% to 9.2%, while the CO_2 selectivity decreased from 52.4% to 48.8%. This was most likely attributed to the dehydrogenation and methanation reactions expedited at higher temperatures [20].

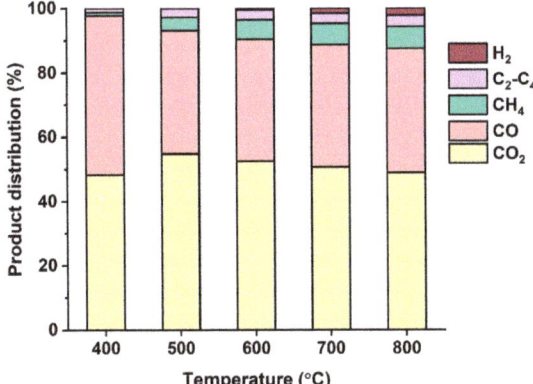

Figure 2. Product distribution of the gas phase pyrolysate made from DJW without Na_2CO_3 as a function of pyrolysis temperature. Average values of triplicate are reported with and standard deviations of the average values of 3–4%.

The DJW-derived liquid phase pyrolysate was a mixture of a wide variety of chemical compounds that could be classified as acids, alcohols, aldehydes, ketones, esters, furans, dioxolanes, hydrocarbons, sugars, and other miscellaneous compounds, as summarized in Table 2. According to the fiber content provided in clothing labels, the average composition of the DJW feedstock was cotton (~90%), polyester (~8%), and elastane (1–2%). Acids, alcohols, aldehydes, ketones, furans, and sugars should originate from natural fiber such as cotton. Esters, dioxolanes, and hydrocarbons were most likely derived from synthetic fibers such as polyester and elastane via thermal cracking of polymeric bonds. N-containing species were associated with blue dye used for manufacturing denim fabric. As found in Table 2, most of the compounds were acids, alcohols, ketones, and sugars. The change in pyrolysis temperature did not markedly affect the product distribution of the liquid phase pyrolysate.

Table 2. Product distribution of the liquid phase pyrolysate made from DJW without Na_2CO_3 at varied pyrolysis temperatures (unit: GC/MS area%).

Component	400 °C	500 °C	600 °C	700 °C	800 °C
Acids	19.6	18.8	19.2	18.8	19.7
Alcohols	13.5	13.7	14.5	13.6	13.9
Aldehydes	1	1.7	1.5	1.4	1.1
Ketones	19.2	21.2	20.8	19.4	19.4
Esters	7.6	7.6	7.8	8.7	7.8
Furans	11.4	12.5	10.9	10.4	10.3
Dioxolanes	2.6	2.6	2.4	2.5	2.4
Hydrocarbons	0	0	0.4	0.9	1.1
Sugars	15.3	12.6	13.6	14	14.7
N-containing species	7.7	7.4	7	8.3	7.8
Others	2.1	1.9	1.9	2	1.8

3.3. Effects of Na$_2$CO$_3$ Addition to Pyrolysis of Denim Jeans Waste

In Figure 3, the yields of three phase pyrolysates obtained with and without Na$_2$CO$_3$ are compared. The addition of Na$_2$CO$_3$ to the DJW pyrolysis increased the yields of gas and solid phase pyrolysates while it decreased the yield of liquid phase pyrolysate at all tested temperatures. For instance, the gas phase pyrolysate yield achieved with Na$_2$CO$_3$ was 34% higher than that achieved without Na$_2$CO$_3$ at 800 °C. As aforementioned, the solid phase pyrolysate yield is highly associated with the release of pyrolytic volatiles from DJW, and the liquid phase pyrolysate yield is highly associated with thermal degradation of the pyrolytic volatiles. Therefore, the increase in the yields of gas and solid pyrolysates and the decrease in the yield of liquid pyrolysate were most likely because Na$_2$CO$_3$ enhanced the thermal degradation of the pyrolytic volatiles released from DJW during the pyrolysis.

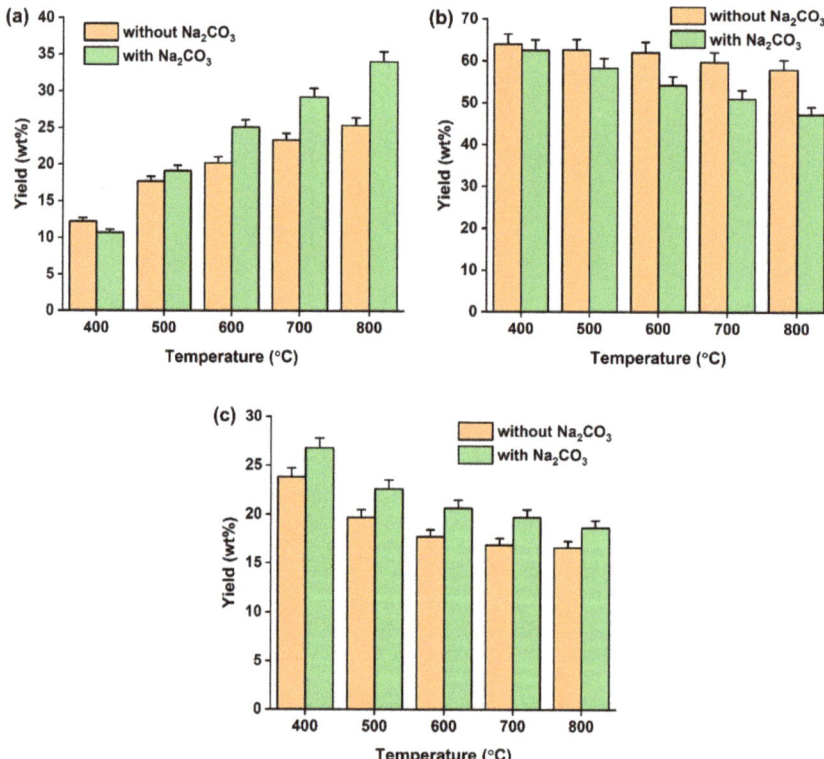

Figure 3. The yield of pyrolysates made from DJW with and without Na$_2$CO$_3$ at varied pyrolysis temperatures: (**a**) gas phase pyrolysate, (**b**) liquid phase pyrolysate, and (**c**) solid phase pyrolysate. Average values of triplicate are reported with standard deviations given as error bars.

The addition of Na$_2$CO$_3$ had considerable effects on the composition of the gas phase pyrolysate. In Figure 4, the yields of non-condensable gases contained in the gas phase pyrolysates obtained with and without the addition of Na$_2$CO$_3$ are compared. It was clearly observed that the addition of Na$_2$CO$_3$ enhanced the yields of H$_2$, CO, and CO$_2$ at 400–800 °C. H$_2$ is formed through secondary decomposition and recombination of C–H groups and aromatic C=C bonds [21]. Alkaline sodium species in Na$_2$CO$_3$ (e.g., Na$^+$) should contribute to the formation of H$_2$ [22]. The addition of Na$_2$CO$_3$ most likely promoted decarboxylation and decarbonylation, resulting in higher CO$_2$ and CO yields, respectively [23]. Furthermore, in the presence of Na$_2$CO$_3$, reactions between Na$^+$ and –COH and –COOH groups leads to phenolic sodium (–CONa) and carboxylate sodium (–COONa), respectively, thereby releasing H$^+$ (Equations (1) and (2)) [24]. The –CONa and –COONa species can react with solid phase carbon, resulting in CO (Equations (3) and (4)) [24]. The extent of the enhancement of H$_2$, CO, and CO$_2$ tended to be greater as the pyrolysis temperature increased, meaning that the above-mentioned reactions were more promoted at higher temperatures. The yield of C$_1$–C$_4$ hydrocarbons was not markedly changed by adding Na$_2$CO$_3$ to the DJW pyrolysis. The higher heating value of the gas phase pyrolysate obtained with Na$_2$CO$_3$ was up to 28% higher than that without Na$_2$CO$_3$, ranging from 8 MJ kg^{-1} to 12 MJ kg^{-1} (calculated based on the heat of combustion of individual combustible gas). In other words, the addition of Na$_2$CO$_3$ to the DJW pyrolysis can help to improve the energy content of the gas phase pyrolysate derived from the DJW feedstock.

$$-COH + Na^+ \rightarrow -CONa + H^+ \tag{1}$$

$$-COOH + Na^+ \rightarrow -COONa + H^+ \tag{2}$$

$$-CONa + C(s) \rightarrow -CNa + CO(g) \tag{3}$$

$$-COONa + C(s) \rightarrow -CONa + CO(g) \tag{4}$$

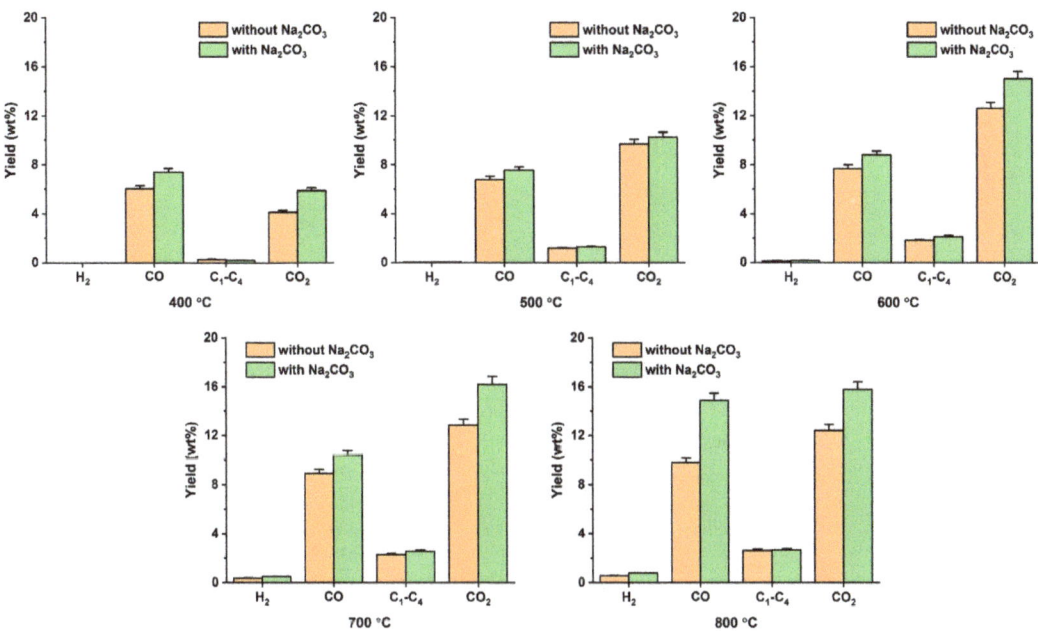

Figure 4. Non-condensable gas yields obtained by the pyrolysis of DJW with and without Na$_2$CO$_3$ at varied pyrolysis temperatures. Average values of triplicate are reported with standard deviations given as error bars.

Figure 5 showed the comparisons of the liquid phase pyrolysates made from DJW with and without Na$_2$CO$_3$. The addition of Na$_2$CO$_3$ decreased the contents of acids, alcohols, esters, and sugars, while it increased the contents of ketones, furans, hydrocarbons, and N-containing species at all the tested pyrolysis temperatures. Among the compounds in the liquid phase pyrolysates, acids can deteriorate the fuel property of pyrolytic liquid because acid species increase the acidity of the pyrolytic liquid; thus, it can cause corrosion issues in engines and boilers [25]. The product distribution of acids in the liquid phase pyrolysate produced without Na$_2$CO$_3$ ranged from 17% to 25%; however, the product distribution of acids in the liquid phase pyrolysate produced with Na_2CO_3 ranged from 12% to 17%. This evidently indicates that the addition of Na_2CO_3 had a deoxidation effect on the liquid phase pyrolysate derived from DJW. As a result, the pyrolysis of DJW in the presence of Na_2CO_3 could improve the fuel property of pyrolytic liquid by decreasing its acid value.

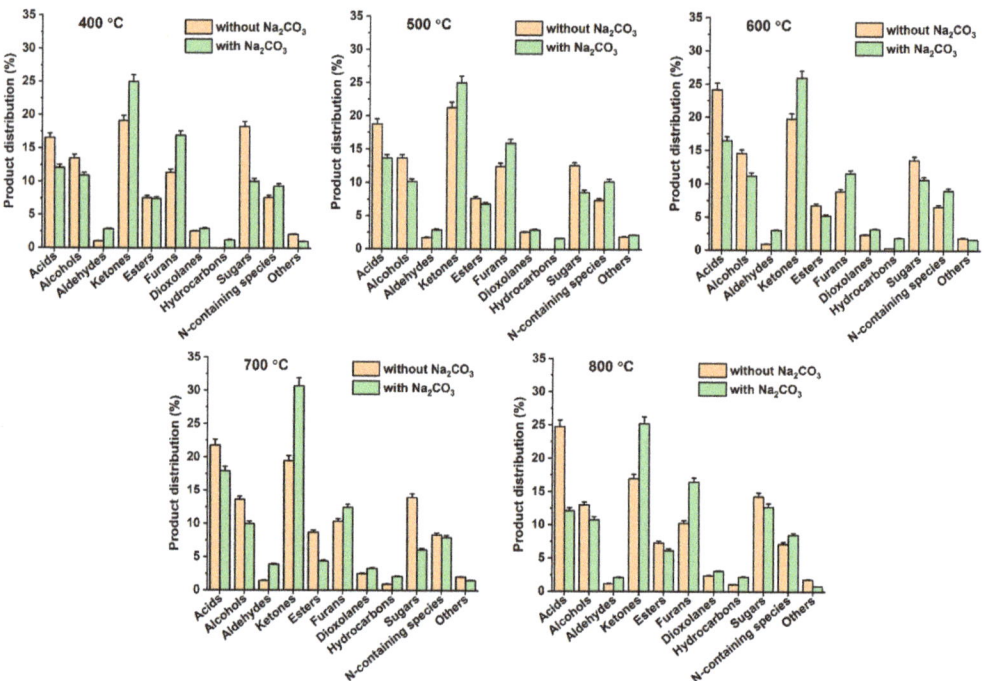

Figure 5. Product distributions of the liquid phase pyrolysates of DJW with and without Na_2CO_3 at varied pyrolysis temperatures. Average values of triplicate are reported with standard deviations given as error bars.

4. Conclusions

Here in this study, the pyrolysis of DJW was carried out in a continuous flow pyrolyser with and without the addition of Na_2CO_3 as a method for the upcycling of textile waste. The increase in pyrolysis temperature increased the yield of gas phase pyrolysate and decreased the yields of liquid and solid phase pyrolysates most likely due to the enhanced thermal cracking of pyrolytic volatiles released during the DJW pyrolysis. The addition of Na_2CO_3 further enhanced the yield of the gas phase pyrolysate reaching 34.1 wt%. The addition of Na_2CO_3 increased the yield of combustible gases (e.g., H_2 and CO) compared with the DJW pyrolysis without Na_2CO_3, thereby increasing the energy content of the gas phase pyrolysate. The major components of the liquid phase pyrolysate were acids, alcohols, aldehydes, ketones, furans, and sugars. In the presence of Na_2CO_3, the content of organic acids was decreased, which potentially improves the fuel property of the pyrolytic liquid by decreasing its acid value.

Supplementary Materials: The following supporting information can be downloaded at: https://www.mdpi.com/article/10.3390/polym14225035/s1, Figure S1. Denim jeans waste used as the feedstock in this study; Figure S2. Scheme of the pyrolyser used for the pyrolysis of denim jeans waste; Table S1. Column information, and analytical conditions for the micro GC; Table S2. Column information, and analytical conditions for the GC/MS.

Author Contributions: Conceptualization, K.-Y.A.L. and J.L.; investigation, J.J. and H.C.; writing—original draft preparation, J.J. and J.L.; writing—review and editing, J.J., H.C., K.-Y.A.L. and J.L.; supervision, J.L. All authors have read and agreed to the published version of the manuscript.

Funding: This work was supported by the National Research Foundation of Korea (NRF) grant funded by the Koran government (MSIT) (No. 2021R1A4A1031357). This work was also financially supported by Korea Ministry of Land, Infrastructure and Transport (MOLIT) as Innovative Talent Education Program for Smart City.

Institutional Review Board Statement: Not applicable.

Informed Consent Statement: Not applicable.

Data Availability Statement: Data available on reasonable request from the corresponding author.

Conflicts of Interest: The authors declare that they have no known competing financial interest or personal relationships that could have influenced the work reported in this paper.

References

1. US EPA. *Facts and Figures about Materials, Waste and Recycling—Textiles: Material-Specific Data*; United States Environmental Protection Agency (US EPA): Washington, DC, USA, 2020.
2. Dissanayake, D.G.K.; Weerasinghe, D.U.; Wijesinghe, K.A.P.; Kalpage, K.M.D.M.P. Developing a compression moulded thermal insulation panel using postindustrial textile waste. *Waste Manag.* **2018**, *79*, 356–361. [CrossRef] [PubMed]
3. Gupta, R.; Shukla, V.K.; Agarwal, P. Sustainable transformation in modest fashion through "RPET technology" and "Dry-Dye" process, using recycled PET plastic. *Int. J. Recent Technol. Eng.* **2019**, *8*, A1432058119. [CrossRef]
4. EEA. *Plastic in Textiles: Towards a Circular Economy for Synthetic Textiles in Europe*; European Environment Agency (EEA): Copenhagen, Denmark, 2021.
5. Lee, J.; Kwon, E.E.; Lam, S.S.; Chen, W.-H.; Rinklebe, J.; Park, Y.-K. Chemical recycling of plastic waste via thermocatalytic routes. *J. Clean. Prod.* **2021**, *321*, 128989. [CrossRef]
6. Lee, N.; Joo, J.; Lin, K.-Y.A.; Lee, J. Thermochemical conversion of mulching film waste via pyrolysis with the addition of cattle excreta. *J. Environ. Chem. Eng.* **2021**, *9*, 106362. [CrossRef]
7. Lee, J.; Lin, K.-Y.A.; Jung, S.; Kwon, E.E. Hybrid renewable energy systems involving thermochemical conversion process for waste-to-energy strategy. *Chem. Eng. J.* **2023**, *452*, 139218. [CrossRef]
8. Park, C.; Lee, N.; Kim, J.; Lee, J. Co-pyrolysis of food waste and wood bark to produce hydrogen with minimizing pollutant emissions. *Environ. Pollut.* **2021**, *270*, 116045. [CrossRef] [PubMed]
9. Lee, S.; Lee, J.; Park, Y.-K. Simultaneous upcycling of biodegradable plastic and sea shell wastes through thermocatalytic monomer recovery. *ACS Sustain. Chem. Eng.* **2022**, *10*, 13972–13979. [CrossRef]
10. Kwon, D.; Yi, S.; Jung, S.; Kwon, E.E. Valorization of synthetic textile waste using CO_2 as a raw material in the catalytic pyrolysis process. *Environ. Pollut.* **2021**, *268*, 115916. [CrossRef]
11. Park, C.; Lin, K.-Y.A.; Kwon, E.E.; Lee, J.; Park, Y.-K. Energy recovery from banner waste through catalytic pyrolysis over cobalt oxide: Effects of catalyst configuration. *Int. J. Energy Res.* **2022**, *46*, 19051–19063. [CrossRef]
12. Kim, S.; Kim, Y.T.; Oh, L.S.; Kim, H.J.; Lee, J. Marine waste upcycling—Recovery of nylon monomers from fishing net waste using seashell waste-derived catalysts in a CO_2-mediated thermocatalytic process. *J. Mater. Chem. A* **2022**, *10*, 20024–20034. [CrossRef]
13. Ding, L.; Zhou, Z.; Guo, Q.; Huo, W.; Yu, G. Catalytic effects of Na_2CO_3 additive on coal pyrolysis and gasification. *Fuel* **2015**, *142*, 134–144. [CrossRef]
14. Yu, D.; Jin, G.; Pang, Y.; Chen, Y.; Guo, S.; Shen, S. Gas characteristics of pine sawdust catalyzed pyrolysis by additives. *J. Therm. Sci.* **2021**, *30*, 333–342. [CrossRef]
15. Zeng, K.; Yan, H.; Xia, H.; Zhang, L.; Zhang, Q. Catalytic pyrolysis of Eupatorium adenophorum by sodium salt. *J. Mater. Cycles Waste Manag.* **2021**, *23*, 1626–1635. [CrossRef]
16. Lin, J.; Cheng, S. Catalytic pyrolysis of crofton weed: Comparison of their pyrolysis product and preliminary economic analysis. *Environ. Prog. Sustain. Energy* **2022**, *41*, e13742. [CrossRef]
17. Sedef Uncu, A.; Cevza, C.; Banu, N.; Neslihan Sebla, Ö. Understanding denim recycling: A quantitative study with lifecycle assessment methodology. In *Waste in Textile and Leather Sectors*; Ayşegül, K., Ed.; IntechOpen: Rijeka, Croatia, 2020; Chapter 4.
18. Kim, S.; Yang, W.; Lee, H.S.; Tsang, Y.F.; Lee, J. Effectiveness of CO_2-mediated pyrolysis for the treatment of biodegradable plastics: A case study of polybutylene adipate terephthalate/polylactic acid mulch film. *J. Clean. Prod.* **2022**, *372*, 133763. [CrossRef]
19. Yang, W.; Kim, K.-H.; Lee, J. Upcycling of decommissioned wind turbine blades through pyrolysis. *J. Clean. Prod.* **2022**, *376*, 134292. [CrossRef]
20. Lee, N.; Lin, K.-Y.A.; Lee, J. Carbon dioxide-mediated thermochemical conversion of banner waste using cobalt oxide catalyst as a strategy for plastic waste treatment. *Environ. Res.* **2022**, *213*, 113560. [CrossRef]
21. Valin, S.; Cances, J.; Castelli, P.; Thiery, S.; Dufour, A.; Boissonnet, G.; Spindler, B. Upgrading biomass pyrolysis gas by conversion of methane at high temperature: Experiments and modelling. *Fuel* **2009**, *88*, 834–842. [CrossRef]

22. Chen, M.-q.; Wang, J.; Zhang, M.-x.; Chen, M.-g.; Zhu, X.-f.; Min, F.-f.; Tan, Z.-c. Catalytic effects of eight inorganic additives on pyrolysis of pine wood sawdust by microwave heating. *J. Anal. Appl. Pyrolysis* **2008**, *82*, 145–150. [CrossRef]
23. Zhao, N.; Li, B.-X. The effect of sodium chloride on the pyrolysis of rice husk. *Appl. Energy* **2016**, *178*, 346–352. [CrossRef]
24. Guo, D.-l.; Wu, S.-b.; Liu, B.; Yin, X.-l.; Yang, Q. Catalytic effects of NaOH and Na_2CO_3 additives on alkali lignin pyrolysis and gasification. *Appl. Energy* **2012**, *95*, 22–30. [CrossRef]
25. Lu, Q.; Zhang, J.; Zhu, X. Corrosion properties of bio-oil and its emulsions with diesel. *Chin. Sci. Bull.* **2008**, *53*, 3726–3734. [CrossRef]

Article

Potent Application of Scrap from the Modified Natural Rubber Production as Oil Absorbent

Anoma Thitithammawong [1,2], Sitisaiyidah Saiwari [1,2], Subhan Salaeh [1,2] and Nabil Hayeemasae [1,2,*]

[1] Research Unit of Advanced Elastomeric Materials and Innovations for BCG Economy (AEMI), Faculty of Science and Technology, Prince of Songkla University, Pattani Campus, Pattani 94000, Thailand
[2] Department of Rubber Technology and Polymer Science, Faculty of Science and Technology, Prince of Songkla University, Pattani Campus, Pattani 94000, Thailand
* Correspondence: nabil.h@psu.ac.th

Abstract: The production of raw natural rubber always ends up with leftover latex. This latex is later collected to produce low grades of rubber. The collection of this latex also depends on the latex's quality. However, reproducing the latex may not be applicable if the latex contains many specks of dirt which will eventually be discarded. In this work, an alternative solution was to utilize such rubber in a processable form. This scrap rubber (SR) from the production of natural rubber grafted with polymethyl methacrylate (NR-g-PMMA) production was recovered to prepare an oil-swellable rubber. The rubber blends were turned into cellular structures to increase the oil swellability. To find the suitable formulation and cellular structure of the foam, the foams were prepared by blending SR with virgin natural rubber (NR) at various ratios, namely 0/100, 20/80, 30/70, 50/50, 70/30, 80/20, and 100/0 (phr/phr). The foam formation strongly depended on the SR, as it prevented gas penetration throughout the matrix. Consequently, small cells and thick cell walls were observed. This structure reduced the oil swellability from 7.09 g/g to 5.02 g/g. However, it is interesting to highlight that the thermal stability of the foam increased over the addition of SR, which is likely due to the higher thermal stability of the NR-g-PMMA waste or SR. In summary, the blending NR with 30 phr of SR provided good oil swellability, processability, and morphology, which benefit oil recovery application. The results obtained from this study will be used for further experiments on the enhancement of oil absorbency by applying other key factors. This work is considered a good initiative for preparing the oil-absorbent material based on scrap from modified natural rubber production.

Keywords: natural rubber; scrap rubber; oil absorbency; foam; waste

1. Introduction

Oil spills in the sea have become a crucial problem for marine environments. It is turning into a more significant issue due to the expansion of the offshore oil sector and the requirement for marine oil transportation [1–3]. This is also not limited to an oil spill in industries or workshops, especially from the machinery and piping systems. In the past, different techniques have been used to clean up oil spills, including in situ burning, mechanical collection of chemical dispersants, bioremediation, and the use of absorbent materials [4,5]. Owing to the economy and efficiency of oil collection and remediation, searching the compromising methods is the most desirable choice. Therefore, it is essential to look for a practical means of producing an absorbent substance for oil cleanup.

Various polymers have been widely used for oil spill redemption. Among them, alkyl acrylate and its derivatives, which have high hydrophobicity, have been attracting much interest in oil spill redemption. Jadhav et al. [6] prepared graft copolymerization of Poly(methyl methacrylate) (PMMA) on the backbone of *Meizotropis pellita* fibers (MPF). A PMMA-grafted MPF by 120% absorbed crude oil at 23.60 g, 17.41 g, and 14.21 g for the first to the third cycle of absorption, whereas diesel oil was absorbed from 13.14 g to 5.25 g for three cycles of absorption. Grafted PMMA onto natural rubber (NR) foam for oil sorbent was prepared by Ratcha et al. [7]. It was claimed that the modified NR foam could rapidly absorb gasoline and petroleum-derived organic solvents like toluene and xylene. The maximum oil absorbency of gasoline, diesel, engine oil, toluene, and xylene was achieved at 9.95 g/g, 8.37 g/g, 6.01 g/g, 11.81 g/g, and 10.96 g/g, respectively. Kanagaraj et al. [8] also prepared the MMA-grafted natural rubber foam with varying PMMA percentages, i.e., 0%, 10%, 20%, 30%, 40%, and 50%. The hardness, compression set, and stiffness of the natural rubber foam increased as the percentage of PMMA increased. The addition of MMA tended to swell more when soaked in toluene.

As stated, PMMA-grafted NR was previously applied for oil-absorbent material and found to be increased over the grafting content of PMMA onto natural rubber. PMMA-grafted NR has been studied and prepared in the last few decades. In the presence of MMA, the polarity of NR increases and can be extensively used in adhesive applications. During the synthesis of NR-g-PMMA, there is abundant rubber waste caught at the end of the manufacturing stage that is particularly not further processed as it contains many specks of dirt. The idea of this work was to utilize the waste as a blending component with NR to prepare oil-absorbent materials based on the scraps from the production of NR-g-PMMA. As previously mentioned, the acrylate-type polymer has been utilized as an oil-absorbent material. It is expected that the NR-g-PMMA could also apply the same solution.

To increase the oil absorbency of the rubber, the appearance of the rubber should be modified by turning the general solid form of rubber into a cellular structure. Cellular rubber, or so-called rubber foam, is an interesting rubber product that consists of rubber and gas phases in one item. Natural rubber foam is lightweight, has excellent thermal insulation, and absorbs sound, making it a popular choice for various applications [9,10]. Further, the porosity of rubber can also increase the penetration efficiency of oil towards the rubber. There are several methods to produce rubber foam, either latex or dry stages. As the collected scraps from the final production are in solid form, it is better to prepare the foam using an additive called a blowing agent. The foam structure can be controlled by the proper selection of blowing agents and curatives, which achieves the correct balance between the gas generated and the degree of curing. There are many types of chemical blowing agents, such as azodicarbonamide (ADC), dinitroso pentamethylene tetramine (DPT), sodium bicarbonate, p-toluenesulphonyl semicarbazide, 5-phenyl tetrazole, and 4,4-oxydibenzenesulphonyl hydrazide (OBSH). Numerous authors have investigated how elastomers are influenced by the content and type of a chemical blowing agent [11–13]. ADC is an attractive chemical blowing agent among these since it produces uniform cell foam and decomposes at low temperatures [14].

This work is a preliminary study on the potential application of scrap rubber (SR) as an oil absorbent material. Here, various blending ratios between NR and SR were designed and their sorption capacities, mainly for crude oil solution, were determined. The results obtained from this study will be used for further experiments on the enhancement of oil absorbency by applying other key factors. This is considered a good initiative for preparing the oil-absorbent material based on scraps from raw rubber production.

2. Experimental Details

2.1. Materials

The natural rubber (NR) used in this study was STR 5L, purchased from Suansom Kanyang, Yala, Thailand. Scrap rubber (SR) was supplied by a local company in the Southern part of Thailand. It was from the production of NR-g-PMMA in their factory. The basic properties of NR and SR are listed in Table 1. It can be seen that the properties were more or less the same except for the acetone extract, original plasticity (P_o), and Mooney viscosity (MU). The higher acetone extract was due to the unreacted chemicals left after the production of SR whereas higher P_o and MU were due to the PMMA hard phase that might affect the hardness and the viscosity of the SR. ADC was used as the blowing agent and was purchased from A.F. Supercell Co., Ltd., Rayong, Thailand. Treated distillate aromatics extract (TDAE oil) was used as processing oil. Processing oil can help to reduce the viscosity and facilitate the diffusion of gas [15]. TDAE oil was purchased from H&R ChemPharm (Thailand) Co., Ltd. Calcium carbonate ($CaCO_3$) was obtained from Krungthepchemi Co., Ltd., Bangkok, Thailand. It is widely known to act as a nucleating agent in the production of foam. Stearic acid was used as a curing activator and was purchased from Imperial Chemical Co., Ltd., Bangkok, Thailand. ZnO was used as a curing activator and foam kicker which was obtained from Global Chemical Co., Ltd., Samut Prakan, Thailand. 2,2,4-trimethyl-1,2-dihydroquinoline (TMQ) and N-cyclohexyl-2-benzo thiazole sulphenamide (CBS) were used as an antioxidant and accelerator, respectively. These were bought from Flexsys America L.P., West Virginia, USA. Sulfur, which was used as a curing agent, was purchased from Siam Chemical Co., Ltd., Samut Prakan, Thailand.

Table 1. Characteristics of NR and SR.

Characteristics	Values	
	NR	SR
Ash content	0.02%	0.04%
Volatile matter	0.52%	0.67%
Acetone extract	2.67%	14.99%
Soluble fraction	98.7%	96.5%
Nitrogen content	0.26%	Not detected
Original plasticity (P_o)	40.3	72.1
Moony viscosity (ML 1 + 4 at 100 °C)	77.4	113.1

2.2. Preparation of the Rubber Foams

Table 2 lists the materials and mixing sequence for compounding the oil-absorbent foam. SR and NR ratios were 0/100, 20/80, 30/70, 50/50, 70/30, 80/20, and 100/0 (phr/phr), respectively. The full amounts of the SR or NR, ADC, TDAE oil, and other additives were prepared using a two-roll mill. Proper control of Mooney viscosity was encouraged for preparing the rubber foam. The compounds were free-blown into specific shapes (12.5 mm in thickness) inside the compression-molded at the temperature of 150 °C. The time consumed was based on the curing times measured by a moving-die rheometer (MDR), as described in the following section.

Table 2. Mixing sequence and compounding ingredients used for preparing the rubber compounds.

Mixing Sequence	Ingredients	Amount (phr)	Mixing Time (min)
1	SR/NR *	100	5
2	Stearic acid	1	1
3	TMQ	1	1
4	ZnO	5	1
5	$CaCO_3$	10	3
6	TDAE oil	10	2
7	CBS	2.5	1
8	ADC	5	1.5
9	Sulfur	0.5	1

* The amount of SR and NR were varied.

2.3. Measurement of Curing Characteristics

Using an MDR (Rheoline, Mini MDR Lite) at 150 °C, the curing characteristics of the rubber were measured according to ASTM D5289. This was utilized to calculate the torque, scorch time (ts_1), and cure time (tc_{90}).

2.4. Measurement of Relative Foam Density and Expansion Ration

The physical properties were investigated, including the relative foam density and expansion ratio. The relative foam density was measured according to ASTM D3575, using Equation (1) as given below.

$$\text{Relative foam density} = D_f/D_c \tag{1}$$

where D_f is foam density (g/cm^3), and D_c (g/cm^3) is compound density.

Equation (2) illustrates how the expansion ratio was calculated by comparing the density of the natural rubber compound specimen to the density of the natural rubber foam.

$$\text{Expansion ratio} = D_c/D_f \tag{2}$$

2.5. Measurement of Hardness

The hardness of the specimens was measured in accordance with ASTM D2240 by using an indentation durometer shore OO, and the readings were taken after a 10-second indentation.

2.6. Oil Absorbency

The sample was prepared by cutting into a dimension of $1.5 \times 1.5 \times 1.0$ cm^3. Then, the sample was weighed prior to immersion in crude oil solution. The swelling lasted for a certain period of time, where the development of swelling uptake or oil absorbency was measured to find the equilibrium swelling. To measure the swollen sample, the sample was tapped with filter paper to remove excess oil and then weighed on a balance. The oil absorbency was calculated by the following formula.

$$\text{Oil absorbency} = W_a/W_b \tag{3}$$

where W_a is the weight of absorbed oil, and W_b is the weight of the sample. The unit of oil absorbency is g/g which is the mass of oil swells per 1 g of sample.

2.7. Diffusion Studies

Diffusion studies were carried out simultaneously as the oil absorbency test but calculated differently. The test piece was weighed using a series of time intervals, beginning with 10 min for the 1st hour, followed by 20 min for the 2nd hour, 30 min for the 3rd hour, and then hourly until the test piece's weight reached equilibrium. The solvent uptake was employed to create a plot of the sorption curve with the mole percent (mol%) against the square root of the time ($t^{1/2}$). The equation is displayed as follows.

$$Q_t = \frac{\left(\frac{m_t - m_o}{M_w}\right)}{m_o} \times 100 \tag{4}$$

where Q_t is the mole percent uptake, m_t is the weight of the swollen sample at a given time, m_o is the initial weight of the sample, and M_w is the molecular weight of the solvent.

The swelling coefficient, which represents the sample's swelling behavior, can then be calculated by inserting the weight into Equation (5).

$$\beta = \frac{m_\infty - m_o}{m_o} \times \rho_s^{-1} \tag{5}$$

where β denotes the swelling coefficient, m_∞ is the weight of the sample at equilibrium, m_o is the weight of the test piece before swelling, and ρ_s is the solvent's density.

Diffusivity (D) is a kinetic parameter that depends on segmental polymer mobility. D can be calculated using Equation (6) (the second Fickian law):

$$D = \pi \left(\frac{h\theta}{4Q_\infty}\right)^2 \tag{6}$$

where h is the thickness of the sample before swelling, θ is the slope of the linear sorption curve, and Q_∞ is the equilibrium solvent uptake.

The action of permeate molecules initially penetrating and dispersing within the polymer matrix is explained by sorption. Equation (7) can be used to calculate the sorption coefficient from the swelling.

$$S = \frac{m_\infty}{m_0} \qquad (7)$$

where m_∞ is the weight of the solvent taken at equilibrium and m_0 is the initial mass of the polymer sample.

Equation (8) can be used to estimate the permeability coefficient (P), which provides data on how much solvent permeates through a uniform region of the sample each minute.

$$P = DS \qquad (8)$$

Yao et al. [16] proposed a practical technique for determining how much solution is released from a slab of time (t) which is in terms of the total amount of solvent uptake as shown in Equation (9).

$$\frac{Q_t}{Q_\infty} = kt^n \qquad (9)$$

The sorption curve's result can be fitted to the empirical data to determine the samples' mode of transport, as indicated in Equation (10).

$$\log \frac{Q_t}{Q_\infty} = \log k + n \log t \qquad (10)$$

where Q_t is the mole percent uptake, Q_∞ is the equilibrium solvent uptake, k is the constant, all of which depend on the polymer's structural properties and how the sample interacts with the solvent. The magnitude of n indicates the type of transport. The linear regression was used to calculate the values of n and k.

2.8. Optical Image and Scanning Electron Microscopy

The physical appearance of natural rubber foams was captured using a mobile phone camera through a default setting. The morphology was screened using a FEI Quanta™ 400 FEG scanning electron microscope (SEM; Thermo Fisher Scientific, Waltham, MA, USA). Each specimen was coated with a layer of gold/palladium to remove the charges that had built up during imaging.

2.9. Thermogravimetric Analysis (TGA)

A PerkinElmer Pyris 6 TGA analyzer was used to perform a thermogravimetric analysis on the samples. The sample was heated at a rate of 10 °C/min in a nitrogen flow while being scanned from 30 °C to 600 °C.

2.10. Fourier Transform Infrared-Spectroscopic Analysis (FT-IR)

Fourier transform infrared spectroscopy (FTIR) was used to examine the functionalities shown in NR and SR via the FTIR spectroscope model TENSOR27 (Bruker Corporation, Billerica, MA, USA). The spectra were captured in transmission mode throughout the range of 4000–550 cm^{-1} at a resolution of 4 cm^{-1}.

3. Results and Discussion

3.1. Characterization of NR and SR

Figure 1 shows the FTIR spectra of NR and SR. These two infrared spectra were different: SR shows intense absorption peaks at 1729 cm^{-1} and 1148 cm^{-1}, which are associated with -C=O and -C-O groups in the PMMA chains grafted onto the NR molecules. The absorbance ratios of the peaks at 1729 cm^{-1} to 837 cm^{-1} might be used to roughly evaluate the amount of grafted PMMA on the NR molecules. The 837 cm^{-1} peak results from the =C-H out-of-plane bending of cis-1,4 polyisoprene, whereas the 1729 cm^{-1} peak is related to the C=O stretching of grafted PMMA. The result clearly shows a higher intensity of 1729 cm^{-1} over the 837 cm^{-1}, suggesting that higher PMMA was grafted to NR molecules. The peaks observed in SR agreed well with the previous works on the preparation of PMMA grafted onto NR molecules [17–19].

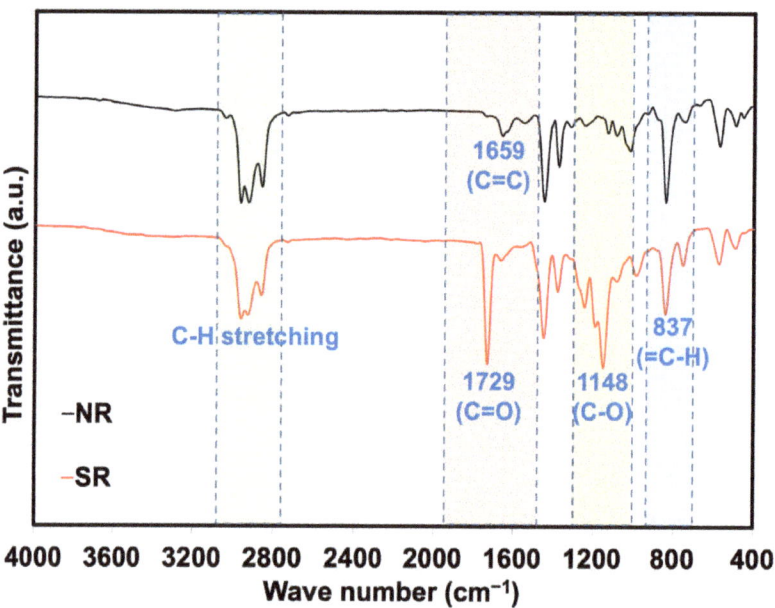

Figure 1. FTIR spectra of NR and SR.

The change in a sample's mass as a function of temperature in a controlled atmosphere is measured by thermogravimetric analysis (TGA). The measurement is primarily used to ascertain the compositional characteristics and thermal and oxidative stabilities of materials. The thermal decomposition behavior of raw NR and SR is shown in Figure 2. The decomposition temperature at 50% weight loss and char residue are also listed and embedded in this Figure. Notably, the decomposition temperature at 50% weight loss of SR was higher than NR. This is simply due to the higher thermal stability of SR as the grafting of MMA onto NR reduces the diene content in the NR. The thermal stability of SR was then improved. Moreover, both NR and SR exhibited low residue which was less than 1%.

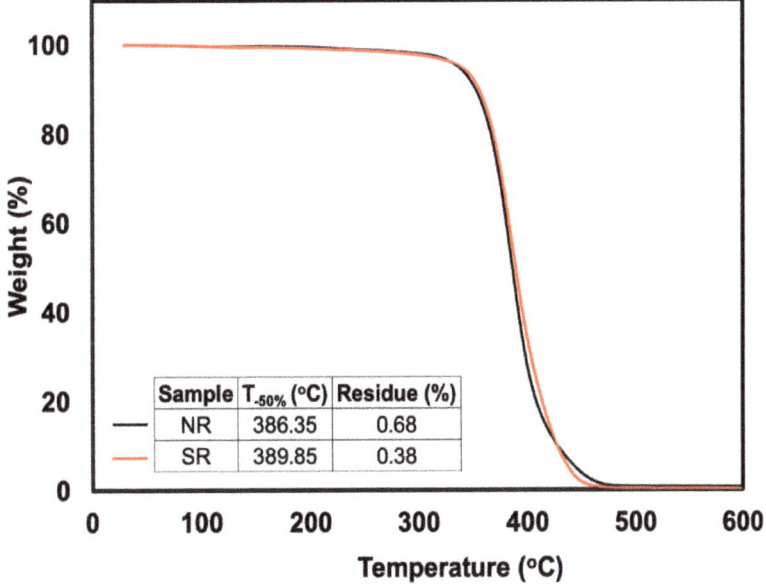

Figure 2. TG curves of NR and SR.

3.2. Cure Characteristics

The rheometric curves of the foams are shown in Figure 3. All the curves showed a marching trend except for the blend at 30/70 phr/phr. The marching curve is present when there is a development of crosslinking [20]. The reversion trend observed for the blend at 30/70 phr/phr may be due to the SR itself. The received SR was from the waste of NR-g-PMMA production, which is uncontrollable material. This phenomenon also occurred in the experiment by Nakason et al. [18], who varied accelerator types to the NR-g-PMMA. They found the reversion of the rubber vulcanizates regardless of accelerator types. The reversion may be associated with the degradation of NR molecules. Separately, the important point to highlight is the increment of the minimum torque (M_L) and maximum torque (M_H) after replacing NR with SR, indicating that the material became stiff over the addition of SR. Grafting MMA onto NR makes the rubber harder due to the presence of the PMMA component as a glassy thermoplastic phase. The scorch time (t_{S1}) and curing time (t_{C90}) were reduced with the addition of SR. t_{S1} is the induction time experienced by a rubber compound before vulcanization is initiated. At the same time, t_{C90} is the time rubber takes to become 90% vulcanized. The decrease in these two values indicated that SR could quicken the vulcanization time of rubber.

According to Harpell et al. [21] and Bhatti et al. [22], the decomposition of ADC produces hydrazodicarbonamide, urazol, and a gaseous mixture of nitrogen (N_2), carbon monoxide (CO), cyanic acid (HNCO), and ammonia (NH_3) through competitive and exothermic chemical pathways. N_2 is the primary source of gas that makes the foam free-flowing. Depending on the circumstances of the process and the status of the result, some paths may be preferred over others. The focal point here is the production of ammonia, which tends to react with PMMA through ammonolysis and results in the formation of primary amide. Such amide derivative may cause the vulcanization of rubber to accelerate, hence decreasing the t_{S1} and t_{C90} in the blends containing a higher content of SR. On the contrary, this kind of phenomenon did not happen for un-foamed specimens. As reported by Nakason et al. [18], they found that the addition of NR-g-PMMA prolongs the t_{S1} and t_{C90}. This was attributed to the polar functional groups of the graft copolymer absorbing certain accelerators as a result of their polarity. Consequently, the accelerator in these quantities was unable to accelerate the vulcanization process. Therefore, a longer cure period was needed to finish the crosslinking process and get the optimum curing capabilities.

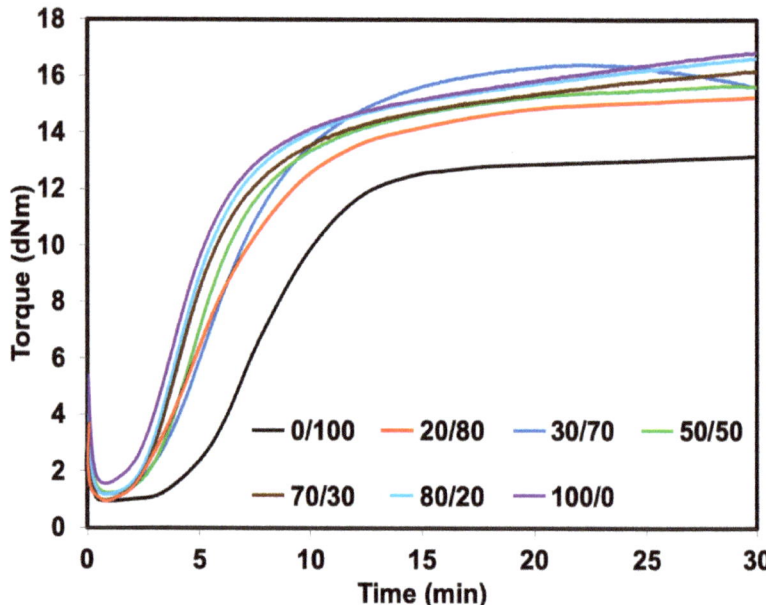

Figure 3. Rheometric curves of SR/NR blends.

3.3. Physical Properties, Appearance, and Morphologies

Table 3 lists the relative foam densities and expansion ratios of the foams with different ratios of blending. As a higher content of SR was used, less gas was subsequently generated due to less flexibility of SR. This increased the relative foam density, which significantly increased from 0.57 g/cm^3 to 0.84 g/cm^3. Higher SR content hardens the rubber matrix, thus restricting the escape of gas through the foam surface. This allowed the foam to have less expansion and, consequently, produce foam with a higher relative density [23]. The relative foam density was directly related to the porosity values and expansion ratio. For this reason, the increased relative foam density when using a high content of SR revealed less expansion of these specimens. The porosity values and expansion ratios are also shown in Table 3. The increase in relative foam density also played a role by decreasing the size of cells per unit volume. The glassy phase of SR may prevent the penetration of gas during compression. Table 3 also lists the hardness of the foams prepared using various SR content. The hardness increased with an increase in the SR content. This was due to the higher foam porosity in the matrix during the formation of the gas phase. Again, an inherent chain stiffness of SR responds to an increase in the hardness of the samples.

Further evidence can be identified from the SEM images shown in Figure 4. The SEM images showed a systematic correlation between the number of cells per unit volume and the average cell size. An increase in the SR resulted in smaller cells and a thick cell wall, indicating the difficulty of gas to diffuse and generate the foams. In this experiment, ADC content was fixed at 5 phr, and the volume of gas generated after the decomposition was assumed to be the same. However, the gas produced could not diffuse or penetrate through the rubber matrix, especially at a higher content of SR. Consequently, the number of cells per unit volume decreased, resulting in a smaller average cell size and a thicker cell wall in the foam [9]. Increasing the levels of SR also affected cell distribution. The foam cell distributed unevenly as the content of SR increased, and random cell size was seen for the sample using a high content of SR. The SEM images are in good agreement with the porosity values and expansion ratios reported in the previous section.

Table 3. Physical properties of SR/NR blends.

SR/NR (phr/phr)	Relative Foam Density (RFD)	Expansion Ratio	Porosity (1 − RFD)	Hardness (Shore OO)
0/100	0.57 ± 0.01	1.75	0.43	58 ± 0.71
20/80	0.59 ± 0.02	1.70	0.41	61 ± 0.44
30/70	0.66 ± 0.01	1.52	0.34	64 ± 0.50
50/50	0.70 ± 0.01	1.44	0.30	69 ± 0.26
70/30	0.72 ± 0.02	1.38	0.28	81 ± 0.56
80/20	0.75 ± 0.01	1.34	0.25	86 ± 0.52
100/0	0.84 ± 0.01	1.19	0.16	91 ± 0.59

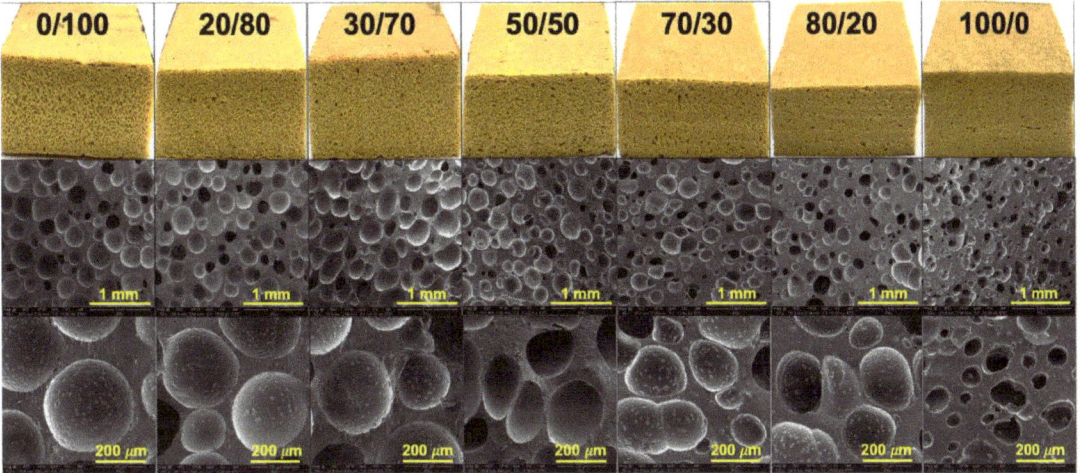

Figure 4. SEM images obtained from the samples' surface of SR/NR blends.

3.4. Oil Absorbency

Figures 5 and 6 depict the oil absorbency over the contact time and the equilibrium absorbency of an oil-absorbent material prepared from various blend ratios. It was observed that the oil absorbency was reduced over the content of SR. Adding SR reduced the oil swellability from 7.09 g/g to 5.02 g/g. The results obtained in this work were found differently compared to previous literature [7,8]. Previous works prepared the foam differently, where the NR and NR-g-PMMA were mixed in the latex stage. The generation of the foam or foam forming was done by Dunlop Process [24]. The foam is generated easier than with the dry rubber method. This has provided a different cellular structure. In this experiment, foaming became more difficult due to the harder phase of SR, where the decomposed gas ineffectively penetrated throughout the matrix. This can be seen from the morphology of the foams (see Figure 4). As mentioned previously, the size of cells decreased, and the thickness of cell walls increased when a higher content of SR was used. As the cell size decreased, the porosity decreased, and the oil could not penetrate easily from one cell to another, leading eventually to a decrease in the swelling uptake. This is further explained in a study by Lee et al. [25], which stated that the swelling of NR foam is influenced by the cell structure and the density, where a lower foam density has caused a higher swelling uptake. This explanation can be clearly understood when correlated with the swelling schematic shown in Figure 7.

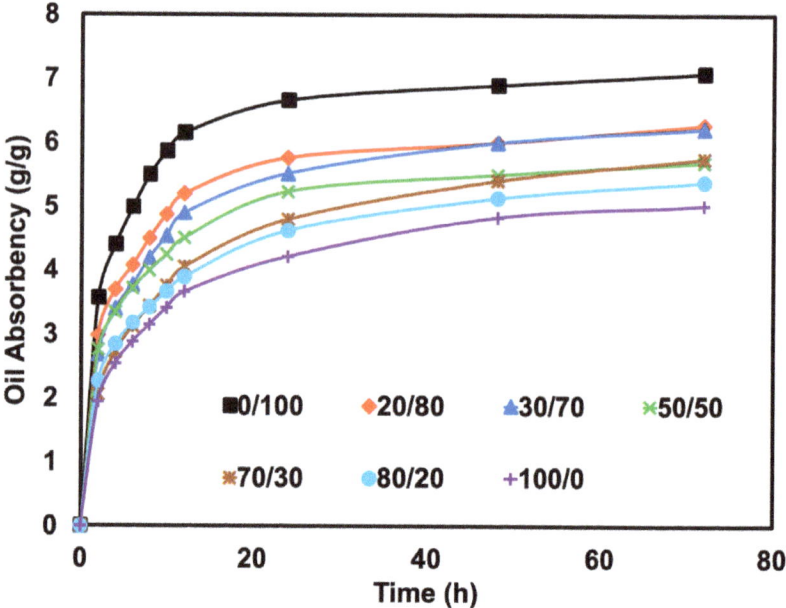

Figure 5. Oil absorbency as a function of contact time of SR/NR blends.

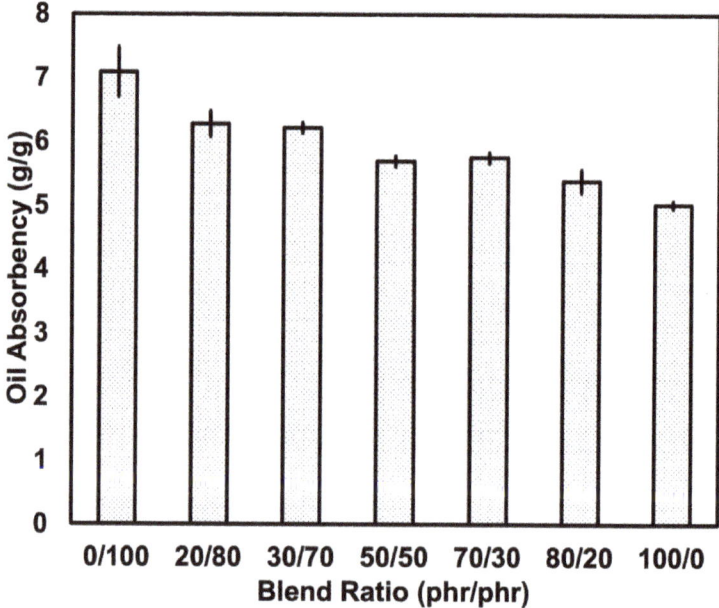

Figure 6. Oil absorbency of SR/NR blends at 72 h uptake.

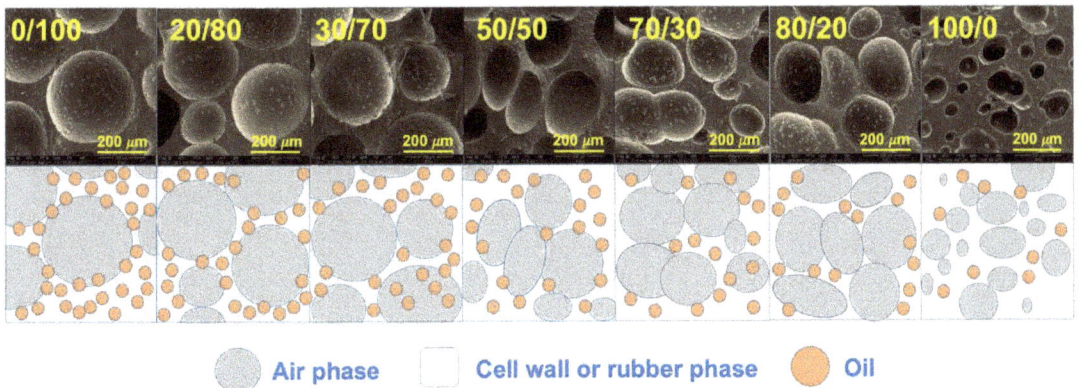

Figure 7. Swelling schematic of oil penetrant through the SR/NR blends.

3.5. Diffusion Study

The most common method of delivering small compounds to polymers is solution diffusion. The penetrant molecules, in this case crude oil, are first absorbed by the rubber before being diffused through it. The sorption data of the crude oil into the rubber at room temperature was determined and expressed as the mole percent uptake (Q_t) against $t^{1/2}$ ($\min^{1/2}$), as shown in Figure 8. The curves show gradual steps of absorption. A significant concentration gradient caused the initial steep zone with a high sorption rate. In contrast, as equilibrium approached, the sorption rate decreased in the later regions. It is well acknowledged that the cross-link density of a network chain and the equilibrium mole percent uptake are correlated [26] and the efficacy of oil or solvent in penetrating rubber molecules since this study was prepared under controlled formulation. It was expected that the cross-linking was more or less the same. The only reason to be regarded with a lower mole percent uptake or penetration into the rubber was due to the efficacy of oil or solvent to penetrate. Thus, the swelling resistance of the blends containing a higher content of SR increased. Similar results were obtained for the rubber's diffusion and swelling coefficients, as shown in Figure 9. As the SR component increased from 0 to 100 phr, the swelling coefficient of the rubber steadily dropped from 8.19 to 5.81 cm^3/g. Therefore, it was more difficult for the oil to penetrate the rubber.

The diffusion coefficient (D) is a kinetic parameter that relies on segmental mobility. Based on the outcome depicted in Figure 9, D was determined using an equation derived from the second Fickian's law. It demonstrates a non-steady decreasing trend, indicating that the oil had difficulty penetrating the rubber matrix with a high SR content. The sorption coefficient and the permeability coefficient are two more characteristics that can be derived from the rubber's diffusion studies (see Figure 10). The initial absorption and dispersion of permeate molecules into the rubber matrix can be explained by the sorption coefficient. In contrast, the amount of penetrant that passes through a consistent area of the sample per minute is shown by the permeability coefficient. As the SR content increased, the rubber's sorption and permeability coefficients gradually declined, indicating that SR segments impede oil diffusion of the rubber foam upon the addition of SR.

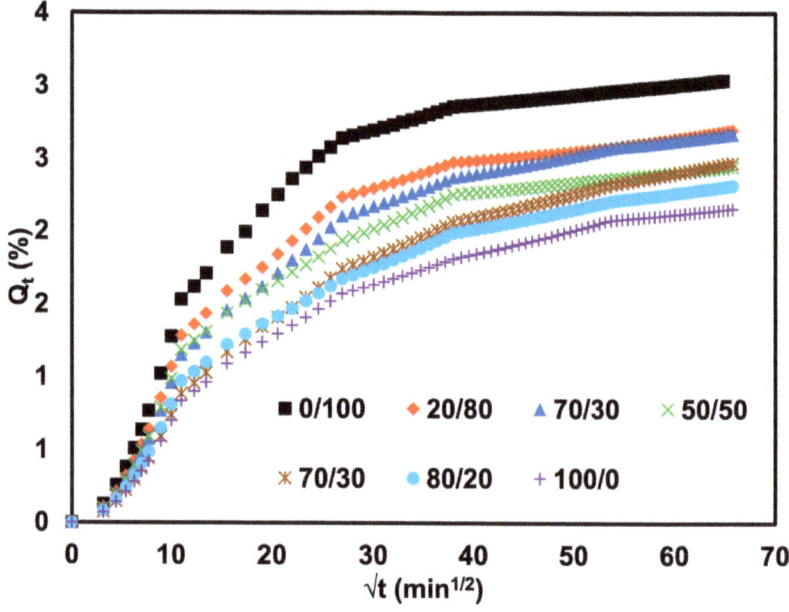

Figure 8. Mole percent uptake as a function of time of SR/NR blends.

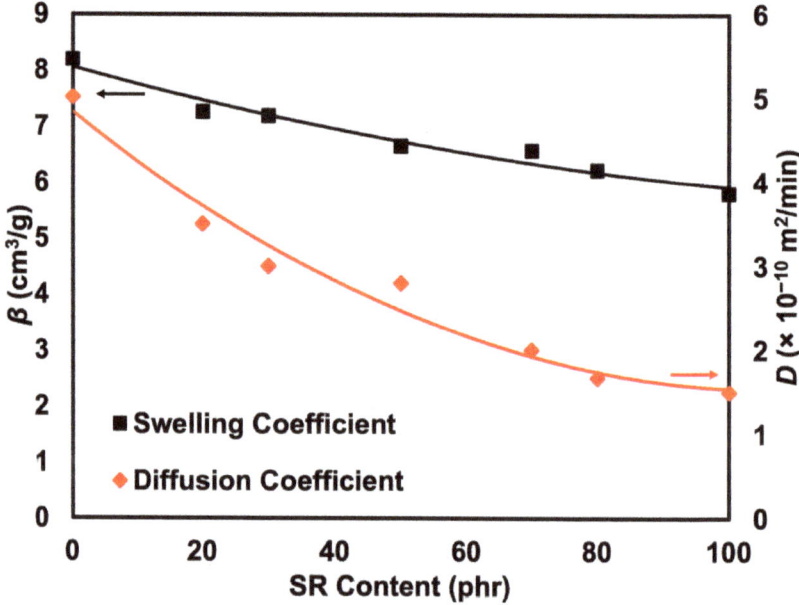

Figure 9. Swelling and diffusion coefficients of SR/NR blends.

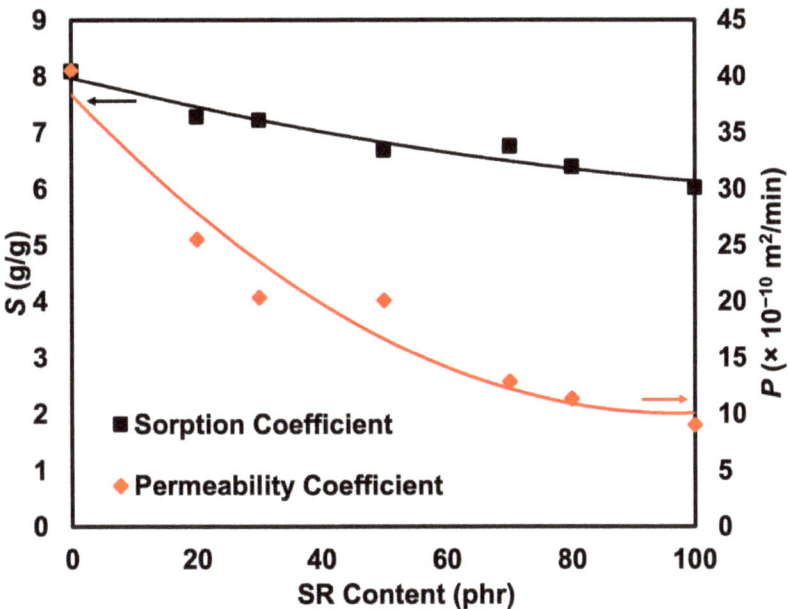

Figure 10. Sorption and permeability coefficients of SR/NR blends.

3.6. Transport Mechanism

Figure 11 displays the results of fitting the oil uptake data into Equation (10) to determine the mode of the transport mechanism. According to Table 4, linear regression analysis of the initial linear slope was used to obtain the values of n and k. Based on the relative mobility of the penetrant and polymer segments, a few categories can be used to classify the transport mechanism, which are: (i) Case I or Fickian diffusion, (ii) Case II diffusion, and (iii) non-Fickian or anomalous diffusion [27]. When n has a value of less than or equal to 0.5, the concentration gradient acts as the main driving factor for diffusion in a Fickian transport mode. The diffusion rate is, therefore, lower than the polymer chain relaxation rate. However, for Case II transport, where n is equal to 1, the diffusion rate is greater than the relaxation process. On the other hand, if the n value is between 0.5 and 1, the transport is anomalous, and the diffusion rate matches the rate at which the polymer chains are relaxing [28].

It is clear from Table 4 that the transport mechanism was anomalous for rubber containing SR between 0 and 100 phr. This result agrees well with the sorption curve in Figure 8, which, over the same time period, showed a modest increase in the mole percent solvent uptake before reaching equilibrium. Polymer chains adapt to a penetrant's presence quickly, but it takes a while for the equilibrium solvent absorption to occur. Additionally, the value of k reflects how oil penetrates rubber or how the penetrant interacts with the rubber matrix. A lower k value represents a lower speed of solvent or oil-penetrating rubber. Here, the calculated results show that the penetrant was more difficult when having more SR. This is simply because the foam contains a thicker cell wall which deactivates the efficacy of oil in penetrating rubber molecules.

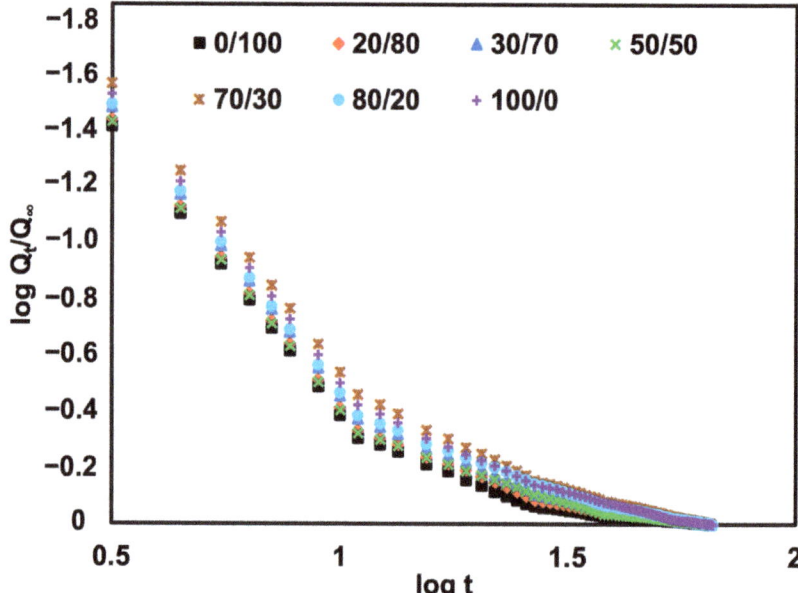

Figure 11. Fitted sorption curves of SR/NR blends.

Table 4. The values of n and k of SR/NR blends.

SR/NR (phr/phr)	n	k
0/100	0.7348	0.0575
20/80	0.7566	0.0517
30/70	0.7587	0.0512
50/50	0.8074	0.0420
70/30	0.8178	0.0394
80/20	0.8524	0.0344
100/0	0.8668	0.0297

3.7. Thermal Stability

Figure 12 depicts the TG curves of the samples. The decomposition temperature at 50% weight loss ($T_{-50\%}$) and the content of char residue are embedded in this Figure. Two regions of degradation of specimens were seen. The initial minor mass loss at around 180–200 °C was due to the presence of volatile matter such as stearic acid and TDAE oil, and the process was complete at about 300 °C [29]. Then, the major step of degradation of the blends (330–450 °C) was caused by the degradation of both SR and NR segments. It is noteworthy that the specimens' thermal stability marginally improved as the content of SR increased. This can be seen from the $T_{-50\%}$, which clearly shows that by introducing SR, the temperature was shifted to a higher temperature. The enhancement in thermal stability can be evidently related to the original stability of SR seen previously in Figure 2. The replacement of MMA onto the backbone of NR has made the NR-g-PMMA more stable against the formation of degradation. A similar observation was also found in the literature [17]. They found that increasing the content of MMA onto NR has enhanced the thermal resistance of rubber. With increased PMMA content in the grafting reaction, stronger chemical interactions between the molecules were speculated to be the cause. Additionally, increasing oxygen compounds exhibited a higher resistance to thermal degradation than that of the original NR. Moreover, the decomposition temperature of foam specimens was slightly lower than raw rubbers (see Figure 2). This is because the foam was heated during processing and vulcanization. Therefore, slight degradation of foam specimens may occur during such a process.

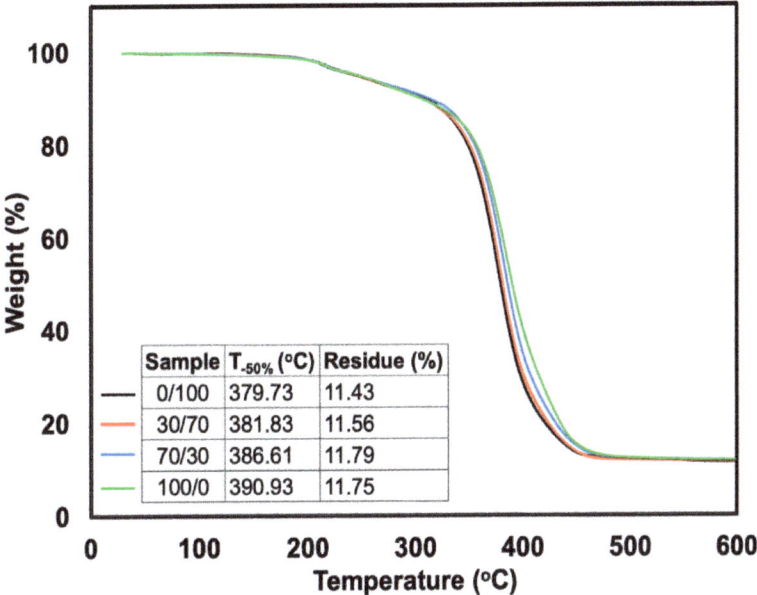

Figure 12. TG curves of SR/NR blends.

4. Conclusions

The aim of this work was to utilize the waste from the production of NR-g-PMMA as an oil-absorbent rubber. The rubber blends were then turned into cellular structures to increase the oil swellability. Results indicated that adding SR quickened the vulcanization process of rubber due to the ammonolysis taking place during the vulcanization of the foam. It was observed that the cellular structure of the foam was difficult to generate when using a higher content of SR. This made the oil swellability reduce from 7.09 g/g to 5.02 g/g. This agreed well with the study of swelling kinetics. It demonstrates a non-steady decreasing trend; the transport mechanism was anomalous for rubber containing SR between 0 and 100 phr. There was a minor degradation of the samples at temperatures of 180–300 °C due to the presence of volatile matters, mainly from TDAE oil. This would not affect the application of oil-absorbent rubber since it was added to facilitate the foaming process. However, it is interesting to note that the prepared oil-absorbent rubber still provided higher thermal stability, particularly at temperatures over 400 °C. This is a very good compromise between oil absorbency and thermal stability. Based on the overall properties, the SR content at 30 phr is suggested. This is considered from oil absorption capacity, yet other factors include blend processability, foam structure, and physical and thermal properties. The results obtained from this study will be used for further experiments on the enhancement of oil absorbency by applying other key factors. This work is a good initiative for preparing the oil-absorbent material based on scrap from modified natural rubber production. At the same time, our work provides an alternative route to fabricate oil-absorbent foam for an oil recovery application.

Author Contributions: Investigation, formal analysis, and writing—original draft preparation, A.T.; methodology, and validation, S.S. (Sitisaiyidah Saiwari) and S.S. (Subhan Salaeh); conceptualization, methodology, validation, writing—review and editing, supervision, N.H. All authors have read and agreed to the published version of the manuscript.

Funding: This research was supported by the National Science, Research, and Innovation Fund (NSRF) and Prince of Songkla University for the Fundamental Funds through grant No. SAT6505037a.

Institutional Review Board Statement: Not applicable.

Informed Consent Statement: Not applicable.

Data Availability Statement: The data presented in this study are available on request from the corresponding author.

Acknowledgments: The authors thank a local company in the southern part of Thailand for supplying the SR.

Conflicts of Interest: The authors declare no conflict of interest in this publication.

References

1. Mardiyati, Y.; Fauza, A.N.; Rachman, O.A.; Steven, S.; Santosa, S.P. A Silica–Lignin Hybrid Filler in a Natural Rubber Foam Composite as a Green Oil Spill Absorbent. *Polymers* **2022**, *14*, 2930. [CrossRef]
2. Passow, U.; Lee, K. Future oil spill response plans require integrated analysis of factors that influence the fate of oil in the ocean. *Curr. Opin. Chem. Eng.* **2022**, *36*, 100769. [CrossRef]
3. Dong, J.; Asif, Z.; Shi, Y.; Zhu, Y.; Chen, Z. Climate Change Impacts on Coastal and Offshore Petroleum Infrastructure and the Associated Oil Spill Risk: A Review. *J. Mar. Sci. Eng.* **2022**, *10*, 849. [CrossRef]
4. Zhang, T.; Li, Z.; Lü, Y.; Liu, Y.; Yang, D.; Li, Q.; Qiu, F. Recent progress and future prospects of oil-absorbing materials. *Chin. J. Chem. Eng.* **2019**, *27*, 1282–1295. [CrossRef]
5. Chin, C.C.; Musbah, N.D.L.; Abdullah, I.; Lazim, A.M. Characterization and Evaluation of Prudent Liquid Natural Rubber-Based Foam for Oil Spill Control Application. *Arab. J. Sci. Eng.* **2018**, *43*, 6097–6108. [CrossRef]
6. Jadhav, A.C.; Jadhav, N.C. Graft copolymerization of methyl methacrylate on Meizotropis Pellita fibres and their applications in oil absorbency. *Iran. Polym. J.* **2021**, *30*, 9–24. [CrossRef]
7. Ratcha, A.; Yoosuk, B.; Kongparakul, S. Grafted Methyl Methacrylate and Butyl Methacrylate onto Natural Rubber Foam for Oil Sorbent. *Adv. Mater. Res.* **2013**, *844*, 385–390. [CrossRef]
8. Kanagaraj, L.; Bao, C.A.; Ying, C.S.; Ing, K. Mechanical Properties and Thermal Stability of methyl methacrylate grafted latex and natural rubber latex foam blends. *J. Eng. Sci. Technol.* **2019**, *14*, 3616–3627.
9. Charoeythornkhajhornchai, P.; Samthong, C.; Boonkerd, K.; Somwangthanaroj, A. Effect of azodicarbonamide on microstructure, cure kinetics and physical properties of natural rubber foam. *J. Cell. Plast.* **2016**, *53*, 287–303. [CrossRef]
10. Lee, E.-K.; Choi, S.-Y. Preparation and characterization of natural rubber foams: Effects of foaming temperature and carbon black content. *Korean J. Chem. Eng.* **2007**, *24*, 1070–1075. [CrossRef]
11. Najib, N.N.; Ariff, Z.M.; Bakar, A.A.; Sipaut, C.S. Correlation between the acoustic and dynamic mechanical properties of natural rubber foam: Effect of foaming temperature. *Mater. Des.* **2011**, *32*, 505–511. [CrossRef]
12. Ramasamy, S.; Ismail, H.; Munusamy, Y. Effect of rice husk powder on compression behavior and thermal stability of natural rubber latex foam. *BioResources* **2013**, *8*, 4258–4269. [CrossRef]
13. Panploo, K.; Chalermsinsuwan, B.; Poompradub, S. Natural rubber latex foam with particulate fillers for carbon dioxide adsorption and regeneration. *RSC Adv.* **2019**, *9*, 28916–28923. [CrossRef]
14. Ariff, Z.M.; Zakaria, Z.; Tay, L.H.; Lee, S.Y. Effect of foaming temperature and rubber grades on properties of natural rubber foams. *J. Appl. Polym. Sci.* **2007**, *107*, 2531–2538. [CrossRef]
15. Baru, F.; Saiwari, S.; Hayeemasae, N. Classification of natural rubber foam grades by optimising the azodicarbonamide content. *Polímeros* **2022**, *32*, e2022014. [CrossRef]
16. Yao, K.D.; Peng, T.; Feng, H.B.; He, Y.Y. Swelling kinetics and release characteristic of crosslinked chitosan: Polyether polymer network (semi-IPN) hydrogels. *J. Polym. Sci. Part A. Polym. Chem.* **1994**, *32*, 1213–1223. [CrossRef]
17. Kalkornsurapranee, E.; Sahakaro, K.; Kaesaman, A.; Nakason, C. From a laboratory to a pilot scale production of natural rubber grafted with PMMA. *J. Appl. Polym. Sci.* **2009**, *114*, 587–597. [CrossRef]
18. Nakason, C.; Pechurai, W.; Sahakaro, K.; Kaesaman, A. Rheological, thermal, and curing properties of natural rubber-g-poly(methyl methacrylate). *J. Appl. Polym. Sci.* **2006**, *99*, 1600–1614. [CrossRef]
19. Thiraphattaraphun, L.; Kiatkamjornwong, S.; Prasassarakich, P.; Damronglerd, S. Natural rubber-g-methyl methacrylate/poly(methyl methacrylate) blends. *J. Appl. Polym. Sci.* **2001**, *81*, 428–439. [CrossRef]
20. Jin, J.; Noordermeer, J.W.M.; Dierkes, W.K.; Blume, A. The origin of marching modulus of silica-filled tire tread compounds. *Rubber Chem. Technol.* **2020**, *93*, 378–394. [CrossRef]
21. Harpell, G.A.; Gallagher, R.B.; Novits, M.F. Use of azo foaming agents to produce reinforced elastomeric foams. *Rubber Chem. Technol.* **1977**, *50*, 678–687. [CrossRef]
22. Bhatti, A.S.; Dollimore, D.; Goddard, R.J.; O'Donnell, G. The thermal decomposition of azodicarbonamide. *Thermochim. Acta* **1984**, *76*, 63–77. [CrossRef]

23. Guan, L.T.; Du, F.G.; Wang, G.Z.; Chen, Y.K.; Xiao, M.; Wang, S.J.; Meng, Y.Z. Foaming and chain extension of completely biodegradable poly(propylene carbonate) using DPT as blowing agent. *J. Polym. Res.* **2007**, *14*, 245–251. [CrossRef]
24. Phomrak, S.; Nimpaiboon, A.; Newby, B.-M.Z.; Phisalaphong, M. Natural Rubber Latex Foam Reinforced with Micro- and Nanofibrillated Cellulose via Dunlop Method. *Polymers* **2020**, *12*, 1959. [CrossRef]
25. Lee, H.-K.; Chung, T.-K.; Kim, S.-C.; Kim, H.-G.; Choi, K.-M.; Kim, Y.-M.; Han, D.-H. Influence of the type of curing agent on swelling behaviour of natural rubber foam. *J. Korea Acad.-Ind. Coop. Soc.* **2008**, *9*, 1775–1781.
26. Sae-oui, P.; Sirisinha, C.; Thepsuwan, U.; Thapthong, P. Influence of accelerator type on properties of NR/EPDM blends. *Polym. Test.* **2007**, *26*, 1062–1067. [CrossRef]
27. Kee, D.D.; Liu, Q.; Hinestroza, J. Viscoelastic (Non-Fickian) Diffusion. *Can. J. Chem. Eng.* **2005**, *83*, 913–929. [CrossRef]
28. Bengfort, M.; Malchow, H.; Hilker, F.M. The Fokker–Planck law of diffusion and pattern formation in heterogeneous environments. *J. Math. Biol.* **2016**, *73*, 683–704. [CrossRef]
29. Samarasinghe, I.H.K.; Walpalage, S.; Edirisinghe, D.G.; Egodage, S.M. The use of diisopropyl xanthogen polysulfide as a potential accelerator in efficient sulfur vulcanization of natural rubber compounds. *J. Appl. Polym. Sci.* **2022**, *139*, e52063. [CrossRef]

Article

Risk Analysis on PMMA Recycling Economics

Jacopo De Tommaso [†] and Jean-Luc Dubois *

ARKEMA France, 420 Rue d'Estienne d'Orves, 92705 Colombes, France; jacopo.de-tommaso@polymtl.ca
* Correspondence: jean-luc.dubois@arkema.com; Tel.: +33-4-7239-8511
[†] Current address: Chemical Engineering, Polytechnique Montréal, Montréal, QC H3T 1J4, Canada.

Abstract: Poly(methyl methacrylate) (PMMA) is a versatile polymer with a forecast market of 4 Mtons/y by 2025, and 6 USD billion by 2027. Each year, 10% of the produced cast sheets, extrusion sheets, or granules PMMA end up as post-production waste, accounting for approximately 30 000 tons/y in Europe only. To guide the future recycling efforts, we investigated the risks of depolymerization process economics for different PMMA scraps feedstock, capital expenditure (CAPEX), and regenerated MMA (r-MMA) prices via a Monte-Carlo simulation. An analysis of plastic recycling plants operating with similar technologies confirmed how a maximum 10 M USD plant (median cost) is what a company should aim for, based on our hypothesis. The capital investment and the r-MMA quality have the main impacts on the profitability. Depending on the pursued outcome, we identified three most suitable scenarios. Lower capital-intensive plants (Scenarios 4 and 8) provide the fastest payback time, but this generates a lower quality monomer, and therefore lower appeal on the long term. On 10 or 20 years of operation, companies should target the very best r-MMA quality, to achieve the highest net present value (Scenario 6). Product quality comes from the feedstock choice, depolymerization, and purification technologies. Counterintuitively, a plant processing low quality scraps available for free (Scenario 7), and therefore producing low purity r-MMA, has the highest probability of negative net present value after 10 years of operation, making it a high-risk scenario. Western countries (especially Europe), call for more and more pure r-MMA, hopefully comparable to the virgin material. With legislations on recycled products becoming more stringent, low quality product might not find a market in the future. To convince shareholders and government bodies, companies should demonstrate how funds and subsidies directly translate into higher quality products (more attractive to costumers), more economically viable, and with a wider market.

Keywords: methyl methacrylate; PMMA recycling; Monte Carlo; economic analysis; regenerated MMA; depolymerization; scenario; net present value (NPV); payback period; risk analysis

Citation: De Tommaso, J.; Dubois, J.-L. Risk Analysis on PMMA Recycling Economics. *Polymers* **2021**, *13*, 2724. https://doi.org/10.3390/polym13162724

Academic Editor: Didier Perrin

Received: 22 July 2021
Accepted: 12 August 2021
Published: 15 August 2021

Publisher's Note: MDPI stays neutral with regard to jurisdictional claims in published maps and institutional affiliations.

Copyright: © 2021 by the authors. Licensee MDPI, Basel, Switzerland. This article is an open access article distributed under the terms and conditions of the Creative Commons Attribution (CC BY) license (https://creativecommons.org/licenses/by/4.0/).

1. Introduction

Poly(methyl methacrylate) (PMMA), known as acrylic or acrylic glass, as well as Perspex, Plexiglas, and Altuglas, is a transparent thermoplastic and an alternative to glass due to its UV resistance and transparency.

The methyl methacrylate (MMA) global market was valued 7.8 billion US dollars in 2018 [1], and the production volume assessed around 3.9 Million tons in 2019/2020 [2,3]. These figures are foreseen to increase steadily in the next years, to reach 5.7 Million tons of MMA worldwide produced by 2028 [2], and an estimated worth of 4.8 USD billion for MMA by 2028 in Europe only. Similarly, the PMMA market volume passed from 2 Million tons/y in 2017 [4], to 2.6 Million tons in 2019 [5], and it is estimated to reach a market value of 6 USD billions by 2027 [6].

The Covid 2019 pandemic had a significant effect on the PMMA market. On one side, the global crisis put a strain on the automotive and construction industry, that accounts for more than 60% of the PMMA application [3,7]. At the same time, sheets production boomed, along with demand for medical diagnostic equipment such as incubators, or medical cabinets, among others. For this reason, although the production of PMMA incidentally decreased in the first two quarters of 2020, spot prices and requests for PMMA increased by 25% in 2020 [7], making it the most sought-after polymer in the world. In particular, the protective sheet market made the feedstock price (MMA) bouncing back after 27 months of pricing decline [8], and therefore ICIS foresees a "strong (+10%) rebound in 2021" (of PMMA market), "to be completed in 2022 returning to pre-crisis demand levels" [3].

Concurrent with the increasing market and production volume trend of the last years, the End-of-Life and post-production waste ramped up as well. The MMAtwo consortium estimated that, in Europe only, out of the 300,000 tons/y produced, only 8000 are recycled in Europe, most of which (70%) coming from post-production streams such as PMMA offcuts, saw dust, production scraps, or off-grades [9].

The market demands of PMMA and MMA represent a relatively small volume compared to other plastics (10–100 times less [10,11]). Therefore, the general public often oversees strength and opportunities of PMMA recycling. Nevertheless, already back in 1945, only 17 years after Otto Röhm first synthesized PMMA, Gem Participations patented a "heat transfer bath process" to depolymerize "Lucite" scrap back to the monomer [12]. Due to its higher price (compared to large volume plastics like polyolefins for instance), several companies and academic groups investigated PMMA chemical recycling to MMA. PMMA thermally depolymerizes to MMA by radical unzipping starting at 350 °C, and pure PMMA fully depolymerizes to MMA at 450 °C [13]. Thermochemical technologies emerged in time, with the principal being dry distillation, molten metal bath, fluidized beds, and extruders [10,14]. Regardless the technology, the economic viability has always hurdled the large-scale industrialization of any recycling process [15]. To guide the future recycling efforts, we investigate the economics of PMMA recycling, detailing the effect of feedstock type, product quality, and capital investment.

The "Dry-distillation" is commonly practiced in several countries, including India, China, Brazil ... In this process, PMMA scraps are loaded in a tank, which is heated with a direct burner most of the time, and when no more cracked products are emitted, the unit is cooled to be reloaded. Crude MMA is condensed, and purified, most often with a rough distillation. The "Rotating-drum" process is a variation of the dry-distillation, which is also popular in Asia. It can be operated continuously, but most of the time it is a batch process. The PMMA scraps are loaded in the rotating cylinder, which provides a kind of mixing of the material, and avoids one of the drawbacks of the dry distillation (the heat transfer limitation through the char layer which forms in the unit). The "Molten Metal" process, uses most often molten lead, but can also use tin or zinc. The molten metal has a high specific heat which is efficiently transferred to the PMMA. When these units are operated properly there is no contamination of the product nor the environment. However, when low quality scraps are processed, a high amount of solid residues are produced, contaminated with the metal used. This generates additional cost to dispose them of properly. This technology is still in operation in several countries, including Italy, Spain, Egypt, and in South East Asia for example. The "Fluid-Bed" process involves a twin fluid bed circulating reactor system. Hot sand is circulated between a depolymerization reactor and a regenerator where carbon deposits are burned and sand is reheated. This process has been piloted at rather large scale, but was not industrialized yet, although several variations have been investigated. Other technologies are under development like the use of microwaves, Joule effect (ohmic heating), and inductive heating. "Stirred-Tank" process is also a variation of the dry distillation, in which a stirrer is used in the reactor for better heat and mass transfer. That requires a molten polymer base to be able to stir the scraps. To the best of our knowledge, a few units are using this type of technology. The technology investigated in the MMAtwo research project is based on a "Twin-screw extruder". This process was used at industrial scale on high-quality cast PMMA scraps, and is investigated in the project for all kinds of PMMA wastes, including post-industrial and post-consumer wastes. In this process, the scraps are continuously fed in the extruder where the residence time is very short, and the crude MMA vapors are condensed, and further purified.

All the processes, in use today, include a scrap pretreatment (crushing, separation of the polyethylene films and PVC parts), a condensation, and a purification section. The level of purification varies greatly between the sites, with product as low as 91 wt % MMA content seen and up to 98.5 wt % seen on the market, or 99.8 wt % achieved in the MMAtwo project. The sequence of purification steps depends on the usual impurities and the target for the regenerated MMA producer. The distillation can be operated in a batch mode with a single column or in a continuous mode with two distillation units, and in some cases with extra purification steps such as washing steps and dehydration steps. So, the capital cost and operating cost are strongly affected by the level of purity expected for the r-MMA. At the same time, the quality of the scraps has also a strong impact on the final r-MMA purity: a clean cast PMMA will give better r-MMA quality as it does not contain co-monomers unlike the injection and extrusion grades PMMA. Therefore, not all the scraps have the same market value.

2. Materials and Methods

2.1. Depolymerization Process

At a first glance, typical PMMA recycling processes have a common backbone (Figure 1):
- Pre-treatment in or off-site (1000);
- Depolymerization section (2000);
- Purification section (3000).

The nature of each process step varies based on the interdependence of feedstock source, the target product quality, and the available technological know-how. Combination of different steps gives the full spectrum of process technologies (dry distillation, molten metal bath, fluidized bed, extruder, etc.,).

Although no process route is dominating worldwide [10], some are preferred when lower quality r-MMA is acceptable, or less capital-intensive plants are required. For instance, dry distillation plants have a low capex, but produce a lower quality monomer (<98% wt.), nevertheless accepted in the local markets (e.g., India, China, Brazil) [10,16,17]. Molten lead baths, instead, are very sensitive to PMMA scraps quality, and have a proven track record with high-quality waste [10,18]. The MMAtwo twin-screw extruder technology depolymerized several post-industrial, and end-of-Life PMMA grades, into a r-MMA as pure as 99.8 wt. % (after distillation) [19]. However, this comes with a higher capital investment, compared to dry distillation, for instance. Theoretically, the ideal technology not only produces the best product from the worst waste, at the lowest capital investment, but it also eases as much as possible its environmental footprint. For this reason, we report the flowsheet of our own PMMA design of recycling process, organized here to be a compromise or average representative of the existing technologies and heavily energy integrated (Figures 1 and 2).

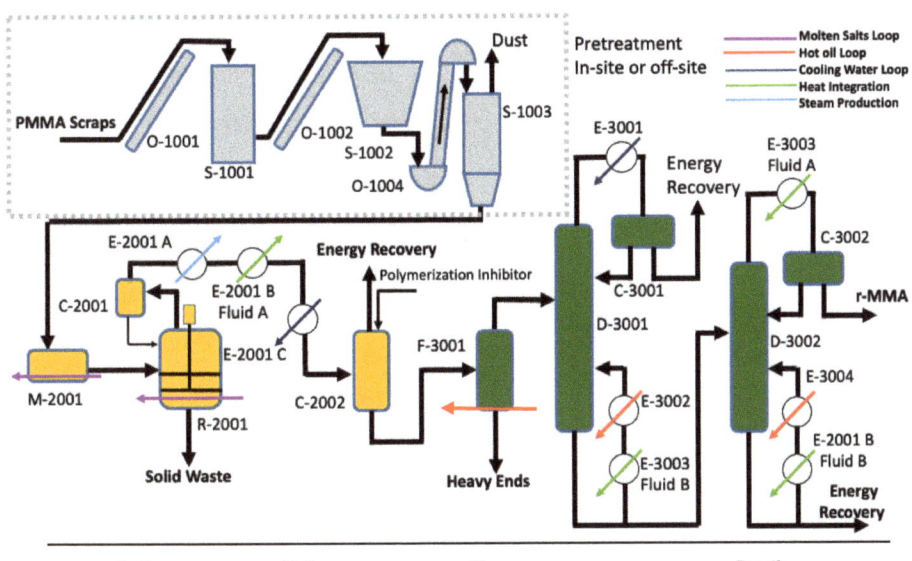

Section	Unit	Name	Details
1000-Pretreatment	S-1001	Impactor	
	S-1002	Granulator	
	S-1003	Bag filter	
	O-1001/1002	Conveyers	
	O-1003/1009	Screw Conveyers	Not shown
	O-1004	Bucket Elevator	
2000-Depolymerization	M-2001	Melter	Extruder premelter
	C-2001	Deduster	
	C-2002	Liquid gas separator	
	V-2001	Skimmer separator	Not shown
	R-2001	Depolymerization Reactor	Stirring tank
	E-2001 A	Cracking gas cooler	Steam generator
	E-2001 B	Cracking gas condenser	Fluid A
	E-2001 C	Cracking gas subcooler	
	F-2001	Molten salts Furnace	Not shown
3000-Purification	F-3001	Evaporator Heavies	Thin-film evaporator
	C-3001	D-3001 Reflux drum	
	C-3002	D-3002 Reflux drum	
	E-3001	D-3001 overhead condenser	
	E-3002	D-3001 reboiler	
	E-3003	Condenser/reboiler Heat integration	C-3001 reboiler/C-3002 Condenser
	E-3004	C-3002 reboiler	
	E-3005	r-MMA air cooler	Not shown
	D-3001	Lights Column	70°C, 0.5 atm
	D-3002	Monomer Column	80°C, 0.5 atm
	F-3001	Hot-oil Furnace	Not shown
	T-2001	Monomer storage tank	Not shown

Figure 1. Simplified flowsheet for our design of a PMMA depolymerization process.

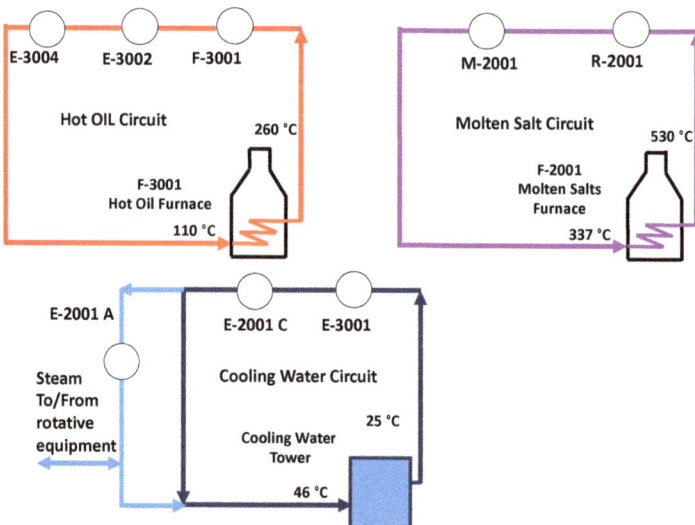

Figure 2. Utilities loop for a typical PMMA depolymerization process.

The PMMA scraps are pre-treated in or off site (section 1000), crushed, granulated, and then conveyed to the reaction section (2000). Because it is nowadays seen more and more often in mixed waste plastic recycling technologies, and because it was explored in the past by Mitsubishi Rayon, [20] we considered a stirred tank reactor for the depolymerization. The PMMA granules are pre-melt (M-2001), for instance in a single screw extruder, and then depolymerized in a stirred tank reactor R-2001 (Figure 1). This combination of melter/reactor is similar to what we see in Agilyx, or plastic energy [21,22] recycling technologies. Compared to a "typical" PMMA process, we minimized the capital and operating expenses (Capex/Opex) at the pre-treatment by not including a washing step.

The depolymerization vapors condense and are then purified to r-MMA in a series of columns. First, the crude is pre-purified from the very heavy ends in the evaporator F-3001, and then two distillation columns fix the purity of the monomer. The plant produces by-products, burnt on site for energy recovery (section 4000—not shown), r-MMA, and solid and liquid waste (heavy ends), which are disposed of outside the battery limits. The pigments, fillers, charred polymer, and other polymer residues tend to accumulate in the solid residue, which might also contain some substances of very high concern, therefore requiring proper disposal protocols. The heavy ends contain dimers and oligomers, but also some other cracked polymers that contaminated the PMMA.

Reaction and purification sections are technology-neutral, meaning that any other reactor or purification choice are valid in lieu of our design, since our goal is to explore the economics of PMMA recycling at large. Same applies for the heating medium choice (molten salt or hot oil), that are interchangeable with any other heating carrier (e.g., steam, flue gases, or direct heating) as long as we assume the same heating energy demand. Other purification technologies are sometimes used by operators such as a washing step with sodium hydroxide solution, but that introduces water in the MMA, and increases the complexity of the distillation due to an azeotrope between water and acrylates and methacrylates. In absence of added water washing step, the losses due to the azeotropes can be minimized.

The by-products combustion provides heat energy and steam for the rotative equipment. Natural gas complements the heating duty required. Natural gas and by-products fuel a molten salt, and a hot oil circuit. The molten salt loop supplies the heat to the reaction section, while a hot oil circuit supplies the purification section. Because of the high energy content of the by-products (comparable to that of MMA—26.2 MJ/kg [23]), the energy recovered from their combustion is high enough to satisfy almost all the energy requirements of the process, with only a little natural gas contribution.

In general, the cast scraps have a better yield in r-MMA, with a lower production of waste, and they require a proportionally higher natural gas consumption compared to the mixed PMMA scraps (Table 1).

Table 1. Mass balance for 5000 PMMA scraps tons/year plant, in case of Cast and Mixed PMMA scraps.

Feedstock/Product/Waste	Cast PMMA Scraps (90 wt % yield)	Mixed PMMA Scraps (78 wt % yield)
PMMA scraps (tons/y)	5000	5000
Natural Gas (tons/y)	106	80
Heavy Ends (tons/y)	119	411
Solid Waste (tons/y)	100	200
r-MMA (tons/y)	4512	3889

The quality of the feedstock affects the energy requirement of the process. High-quality waste, such as clean cast PMMA scraps, are more expensive, require less energy for the purification, and produce less by-products. Regarding the net energy recovery, cast scraps will require slightly more external energy source than mixed PMMA waste scraps. Although the MMAtwo consortium alone has committed to collect and treat up to 27 000 tons/y of PMMA by 2023, the current European plant size is around 5000 tons/y, as the experience of Madreperla, Monomeros del Valles, and in the past Evonik, Arkema, or Du Vergier [9,24] suggests. While sorting and collection volumes ramp up, especially for end-of-life waste, highly industrialized and densely populated areas can easily have access to similar volumes. Therefore, we believe that 5000 tons/year of scraps is what any new plant should realistically aim to start with, especially while establishing a sound market.

2.2. PMMA—Our Case Studies

Since the nature of scraps affects the mass and energy balance of the process, to investigate the PMMA recycling options, we benchmarked cast and mixed scraps, on a common final r-MMA purity. In general, for any given r-MMA grade, mixed scraps present a lower global (depolymerization + purification) r-MMA yield, compared to cast (78 wt.% vs 90 wt.% respectively—Table 1). At the same time, mixed scraps produce proportionally more waste, but require less external source of energy. In Table 1, we report the mass balance for both cast and mixed scraps processes.

2.3. Economic Analysis

When a portfolio of potential projects is available, companies or public investors need to evaluate the potential profitability of each project, to assess when and if to take a risk. The profitability of a project can be assessed by different methodologies, such as the payout period plus interest (POP), the net present value (NPV), the discounted cash flow rate of return (DCFROR), and the uniform annual cost (UAC), among others [25]. The NPV, also known as net present worth (NPW), is the most common quantitative methodology to assess the profitability, since it considers the effect of the time value of money on the profitability [25], and the annual variation in expenses and revenues [26]. The NPV of a project is the sum of the present values of future cash flows [26]:

$$\text{NPV} = \sum_{n=1}^{n=t} \frac{CF_n}{(1+i)^n} \tag{1}$$

where CF_n represents the annual operational cash flow for year n, i is the interest rate, and t the project life in years.

When the NPV is positive, the project earns more than the interest (discount) rate selected, or the best alternative project [25]. The annual operational cash flow CF_n is the sum of the annual net profit after taxes (ANP), and the depreciation D:

$$CF_n = \text{ANP} + D \qquad (2)$$

With ANP being the annual gross profit—the taxes (TR). The annual gross profit (AP), is the sum of sales S, minus expenses E, minus depreciation D, so that the annual operational cash flow CF_n becomes:

$$CF_n = (S - E - D) * (1 - \text{TR}) + D \qquad (3)$$

The sales are the revenue generated from the r-MMA, the expenses are the total expenses of the plant, the tax rate is set at 35% as for France, and the depreciation follows the straight line depreciation on 10 years [25]. In our assessment, we divided the expenses in *Fixed* and *Variable*. The *Fixed* expenses are:

- Direct labor;
- Operating supervision, assumed as 18% of the direct labor;
- Laboratory charges, assumed as 18% of the direct labor;
- Plant overhead, assumed as 60% of direct labor;
- Administration, assumed as 20% of direct labor
- Maintenance and repairs, assumed as 2% of the total investment, but which could be higher for a cheap equipment;
- Operating supplies, assumed as 1% of the total investment;
- Financial interests, assumed as 2% of the total investment;
- Property taxes, assumed as 2% of the total investment; and
- Insurance, assumed as 2% of the total investment.

The *Variable* expenses are:

- Raw Materials (PMMA scraps, natural gas);
- Waste disposal (Solid waste, heavy ends);
- Other utilities (water, electricity, etc.) assumed as 3% of total raw material and waste disposal because we did not go into that level of details;
- Distribution and selling, as 5% of the sales, because we assumed here to have a limited pool of customers, that therefore minimizes the distribution related costs;
- R&D and Royalties, as 3% of the sales in total. This way R&D budget depends on the sales, but remains in the range of what is seen for commodity chemicals.
- We also made the additional general assumptions for the plant:
- Plant life is 20 years;
- The plant is located in France;
- Operating time is 8000 h, or 330 days;
- Depreciation is in 10 years;
- Internal rate of return (IRR) is 10%;
- Labor cost is calculated for 5 shifts per day, continuous process;
- The plant needs 2 operators per shift, at a cost of 60 000 USD/y per operator;
- The capital investment is spread over two years (year 0 and year 1), where 2/3 of the capital is invested in the first year, and 1/3 in the second;
- The plant starts to operate in the third year, at 50% production capacity, then it ramps up to 75% in the fourth year, 90% in the fifth, and finally 100% starting from the sixth year (year 5 in following tables and figures).

At this early stage, the capital investment can be estimated in different ways.

In 2017 Tsagkari et al. reviewed cost estimation methods for biomass conversion processes at class 4–5 AACE (Association for the Advancement of Cost Engineering) [27,28], pointing out that there are several methods for early stage estimation. In our case, we can rely on the Petley [29], and Lange methods [30]. The Petley method estimates the ISBL of the plant starting from the number of functional units, the capacity of the plant, the maximum pressure and temperature. In 2001, Lange proposed two capital cost estimation methods, one based on the plant energy losses, and one on the plant energy transfer. Both methods are appropriate for highly exothermic, or endothermic processes and start to be reliable for energy transfer/losses higher than 10 MW.

The original Petley's correlation (from 1988) updated to 2019, and relocated to France is:

$$ISBL\,(2019) = 55882 Q^{0.44} N^{0.486} T_{max}^{0.038} P_{max}^{-0.22} F_m^{0.341} \frac{CEPCI(2019)}{CEPCI(1988)} \cdot Fl \quad (4)$$

where $ISBL$ is the Inside Battery Limit investment, Q is the plant capacity in tons/y of scraps, N the number of process steps (4 here), T_{max} the maximum process temperature in K (723 K), P_{max} the maximum pressure in bar (1.5 bars), and F_m the material construction factor (1.5). The Chemical Engineering Plant Cost Index (CEPCI) updates the plant cost from 1988 to 2019, and the relocation factor (Fl) relocate it to France. In 1988 the CEPCI was 342.5, while in 2019 it was 607.5 [31,32]. The Peters *and* Timmerhaus handbook suggests that the Outside Battery Limit (OSBL) investment, is 25–40% of the ISBL [33], so we selected 40% to be conservative.

Lange correlated the ISBL with the energy losses in the plant, calculated as $LHV_{(feed+fuel)} - LHV_{(product)}$:

$$ISBL + OSBL\,(2019) = 3.0 \cdot (\text{energy losses [MW]})^{0.84} \cdot \frac{CEPCI(2019)}{CEPCI(1993)} \cdot Fl \quad (5)$$

Again, we updated the investment with the CEPCI of 1993 (343.5), and the relocation factor for France.

Alternatively, he correlated the ISBL with the energy transferred in within the process, as:

$$ISBL(2019) = 2.9 \cdot (\text{energy transfer [MW]})^{0.55} \cdot \frac{CEPCI(2019)}{CEPCI(1993)} \cdot Fl \quad (6)$$

A fourth method is by expert judgment adjusted on existing plant capital costs. We know that the PMMA recycling process does not differ that much from a plastic to oil pyrolysis plant, some pyrolysis-based plastic to fuel, or some existing plastic to plastic recycling plants. In the last 10 years, several plants suitable for comparison have been built, in Europe or US, and even more have been announced in the near future (Table 2). Reviewing such plants, along with process reviews from expert companies (e.g., IHS [34], or Nexant [35]), we gave a best educated guess on the capital cost of the plant.

In Figure 3, we report the capital cost versus capacity of some selected projects scaled to 2019 in France, as well as the estimated capital cost. For a 5000 tons/y plant, the two Lange methods reach their validity limit, (under)estimating a capex of 3.3 M USD, and 7.2 M USD respectively. The Petley method probably overestimates the capex at 18 M USD. Both methods are indeed "Class V" for process engineers, or technology readiness level "TRL 3–4" for chemists, and they carry significant uncertainties. For comparison, for similar plant sizes, announced investments for recycling plant were between 11 M USD, and 25 M USD for the first of a kind (FOAK) Plastic energy Plastic to Oil plant [36]. Obviously, FOAK plant are more expensive than n^{th} of a kind (NOAK) for the same technology/company, because of learning curve and technology maturity [37]. For instance, this is the case of Plastic Energy, Quantafuel, or QM Recycled Energy, in Table 2.

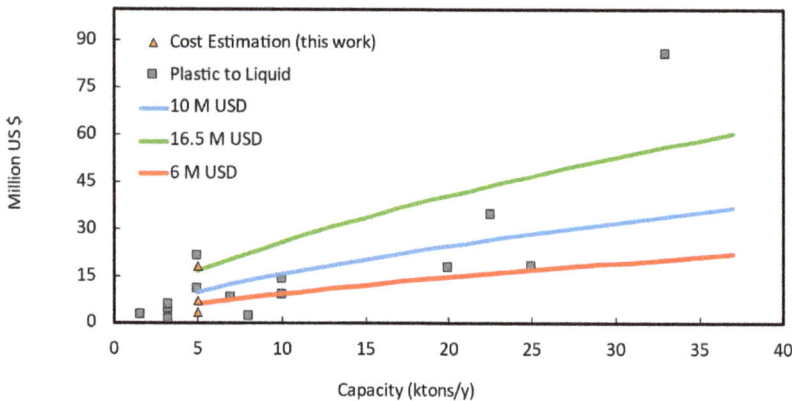

Figure 3. Scaled Investment (2019, France) versus original plant capacity. Historical data of plastic to liquid (oil or fuel) plants (grey squared marks), and early estimation methods (triangular marks) for a 5 kt/year scraps Mixed or Cast scraps PMMA depolymerization plant. The blue line represents the power-law with a reference investment of 10 M USD, while the green and the red with 16.5 M USD, and 6 M USD, respectively.

Table 2. Historical data of first of a kind (FOAK) or nth of a kind (NOAK) plastic to liquid plants.

Company	Capacity (ktons/year)	Investment (M US$)	Year	Project Type	Process Type
Cyclix [38]	33	80	2020	NOAK	Plastic to Oil
Plastic Energy [36,39]	5	20	2014/15	FOAK	Plastic to Oil
Plastic Energy [36]	20–25	35	2019	NOAK	Plastic to Oil
Renewlogy [40]	3.3	4	2018	NOAK	Plastic to Fuel
Renew Phoenix [41,42]	3.3	5.5	2019	NOAK	Plastic to Fuel
QM Recycled Energy [43]	10	9	2018	FOAK	Plastic to Oil
QM Recycled Energy/Biofabric [44]	1.6	2	2020	FOAK	Plastic to Oil
Quantafuel [45]	20	17.5	2018	FOAK	Plastic to Fuel
Quantafuel [46]	25	18	2020	NOAK	Plastic to Fuel
Quantafuel/Vitol [45,47]	100	75	2020	NOAK	Plastic to Fuel
Recycling Technologies [48]	7	7.7	2018	FOAK	Plastic to Oil
Carbo Hydro Transformation (seems to have now disappeared)	8	2	2018	FOAK	Plastic to Fuel

To select the most appropriate investment at this early stage, we extrapolated the data of (Figure 3) with the power law method, scaling everything to 5000 tons/y:

$$\frac{C_1}{C_2} = \left(\frac{S_1}{S_2}\right)^{0.65} \tag{7}$$

where C is the Cost (ISBL or OSBL, etc.) and S the plant size, the scaled 5000 tons/y would therefore be in the 6 to 16.5 M USD range. On the basis of different evaluation methods, we assumed a reference capital investment of 7.5 M USD, that translates into a median (p50) CAPEX of 10 M USD (p10 at −20% reference Capex (i.e., 6 M USD); and p90 at +120% (i.e., 16.5 M USD)—Figure 3), and which also includes working capital and start-up costs. When a plant starts, it always needs raw material/product stocks, and some budget allocated for unforeseen contingencies.

2.4. Monte-Carlo Simulation

To factor-in all the variables uncertainties at once, as opposed to inspect only one variable at the time (sensitivity analysis), we performed a probability risk analysis, also known as Monte-Carlo simulations. The way we described it so far, the NPV method is a deterministic method, meaning that we obtain a single output when plugging in a fixed input. However, if we input probabilistic distributions instead, we obtain an output with its own probabilistic distribution. In the Monte-Carlo simulation technique, each input has its own statistical distribution (normal, log-normal, gaussian, gamma, etc.,), from which the calculator selects one random value. From the random input series, the simulation calculates the output variable, for instance the NPV. The process is then repeated a high enough number of times so that the output becomes reliable, for instance in the order of thousands. Eventually, out of the frequency of recurring of each outcome, we can calculate the probability of happening of a certain outcome (e.g., NPV on a certain year). The Perry's Handbook, or our previous paper on Risk assessment using Monte-Carlo simulation is a good starting reference on this technique [25,49,50].

In layman's terms, the risk assessment analysis consists in:
- Based on production price index-adjusted historical data, set a statistical distribution for the future price of raw materials (PMMA, natural gas), waste (solid wastes and heavy ends), and product (r-MMA);
- Based on the combination of cost estimation techniques, plant historical investments, future announced investments, and expert judgement, set a statistical distribution for the capital cost;
- Based on the inflation adjusted historical prices, find the historical correlation between raw materials, waste, and product price;
- Based on the expert judgment, give the best educated guess on the raw materials, waste, and product future prices correlation;
- For each input (capital cost, raw materials, waste, and product prices) adjust the random variables for the correlation matrix. In this way, the inputs do not vary independently, but we force their variation according to the correlation matrix;
- Determine the profitability (i.e., NPV, production cost);
- Repeat the last steps 3000 times;
- Evaluate the probability of each occurrence to happen, and look at the median probability, the p10 and p90 (10 % and 90 % probability) for the variables of interest.

The Monte-Carlo simulation copes with the uncertainties on raw material price, waste price, product price, and investment. This approach is very appropriate, as a risk analysis, in this very early stage evaluation of processes, to identify the most viable scenario leading to a successful project. We kept all the other values as a fixed deterministic variable (labor cost, other utilities cost, depreciation, taxes, all the other fixed costs).

2.4.1. Raw Materials

The main raw materials are the PMMA scraps, and the natural gas to compensate for the extra heat needed.

The MMAtwo consortium estimates that before the Chinese trade ban, Europe was exporting between 15 000 tons/y and 30 000 tons/y to India and China [10], with around 8 000 traded within Europe [9,10]. Hence, we collected some market prices for PMMA scraps for the 2013–2016 period in the major ports of India (Figure 4) from Zauba [51], and Eximpulse [52]. What we see here are relatively good quality scraps, most likely cast materials, traded as "Acrylic scraps HS39159030". Lower quality scraps are difficult to trace, also because they might be sourced from the local market. For this reason, we assumed two different statistical distributions for the PMMA scraps (Figure 5):

- The **"Cast" PMMA scraps**, more expensive because they are a higher purity kind of scrap; and
- The **"Mixed" PMMA scraps**, a lower quality, cheaper feedstock.

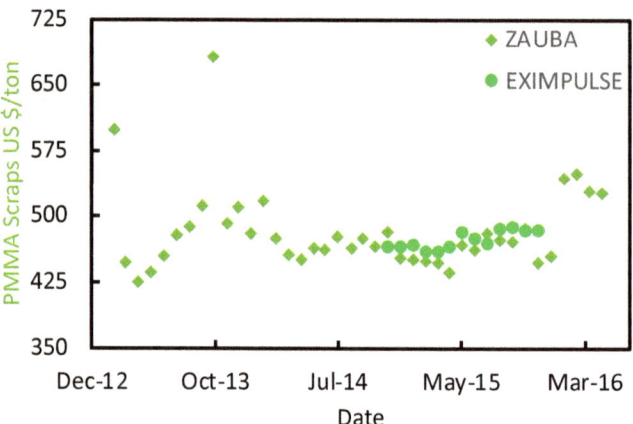

Figure 4. PMMA scraps import prices in India between 2013 and 2016, from Zauba [51], and Eximpulse [52].

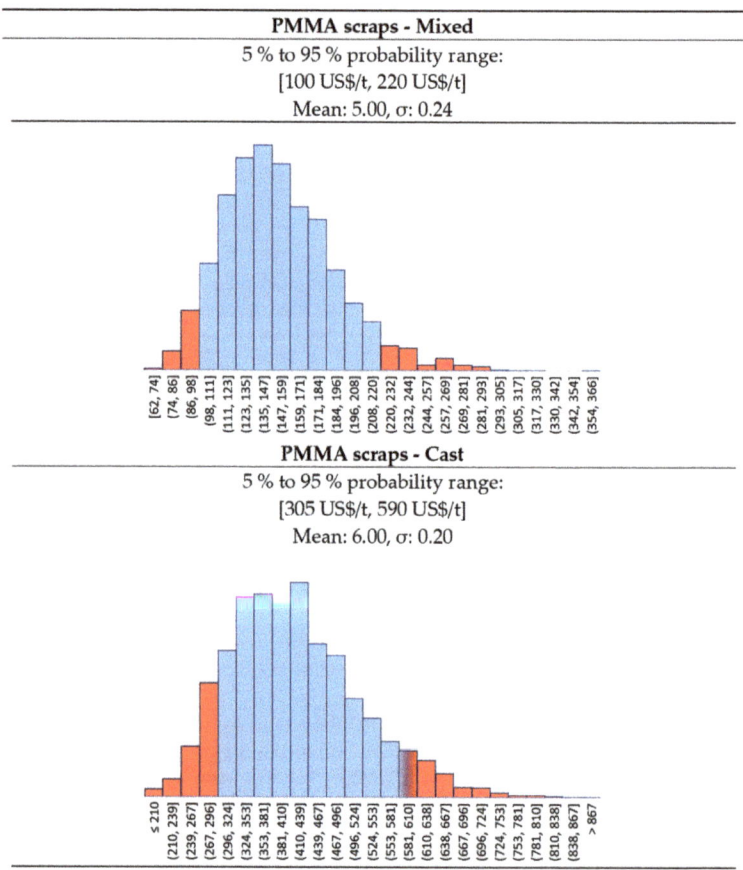

Figure 5. Log-Normal distribution for Mixed and Cast PMMA scraps prices.

In both cases, we assumed a log-normal distribution for the future price of the scraps. The "Cast" scraps have a {p10, p90} interval in between {305, 590} USD/t, while for the "Mixed" we selected {100, 220} USD/t. There is also a **third**, extreme **scenario**, where the scraps quality is so low, that they become available at zero price.

For the European Natural gas, we collected the monthly price in USD between 2000 and 2020 from FRED (Federal Reserve Economic Data) [53] (Figure 6). Then, we adjusted for inflation, and we found out that the prices can be fitted best with a log-normal distribution.

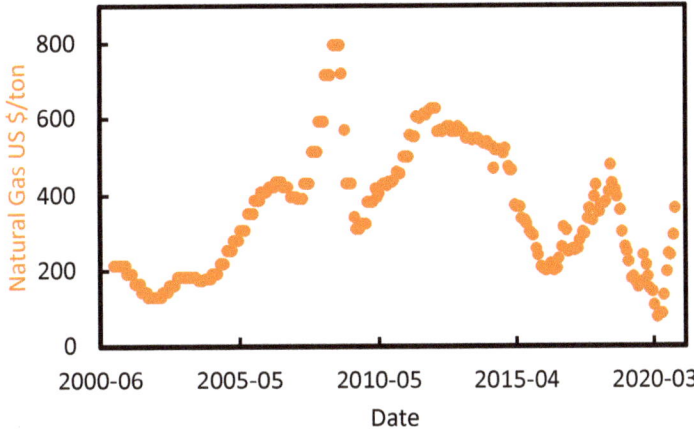

Figure 6. Historical natural gas prices in EU from [53].

We do not see any reason why natural gas should change its statistical distribution over time in the near future, nor we think that its value should deviate much from last years' trend. Therefore, we assumed a log-normal distribution for the natural gas price, with a {p10, p90} interval of {190, 705} USD/t, with the mean lying around 300 USD/t (Figure 7).

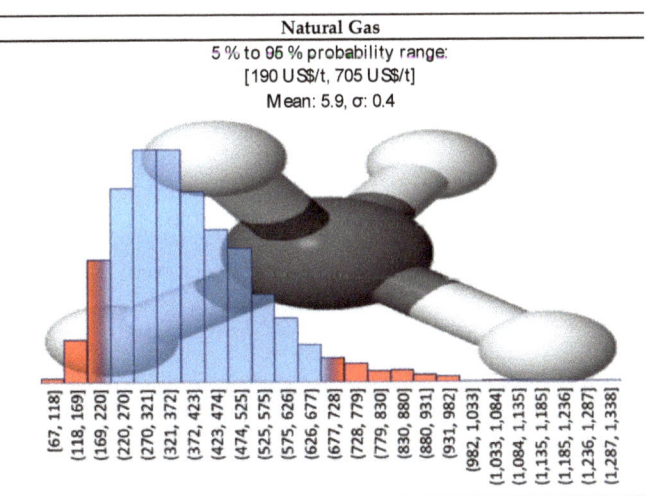

Figure 7. Log-normal distribution for natural gas prices.

2.4.2. Products—r-MMA

The recycled MMA (r-MMA) price depends strongly on the product quality. PMMA producers aim at the very best monomer quality that satisfies their needs, which could be up to that of the virgin material (>99.8% wt. purity [54]). At the same time, the western world, and in particularly Europe, is pushing toward sustainable development, and promoting circular economy models. For instance, Europe recently updated its circular economy action plan (March 2020 [55]), toughening the policy on waste export outside Europe, and it imposed a contribution of about 1000 USD/t for non-recycled plastic packaging to curb plastic waste [56]. In this framework, we foresee how recycled MMA will play an even bigger role hereafter. We collected the virgin MMA spot historical prices in EU (Figure 8), and we adjusted them for inflation with the PPI (producers price indices) index. The PPI index, available online on the organization for economic co-operation and development (OECD), is an "advanced indicator of price changes throughout the economy" that measures the average price change over time in selling price from the sellers' perspective [57]. In this way, the only factor playing a role in the historical prices' trends is geo/political.

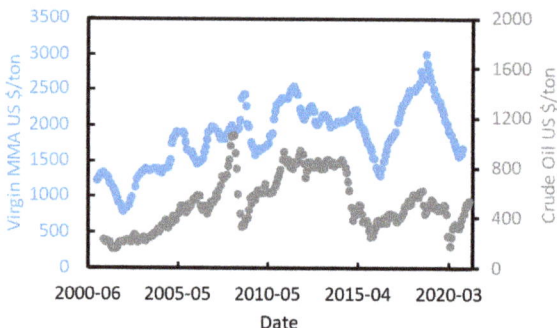

Figure 8. Virgin MMA spot prices from [8,58,59], and crude oil [60].

Virgin MMA price is highly correlated to crude oil price (Figure 8), and we believe that this will not change in the near future.

We also collected some prices for exported "regenerated black MMA" HS 29161400 in the ports of India between 2013 and 2016 [61] (Figure 9), a period during which crude oil prices varied a lot. All shipments were destined to Asia (e.g., China, Bangladesh, or Pakistan), for an average tonnage of 400 ton/y. In the period of interest, this lower quality crude MMA product was selling at an average of 30% discount over the virgin material.

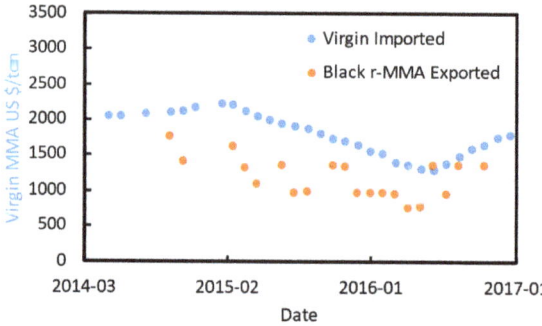

Figure 9. Crude exported MMA price [61], and virgin imported spot price between 2014 and 2017 in India.

Again, the best fit for the PPI adjusted historical prices of virgin MMA was the lognormal distribution. For our future prediction, we imagined different r-MMA prices, for three different product's qualities, each with a log-normal distribution (Figure 10):

- **Virgin-like MMA** price;
- **20% discount MMA** price. In western markets it will be very unlikely to see very low-quality products. Requests for r-MMA will increase due to stricter policies and increasing environmental awareness of the final customers. Therefore, we expect that recycling companies and PMMA producers will settle midway for a fairly good enough "low-quality" r-MMA;
- **20% premium MMA price**. Either for increasing requests, superior quality, or market fluctuations, we believe that a scenario where PMMA (or other MMA derivatives) producers will pay a premium for their r-MMA is plausible. It could also correspond to subsidies given to recycled products based on their quality.

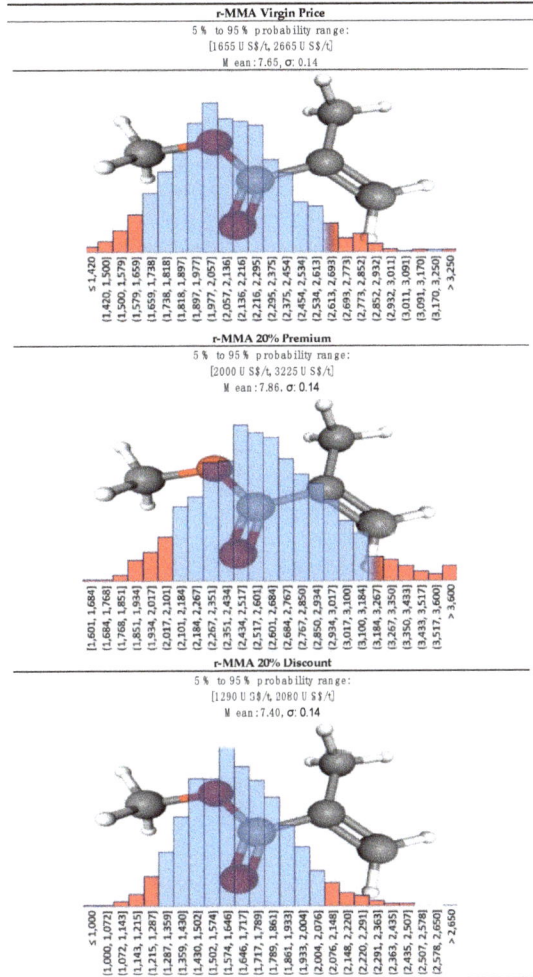

Figure 10. Log-normal distribution for r-MMA at virgin price, 20% premium, and 20% discount.

2.4.3. Waste—Heavy Ends and Solid Waste

The depolymerization step produces char and dust as cracking by-product, while the purification train separates the heavy ends from the recycled monomer. Carbonaceous waste from the depolymerization is collected and either sold as charcoal to heavy industries (e.g., steel), or landfilled. Heavy ends can be either burnt on site for heat recovery, or if their quality as combustible is too low, they must be disposed of. In our case, we assume to remove the very heavy ends in the evaporator V-3001 (Figure 1), in form of a syrup-like fluid. Then, the distillation fixes the purification and removes some slightly lighter heavy cuts that are burnt on site for energy. In the mass balance (Table 1), we assumed the very heavy ends to leave the plant as by-product. In the economic assessment, this oil could either be sold as a heating oil equivalent or disposed of. To be conservative, we assumed that there is no market for these very heavy ends, and they must be disposed of.

Same applies for the solid wastes. Because of the increasing public awareness, landfill taxes and gate fees have been swiftly increased in Europe in the last 20 years, as shown for instance by ADEME (French government agency) in a comparative report of 2017 [62]. As a reference, Finland had a landfill tax of 25 USD/ton in 2003 [63], which rose to 85 USD/ton in 2020 [64]. As per 2020, 23 EU members have a landfill tax, that varies from 10 (Latvia) to 150 USD/ton (Belgium) [65], as well as Switzerland and UK (150 USD/ton) [64].

To this tax, we still need to add the gate (or tipping) fee, at the entrance of the landfill. Gate fees depends again on the country, but they roughly are comparable to landfill taxes [64]. We believe that in the near future, landfill taxes will keep increasing in EU, leaning toward the highest rates in the continent (i.e., central and northern Europe countries such as Belgium, Luxemburg, Denmark, Sweden, etc.,).

For both heavy ends and solid wastes, we assumed a gaussian distribution. For solid wastes we assumed a median cost (landfill tax + gate fee) of 250 USD/t and a deviation of 50 USD/t, and for the heavy ends a median cost of 150 USD/t (Figure 11) and a deviation of 50 USD/t. In this way, we assume to treat the plant by-products according to the highest European standards in terms of waste disposal, as opposed of what happens in existing lower quality PMMA recycling plant in developing countries (e.g., India, Brazil, etc.,).

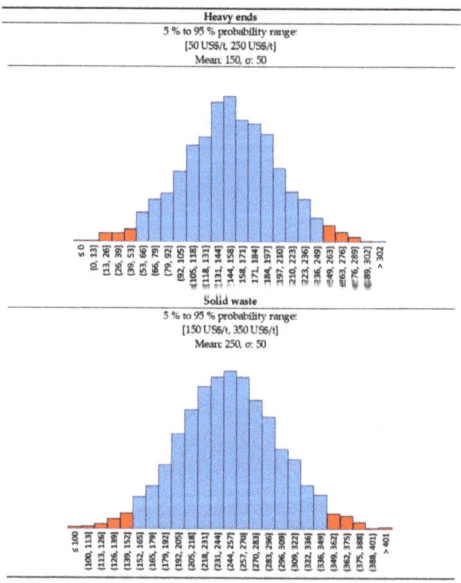

Figure 11. Solid waste and heavy ends statistical distribution.

2.4.4. Investment

In a study of 2012, J.K. Hollmann queried over 1000 projects in different industries, and demonstrated how the p10/p90 accuracy range of the capital investment, is close to −20%/+120%, in a log-normal statistical fashion [66]. Thus, in our Monte Carlo simulation, we assumed a log-normal distribution for the capital investment, where p10 corresponds to −20% of the capital cost, and p90 to the +120%.

For the analysis, we selected four conditions:

- **Full CAPEX at 10 M USD** (Figure 12), for a higher capital-intensive plant. This is the case when we have lower quality scraps (MIXED or 0 price scraps);
- **30% subsidies on Full CAPEX** for a final CAPEX of **7 M USD** (Figure 13). It is again a higher capital-intensive plant treating lower quality scraps. Local authorities are more willing to provide subsidies when they see the benefit of the money they spent. In this case, lower quality scraps might need higher cost at pre-treatment or purification, and subsidies could cover that;
- **Full CAPEX at 7 M USD** (Figure 13), for a slightly lower capital-intensive plant. We imagine this to be the case when treating only Cast scraps. The feedstock is relatively "clean", and a cheaper plant is expected; and
- **Full Capex at 1.3 M USD.** This is the case of a cheap technology, or a plant made of second-hand equipment. In both cases, the lower capital expenditure is likely to be accompanied by a shorter plant life (e.g., 10 years instead of 20), an overall lower quality of the final product, and so a lower price.

Figure 12. Capital cost distribution for 5000 tons/year PMMA scraps plant, at 10 M USD CAPEX—Capital Cost −20%/+30% (Green Area) corresponds to Class 3 confidence interval of the Association for the Advancement of Cost Engineering (AACE) [28].

Figure 13. Capital cost distribution for 5000 tons/year PMMA scraps plant, at 7 M USD CAPEX—Capital cost −20%/+ 30 % (Green Area) corresponds to Class 3 confidence interval of the Association for the Advancement of Cost Engineering (AACE) [28].

3. Results

3.1. Scenarios

To study the challenges and possibilities of the recycling industry at large, we compared the profitability for different scenarios (Table 3). The scenarios differ for kind of feedstock (Mixed or Cast), final product quality and selling price (20% discount, virgin or 20% premium price), or CAPEX. In Table 3, we summarize the scenarios, and the prices distribution for raw materials, waste/by-products, and products. We ended up outlining nine different scenarios (Table 3):

- **Scenario 1**—The plant treats mixed scraps, with a full Capex at 10 M USD, and the r-MMA sells at virgin price. This is our base case. Processing mixed scraps needs a relatively higher capital-intensive plant to reach an acceptable purity for the EU market.
- **Scenario 2**—Like Scenario 1, but the project obtains external funding as grants or subsidies (30% is an accessible value for an European plant);
- **Scenario 3**—Like Scenario 1, but the r-MMA purity requested is higher, and it sells with a premium of 20%. This is more representative of what could happen in the long-term;
- **Scenarios 4** and **8**—The plant treats mixed scraps, with a second-hand, or a cheap process technology (e.g., dry distillation). Therefore, the r-MMA produced is of a lower quality and sells at 20% discount compared to the virgin. In Scenario 4, the plant life is still 20 years, while in Scenario 8 the plant is replaced after 10 years.
- **Scenarios 5** and **6**—The plant processes Cast scraps, more expensive than the mixed, but requiring less/cheaper equipment. The r-MMA sells at virgin or premium price, in Scenarios 5 and 6, respectively;
- **Scenarios 7** and **9**—PMMA scraps are very low quality, and therefore available for free. However, the plant still needs a relatively high investment (10 M USD) to purify the MMA to a grade pure enough to at least enter the market. Nonetheless, r-MMA sells at a 20% discount price. This scenario is representative of very polluted end-

of-life streams, where sorting is difficult due to poor infrastructure. For instance, it could apply to the automotive or the construction industry; Scenario 7 is at full capex (10 M USD), while Scenario 9 is with 30% subsidies. Because the project repurposes waste that would otherwise inevitably end in landfill, local bodies grant subsidies for environmental reasons rather than economic reasons only. Our guess is that for the most part, these two scenarios are probable to be transient. The public opinion calls for more sustainable materials, and the government agencies legislate accordingly. Hence, the collecting and sorting infrastructure will eventually keep up with the demands for scraps. When that will happen, PMMA producers in EU will not easily settle for a poor-quality monomer.

Table 3. Scenarios and prices overview.

Scenario	1	2	3	4	5	6	7	8	9
Scraps	Mixed	Mixed	Mixed	Mixed	Cast	Cast	Mixed 0 price	Mixed	Mixed 0 price
r-MMA (price)	Virgin	Virgin	20% premium	20% discount	Virgin	20% premium	20% discount	20% discount	20% discount
Capital Investment	10 M $	7 M $ (30% subsidies)	10 M $	1.3 M $ 20 years lifetime	7 M $	7 M $	10 M $	1.3 M 10 years lifetime	7 M $ (30% subsidies)

Feedstock/Product/Waste	Distribution Law	Mean, Deviation	[5%, 95%] Probability ($/ton)
PMMA Mixed	Log-Normal	[5.00, 0.24]	[100, 220]
PMMA Cast	Log-Normal	[6.00, 0.20]	[305, 590]
Natural Gas	Log-Normal	[5.90, 0.40]	[190, 705]
Heavy Ends	Gaussian	[150, 50]	[70, 240]
Solid Waste	Gaussian	[250, 50]	[165, 300]
r-MMA 20% discount	Log-Normal	[7.40, 0.14]	[1290, 2080]
r-MMA Virgin	Log-Normal	[7.65, 0.14]	[1655, 2670]
r-MMA 20% premium	Log-Normal	[7.84, 0.14]	[2000, 3225]

3.2. Correlation Matrix

In 2019, the OECD (Organization for Economic Co-operation and Development) demonstrates that, with few exceptions, waste generation "is still very much linked to economic growth" [67]. Therefore, although prices for feedstock and products/waste vary with time, such variations are not completely random. When the economy is booming, people get better salaries, eat more (or better food), and their purchasing power increases. The downside is that energy and food prices increase too, as well as waste production, especially plastic waste [68]. For instance, Germany has a GDP per capita of around 40 000 USD, and produces 0.5 kg of plastic waste per capita, same applies for the Netherlands (Figure 14). On the contrary, Albania has a GDP per capita of 10 000 USD, and produces less than 0.1 kg plastic per person (Figure 14).

Figure 14. Per capita plastic waste vs. GDP per capita (2017), measured in 2011 international USD—From Our World in Data [68]—World Bank [69] and Jambeck et al. [70].

Moreover, in our case, PMMA scraps price is highly correlated to r-MMA price, and r-MMA price is correlated to price of energy (Natural Gas for us). Same applies to the price of waste disposal that are expected to be linked. To be conservative, we decided to consider the heavy ends as a waste. However, we know that its properties are like those of a heating oil [71], so historically it has been correlated to natural gas, as well as MMA. For PMMA scraps very scarce historical data are available, or in a very narrow timespan (e.g., Zauba imports 2013–2017). Based on the historical data (if any), and our expert judgment, we represented our vision of the future in a correlation matrix, that mimic prices trends fluctuations at the best of our knowledge (Table 4). Once correlated, the prices for the 3000 Monte-Carlo cases appear as in Figure 15.

Table 4. Proposed correlation matrix for the future, based on historical data and expert judgement. Green stands for highly correlated, while red for lowly correlated.

	PMMA Scraps	Natural Gas	Heavy Ends	Solid Waste	r-MMA
PMMA Scraps	1				
Natural Gas	0.2	1			
Heavy ends	0.05	0.5	1		
Solid Waste	0.05	0.05	0.8	1	
r-MMA	0.8	0.4	0.4	0.05	1

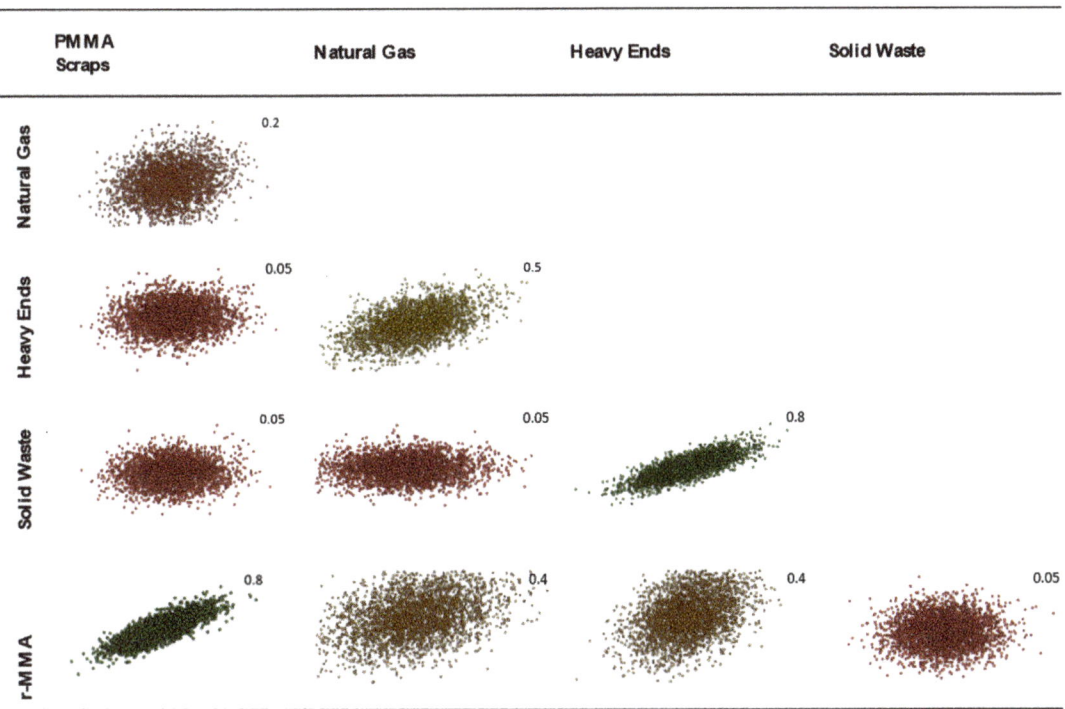

Figure 15. Visualization of the correlated prices of raw materials and products based on our vision of the future.

3.3. Profitability Analysis by Monte Carlo Simulation

Under our assumptions, the base case (Scenario 1), where the plant is built without subsidies, treating 5000 tons/y of mixed scraps to produce 3889 tons/y of r-MMA selling at virgin price is not very appealing. Over 3000 simulations, the project has a (median) payback time of 8 years (Figure 16) with 30% probability to make no money after 10 years of production (Figure 17). To understand where assumptions improvements would be most effective, we studied the impact of the individual uncertainties of feedstock, waste, product, and investment, on the median NPV. In Scenario 1, when the parameters vary between their 10% and 90% probability of their statistical distribution, the NPV changes accordingly to the impact that the single uncertainty has on it. The r-MMA price has the largest influence on the NPV, comparable to that of the investment cost (Figure 18). This means that improving the assumptions on the statistical distribution of either, should improve the uncertainties on the NPV, and lower the risk of losing money. For instance, if the investment uncertainties reduce as low as the equivalent of a Class 3 AACE (extreme values of the tornado at p29 and p75), the investment has an impact on the NPV comparable to that of PMMA scraps (Figure 19). To reach the confidence of a AACE class 3, we need a well identified list of equipment. After class 3, the improvement is marginal (at this stage) (Figure 19). In the Tornado graph (Figure 18), the investment varies in between −20% and +120% (p10, p90), of the selected investment (7.5 M USD), whose median is 10 M USD. Instead, the tornado capex plot (Figure 19) varies from the median. For instance, Class 5 in Figure 19 corresponds to a [p10, p90] of −50% and + 100% of the median, while the [p10, p90] in Figure 18 are the 10th and 90th percentile of that log-normal distribution having 10 M USD as median. The Tornado graph was generated to illustrate the impact of the definition of the project (Capex) on the dispersion of the NPV values.

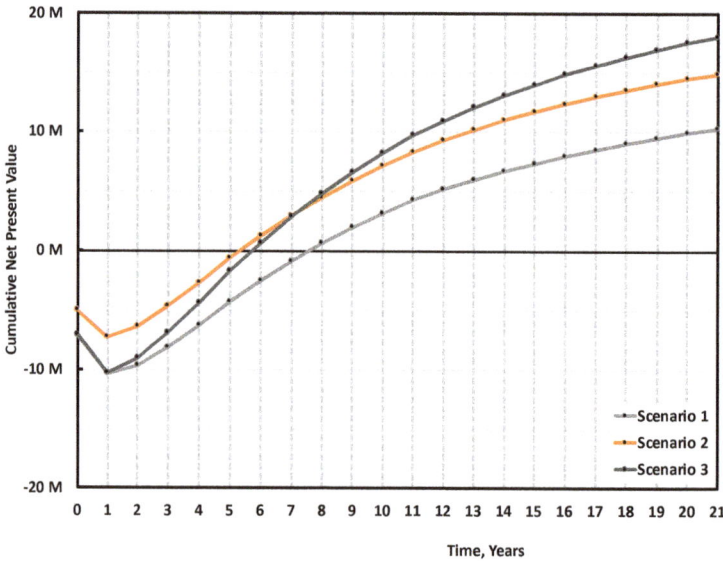

Figure 16. Median cumulative cash flow (NPV) over 20 years of production for 5000 tons/y plant, Scenario 1, 2, and 3 (Table 3). Scenario 1 is the base case, mixed scraps, Full 10 M USD Capex, and r-MMA sold at virgin price. Scenario 2 is when such a project gets 30% subsidies on the Capex, while Scenario 3 is when the r-MMA sells at 20% premium price.

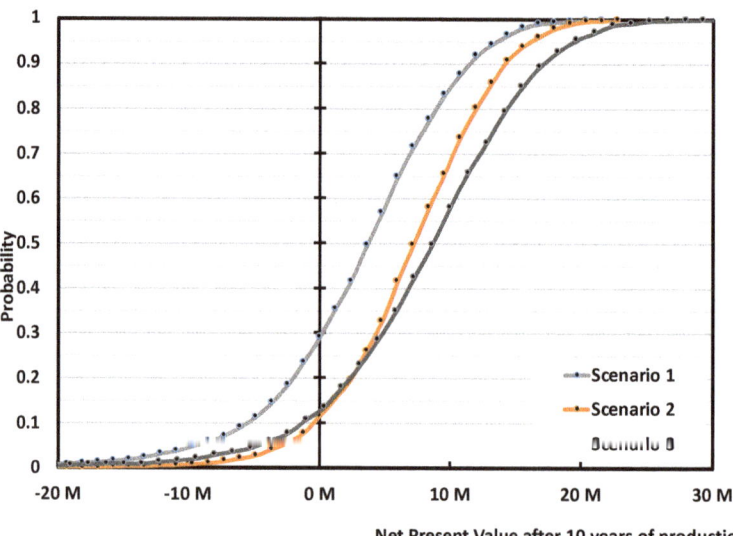

Figure 17. Probability vs Cumulative NPV over 10 years of production 5 000 tons/y plant. Scenario 1, 3 and 3. Scenario 1 is the base case, mixed scraps, Full 10 M USD Capex, and r-MMA sold at virgin price. Scenario 2 is when such a project gets 30% subsidies on the Capex, while Scenario 3 is when the r-MMA sells at 20% premium price.

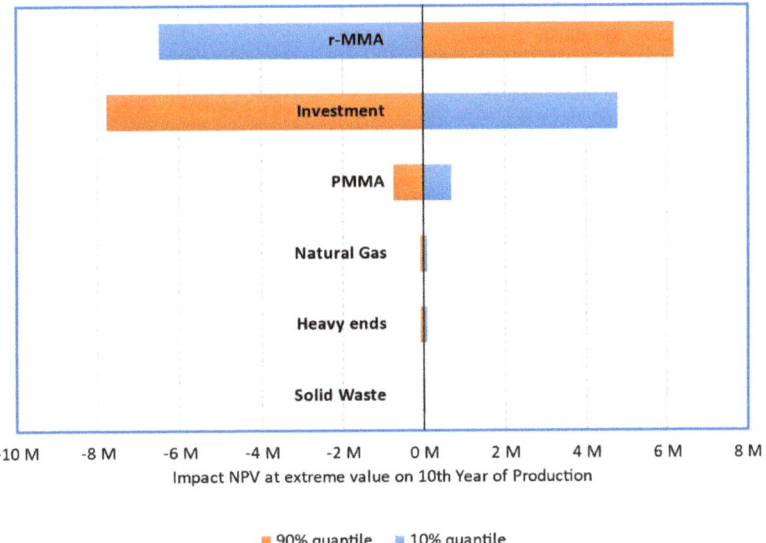

Figure 18. Tornado Plot for 5000 tons/y plant, **Scenario 1**. Effect of Raw-materials, waste, products, and investment variation (10% and 90% quantile), on the median NPV.

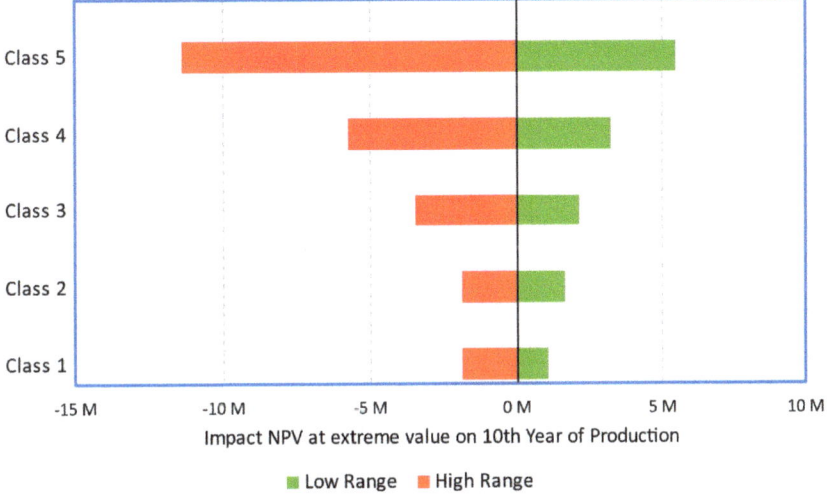

Figure 19. Scenario 1, impact of Capex class on the median NPV. Class 5 AACE corresponds to extremes at p4–p96 (or −50%, +100% of the median); Class 4 AACE corresponds to extremes at p85–p19 (or −30%, +50% of the median); Class 3 AACE corresponds to extremes at p29–p75 (or −20%, +30% of the median); Class 2 AACE corresponds to extremes at p34–p64 (or −15%, +15% of the median); Class 1 AACE corresponds to extremes at p40–p64 (or −10%, +15% of the median).

When Scenario 1 gets either 30% subsidies on the Capex (Scenario 2), or sells r-MMA at 20% premium price (Scenario 3), the project becomes much more viable (Figure 16). Both solutions reduce the median payback period to 5/6 years; while the 20% premium solution (Scenario 3) becomes more profitable on the long-term (Figure 16). Both Scenarios (2 and 3) are equally attractive, but they still have a 10–15% probability of not earning any money after 10 years of production (i.e., probability at 0 NPV) (Figure 17).

In the current political climate, we cannot really give an excessive premium on the r-MMA price, but we do have a good case to present to local authorities when asking for subsidies. Scenarios 4 and 8 are representative of a second-hand plant, or a very cheap technology (Capex 1.3 M USD), that treats mixed scraps to produce low-quality r-MMA selling at 20% discount price. When operating a low capital-intensive plant, provided that the r-MMA quality is still acceptable to enter the market, we cover the plant cost after a little more than 1 year of production (Figure 20). The fast payback time makes this a low-risk alternative. However, with time, the higher r-MMA selling price overcomes the effect that a low depreciation has on the expenses. At the end of the plant life Scenario 2 has the same NPV as Scenario 4. Moreover, if we expect to replace the second-hand equipment/low-quality plant after 10 years, eventually Scenario 8 has a lower NPV than Scenario 2, almost comparable to that of Scenario 1. With the difference that while Scenario 1 has still room for improvement (subsidies, higher r-MMA price due to EU policies), Scenario 8 has limited growing potential. This suggests that public or private investors should look at solid technologies, which are able to provide a good quality product, rather than very cheap plants, even if the former are more expensive up-front.

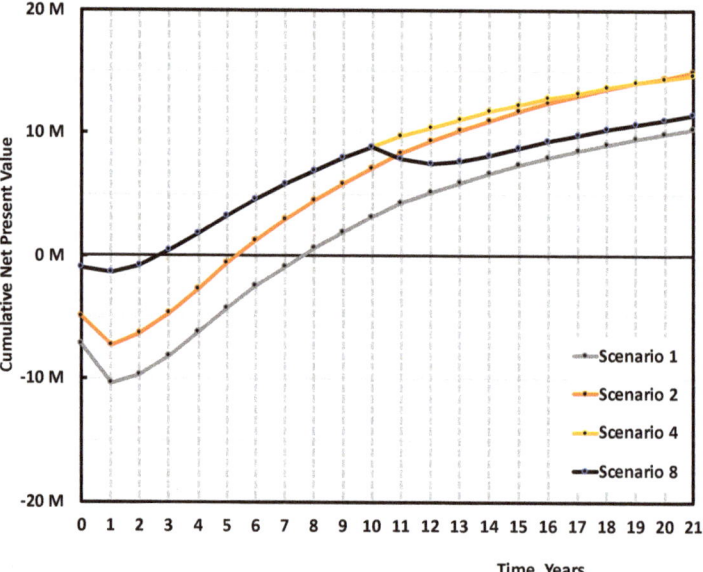

Figure 20. Median cumulative Cash Flow (NPV) over 20 years of production for 5000 tons/y plant, Scenario 1, 2 and 4, 8 (Table 3). Scenario 1 is the base case, mixed scraps, Full 10 M USD Capex, and r-MMA sold at virgin price. Scenario 2 is when such a project gets 30% subsidies on the Capex. Scenarios 4 and 8 are the case of a low-quality plant (1.3 M USD Capex), treating mixed scraps and selling r-MMA at 20% discount price. In Scenario 4 we operate the plant for the full 21 years, while in Scenario 8 we must replace it after 10 years.

When comparing Cast and Mixed scraps, Cast scraps makes for an overall better case. Cast scraps are more expensive, but they can be used in a cheaper plant, and have less carbon losses along the process, making for a better r-MMA global yield. We compared the base case (Scenario 1), with the case of a 5000 tons/y cast scrap 7 M USD plant, producing 4512 tons/y of r-MMA selling at virgin price (Scenario 5), or 20% premium price (Scenario 6) (Figure 21). At full investment cost, selling at virgin price, the mixed scrap plant (Scenario 1), falls short compared to its Cast equivalent (Scenario 5). Over the plant life, the cumulative NPV plot slope is the same in these two cases (Scenario 1 and 5). Namely, the better r-MMA cast global yield compensates for the higher feedstock price, with respect to the Mixed scrap case. Similarly, Scenario 5 and Scenario 2, i.e., the case where we get 30% subsidies on the mixed scrap plant, have NPVs which basically overlap. This means that the investment plays here a fundamental role. At parity of investment and selling price, the economic viability is not influenced by the kind of scraps. However, mixed scraps consist in post-production, but end-of life PMMA as well. Should PMMA producers be willing to pay a premium for r-MMA from end-of-life products only (the only ones that have made a full cycle in the economy), for instance to better advertise their product, the mixed plant would have a more positive case than the cast (Scenario 3 vs 5), even at a higher Capex.

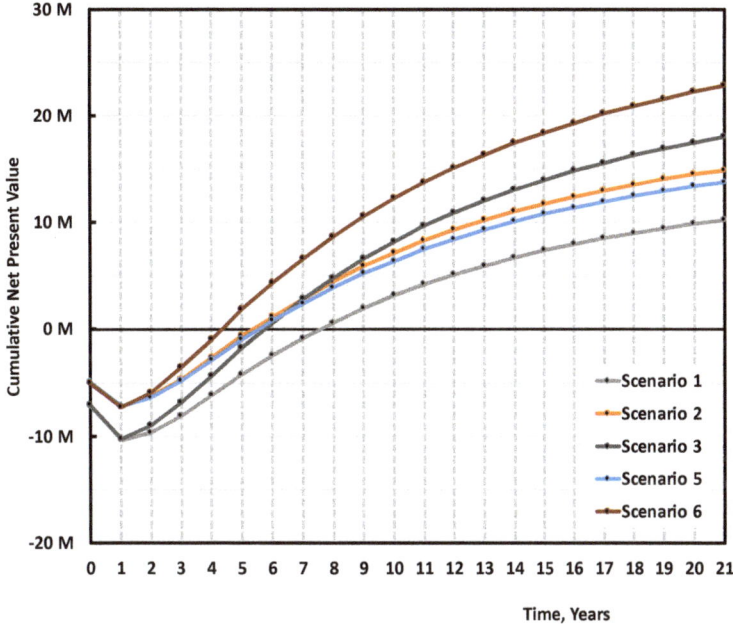

Figure 21. Median cumulative NPV over 20 years of production for 5000 tons/y plant, Scenario 1, 2, 3 and 5, 6 (Table 3). Scenario 1 is the base case, mixed scraps, Full 10 M USD Capex, and r-MMA sold at virgin price. Scenario 2 is when such a project gets 30% subsidies on the Capex, and Scenario 3 when r-MMA is sold at 20% premium price. Scenarios 5 and 6 are the case of a Cast scrap plant, that sells at either virgin (Scenario 5), or 20% price (Scenario 6).

The overall superiority of the cast over the mixed scenarios appears also on the NPV probability curve at the 10th year of production (Figure 22). Scenarios 5 and 6 have a probability of making zero money of 15% and 5% respectively, while Scenario 1 lost money around 30% of the times. Again, Scenarios 2 and 3 are equivalent to Scenario 5 in terms of cumulative NPV equal to zero at year 12 (10th year of production).

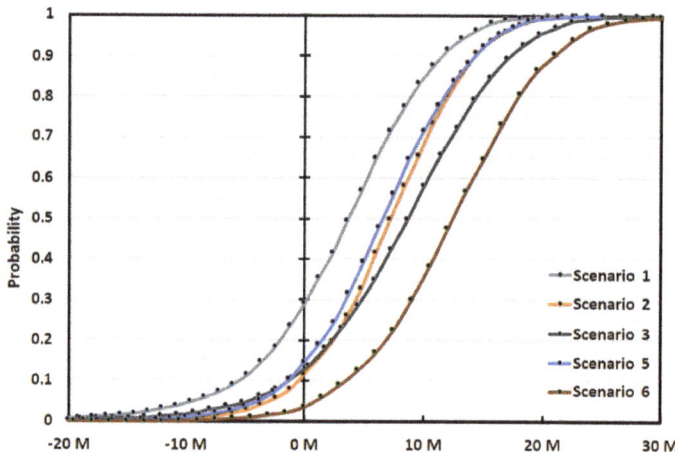

Figure 22. Probability vs Cumulative NPV over 10 years of production 5000 tons/y plant, Scenario 1, 2, 3 and 5, 6 (Table 3). Scenario 1 is the base case, mixed scraps, Full 10 M USD Capex, and r-MMA sold at virgin price. Scenario 2 is when such a project gets 30% subsidies on the Capex, and Scenario 3 when r-MMA is sold at 20% premium price. Scenarios 5 and 6 are the case of a Cast scrap plant, that sells at either virgin (Scenario 5), or 20% price (Scenario 6).

Nowadays, we see more and more techno-economic assessment on plastic recycling setting the feedstock price at zero [72], or even at a negative value [73]. PMMA is a relatively expensive plastic, compared to higher volume materials like PP, PVC, or PE. Therefore, as demonstrated by the few available historical import data, PMMA scraps have a market, nonetheless. However, the socio-political climate around the plastic recycling world is mercurial. Therefore, we wanted to investigate what would happen if our scraps were of such a low quality (e.g., end-of-life not completely sorted) to be available free of charge.

In this case, we can expect to either pay an extra for the pre-treatment (e.g., additional sorting, crushing, washing, and drying), lose in global yield to reach the target quality, or accept a lower quality monomer but with the same global r-MMA yield. We decided for this last option, and assessed the economics of a mixed scrap plant, treating unsorted scraps at 0 cost, selling low quality (e.g., 97–98 wt.% pure or less) r-MMA at 20% discount. We assumed such a plant to require the same capital expenditure of our base case (Scenario 1 at 10 M USD), and we evaluated it with (Scenario 9) and without (Scenario 7) subsidies (Figure 23). The raw materials contribution to the total expenses is now (Scenarios 7 and 9) much lower. In the global model, the other utilities are expressed in terms of a percentage of raw materials cost. In the case of the scrap at 0 price, with the mass yield of the mixed feedstock, we expressed the utility cost as a fixed value, equal to that of the base case.

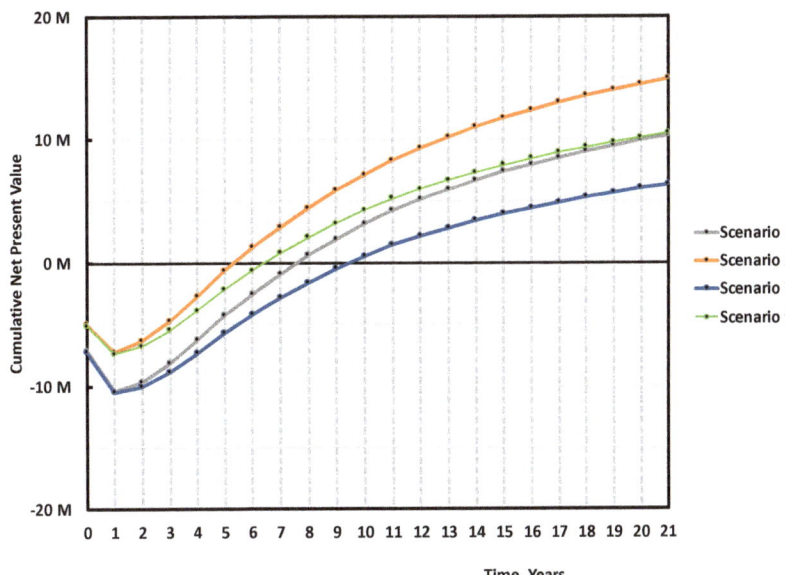

Figure 23. Median cumulative NPV over 20 years of production for 5000 tons/y plant, Scenario 1, 2, and 7, 9 (Table 3). Scenario 1 is the base case, mixed scraps, Full 10 M USD Capex, and r-MMA sold at virgin price. Scenario 2 is when such a project gets 30% subsidies on the Capex. Scenarios 7 and 9 are the case of a mixed scrap plant, with (Scenario 9) or without (Scenario 7) 30% subsidies, processing 0 cost low, quality PMMA, and selling r-MMA at 20% discount price.

Once more, the r-MMA quality differentiates between positive and negative cases. At first glance, scraps at zero cost may seem more attractive, but it is not correct. Scenario 7, the equivalent of Scenario 1, has a 40% probability of nullifying the NPV at the tenth year of production, as opposed to 30% in Scenario 1 (Figure 24). Besides, Scenario 7 cumulative NPV turns positive only after 9/10 years, 2 years later than Scenario 1 (Figure 23). The subsidized Scenarios 2 and 9, follow the same trend, with Scenario 2 being more profitable (Figures 23 and 24). Even if, for environmental reasons, public bodies granted funds to the 0 scrap price plant only, this case (Scenario 9) gains the same profits (in median) than Scenario 1 at the end of the plant life, with the risk that a zero cost feedstock may see an increase of its price when alternatives develop (such as fuel users). To target high-quality product is just overall more viable.

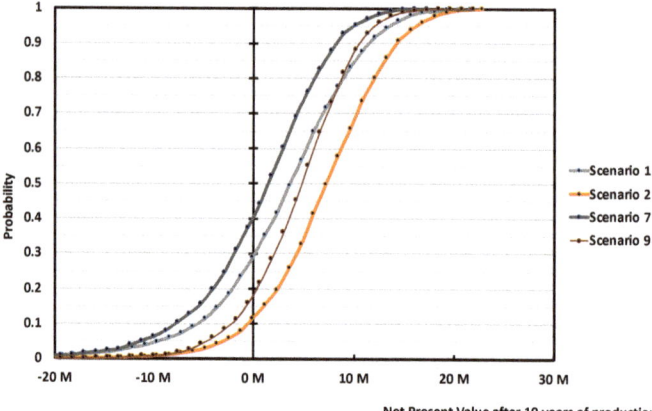

Figure 24. Probability vs Cumulative NPV over 10 years of production 5 000 tons/y plant, Scenario 1, 2, and 7, 9 (Table 3). Scenario 1 is the base case, mixed scraps, Full 10 M USD Capex, and r-MMA sold at virgin price. Scenario 2 is when such a project gets 30% subsidies on the Capex. Scenarios 7 and 9 are the case of a mixed scrap plant, with (Scenario 9) or without (Scenario 7) 30% subsidies, processing 0 cost, low quality PMMA, and selling r-MMA at 20% discount price.

Ideally, the best process technology can process all the scraps grade, yielding to the top quality r-MMA product, regardless the nature of the scrap itself. In the MMAtwo experience we found out how, on some extent, all the technologies tend to be feedstock sensitive (some more than others). These results suggest that any new player in the PMMA recycling field must demonstrate the maturity of their process, to produce high-quality r-MMA. Getting low quality scraps as cheap as possible, is not only difficult, but can be negative for the profitability of the plant. The only case where this could make sense, is if these scraps come with an important negative price.

3.4. Cost of Production

When the input variables are probabilistic functions, the cost of production is a probabilistic function as well. For all the cases, the cost of production before depreciation distribution fits well a normal, or log-normal distribution. For instance (Figure 25), Scenario 1 has a cost of production before depreciation between 784 and 1176 USD/ton for the interval of probability p5–p95 (green interval—Figure 25). Labor cost, the "other fixed", and "other variable" costs are the most important contributors to the cost of production (Figure 26).

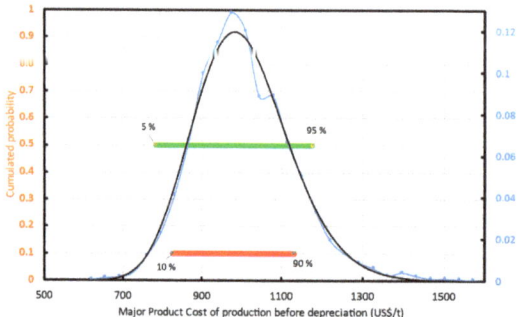

Figure 25. r-MMA cost of production before taxes and depreciation. Distribution for Scenario 1. The Blue line is the simulated distribution that fits a log-normal distribution (black line).

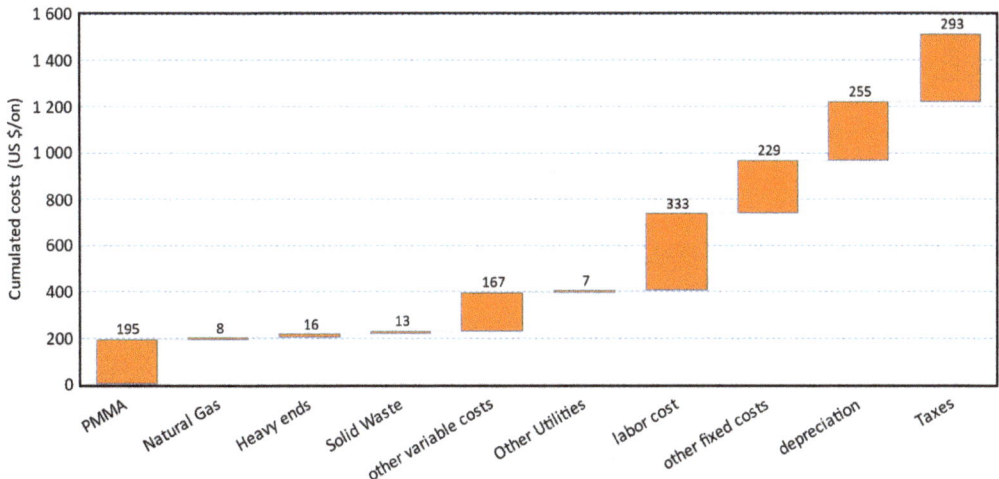

Figure 26. Median production cost cascade for Scenario 1.

The "other fixed" costs comprehend (Section 2.2) maintenance and repairs, operating supplies, property taxes, financing interests, and insurance. The "other variable" costs are instead (Section 2.2) expressed as a % of the sales, includes R&D and Royalties, and distribution and selling.

We already assumed to run a lean plant, with only two operators per shift, which is the minimum for safety reasons. The only way to improve that, would be erecting the plant as a brownfield. In this way, we could run the plant with only 1 operator per shift, while there would still be somebody else on the industrial site.

The other fixed costs (as well as the depreciation) are linked to the investment. The only way to reduce that is to keep the plant layout as simple as possible. An idea could be to reduce the number of spare equipment, or minimize the side streams that have then to be treated on-site.

Classically, savings are made on the other variable cost, but not much can be done in our case. With only a total of 8% of the sales, there is a small room for improvement. Since the plant is in the middle of Europe, close to PMMA producers and scrap collectors, we already imagined keeping the distribution and selling costs quite low. Moreover, because there are numerous running technologies, we also assumed a low R&D and royalty contribution. Most of the processes have already demonstrated to be mature enough to produce sellable r-MMA. R&D should focus on tuning the right technology with the right feedstock available, or target product, rather than finding novel routes.

4. Discussion and Conclusions

To highlight challenges and opportunities of PMMA recycling we performed a Monte Carlo simulation with a total of 3000 iterations. The risk analysis debunks some preconceptions, and it is a support for any future work in the field.

Based on the results of our simulation, we ranked the Scenarios according to different selected outcomes: (i) Pay-back time; (ii) NPV after 10 y of operation; (iii) probability of losing money after 10 y of operation (Table 5).

Table 5. Ranking of Scenarios according to different criteria/expected outcome.

Criteria	Ranking (Scenarios)	Conclusions
Pay-back time (from short to long)	1. **Scenarios 8/4** 2. Scenario 6 3. Scenarios 2/3/5 4. Scenario 9 5. Scenario 1	Cheap technology/second-hand equipment plant pay back in the shortest time.
NPV after 10 y of operation (highest first)	1. **Scenarios 3** 2. Scenario 2/5 3. Scenarios 8 4. Scenario 9 5. Scenario 1	After 10 y of operation, high-quality r-MMA, coupled with expensive technology pays off more than cheap plants, or low-quality scraps at zero price. If the market can accept both high and low purity product, companies should target high-quality r-MMA.
Probability of losing money after 10 y of operation (from low to high)	1. **Scenario 6** 2. Scenario 2/3/5 3. Scenario 9 4. Scenario 1 5. Scenario 7	Despite feedstock available for free, Scenario 7 has more probability of losses than the base case Scenario 1. The better the quality of the scraps/product, the lower the probability to lose money becomes.

The 5000 tons/y, 10 USD M plant of Mixed PMMA scraps, producing 3889 ton/y of r-MMA selling at virgin price, is the most representative case, and acts as baseline for comparison. We demonstrated how:

- A highly energy integrated process minimizes waste streams, utility consumptions, and environmental burden, as well as contributing to make a positive economic case;
- A median Capex of 10 M USD (or lower) is what companies should aim at for a 5000 tons/y PMMA depolymerization plant;
- NPV is most sensitive to the uncertainties on r-MMA price and investment. Expert judgment, experience on similar plants, subsidies, reduce the risk on the investment. Securing a deal with the r-MMA final buyers narrows down the r-MMA statistical distributions and improves the reliability of the business plan;
- New players should not compromise on product quality, even if this means higher capital plants, or "cleaner" and more expensive scraps. Whenever capital cost (Scenarios 4 and 8) and scrap composition (Scenarios 7 and 9) diminish the product quality, the overall economics worsens. Cheap plants allow for a faster pay-back time but are less profitable overall compared to the base case. The lower profits in the zero price waste scraps plant outweigh the reduced cost of operation.
- If we are in a market with a sufficiently high demand for low quality product (e.g., India, or Brazil), there is no need to over-purify the regenerated monomer, the plant is already economic as it is. However, the evolving EU legislation might hurdle the entrance of low quality r-MMA in the European market;
- Currently, Cast scraps plants are more viable than Mixed scrap plants, under the hypothesis that the first are less capital-intensive. However, when a plant is designed and built to treat both, it can switch with no effect on the cumulative NPV. The better global yield of Cast scraps counterbalances the lower feedstock price of Mixed scraps;
- The r-MMA cost of production follows a log-normal distribution, and labor cost, investment, and "other variable cost" are the main contributors.

Therefore, regardless the process technology, we outlined what are the minimum requirements that a plant should check in terms of mass balance, capital investment, feedstock and product prices, feedstock quality, and economic figures, in order to be competitive.

A first improvement for the model could be to investigate further waste streams. For instance, PMMA/fiber glass composites that might be employed in boats and wind blades, PMMA/ATH (aluminum trihydroxide), or solid surface materials, which are sold as a marble substitute. Future research is needed to understand how to best valorize each part of the composite, while still having a positive case.

An equally interesting scenario to investigate, would be when local governments grant subsidies on the product made, rather than the plant. This would be a variation of the 20% premium scenario we analyzed. Because of the volatile nature of the plastic recycling world, it would be wise for governmental agencies to subsidize only the winning horse, i.e., only when a company has already demonstrated production. At the beginning, the company takes all the risk, because they put money up-front. However, if the plant works, and it demonstrates production, subsidies on the product should guarantee a higher return than subsidies on the Capex.

Furthermore, r-MMA is currently sold in Europe under Reach registration exemption for recycled materials. In the framework of this study, it means that the regenerated monomer can directly enter the market. This might change in the future. As pointed out, r-MMA quality is feedstock sensitive and process sensitive, to some extent. Standards to select properly the PMMA scraps are then needed, in order to assign them the best value and address them to the different depolymerization technologies.

Author Contributions: Conceptualization, J.-L.D.; methodology, J.-L.D. and J.D.T.; software, J.-L.D. and J.D.T.; validation, J.-L.D.; investigation, J.-L.D.; resources, J.-L.D.; data curation, J.-L.D. and J.D.T.; writing—original draft preparation, J.D.T.; writing—review and editing, J.-L.D. and J.D.T.; visualization, J.D.T.; supervision, J.-L.D.; project administration, J.-L.D.; funding acquisition, J.-L.D. All authors have read and agreed to the published version of the manuscript.

Funding: This project has received funding from the European Union's Horizon 2020 research and innovation program under grant agreement No 820687.

Data Availability Statement: The authors confirm that the data supporting the findings of this study are available within the article.

Acknowledgments: We acknowledge Paul Masih and Emma Brevot, trainee students in Arkema who helped in the debugging of the Monte-Carlo simulation and worked on the Economic spreadsheet. This publication reflects only the author's view and the Commission is not responsible for any use that may be made of the information it contains.

Conflicts of Interest: The authors declare to have no conflict of interest. The EU project MMAtwo, in which they are involved, aims to promote the recycling of PMMA and MMA regeneration.

References

1. Global Market Study on Methyl Methacrylate (MMA)—Increasing Application in PMMA Production to Account for Significant Revenue Generation Opportunities. Available online: http://www.persistencemarketresearch.com/market-research/methyl-methacrylate-market.asp (accessed on 24 February 2021).
2. Methyl Methacrylate Market to Reach ~5.7 Million Tons by the End of 2028—Business. Available online: https://ipsnews.net/business/2020/06/09/methyl-methacrylate-market-to-reach-5-7-million-tons-by-the-end-of-2028/ (accessed on 24 February 2021).
3. PMMA Faces Long Road to Recovery. ICIS Interactive Insight. Available online: https://www.icis.com/explore/services/pmma-faces-long-road-to-recovery/ (accessed on 24 February 2021).
4. PMMA Market to See 6.2% Annual Growth Through 2023. Available online: https://www.bccresearch.com/pressroom/chm/pmma-market-to-see-62-annual-growth-through-2023 (accessed on 24 February 2021).
5. PMMA Market For Construction Application Growth Statistic Report 2024. Available online: https://www.gminsights.com/industry-analysis/pmma-market-for-construction (accessed on 24 February 2021).
6. ReportLinker Global Polymethyl Methacrylate (PMMA) Industry. Available online: http://www.globenewswire.com/news-release/2020/09/25/2099510/0/en/Global-Polymethyl-Methacrylate-PMMA-Industry.html (accessed on 24 February 2021).
7. Sale, K. Coronavirus Pandemic Creates New Application for PMMA Sheet Market. Available online: https://www.icis.com/explore/resources/news/2020/05/08/10505498/coronavirus-pandemic-creates-new-application-for-pmma-sheet-market (accessed on 24 February 2021).

8. Sale, K. Europe MMA July Prices Increase as Oversupply Eases, Feedstocks Firm. Available online: https://www.icis.com/explore/resources/news/2020/07/31/10536612/europe-mma-july-prices-increase-as-oversupply-eases-feedstocks-firm (accessed on 24 February 2021).
9. van der Heijden, S. Perspectives for PMMA Recycling—The Relevance and Societal Impact of the MMAtwo Project. In Proceedings of the MMAtwo Virtual Workshop on Polymer Recycling, Virtual, 15 September 2020; European Union: Brussels, Belgium, 2020. [CrossRef]
10. Dubois, J.-L. Dubois—MMAtwo Workshop—PMMA Depolymerization: Scale-up and Industrial Implementation. In Proceedings of the MMAtwo Virtual Workshop on Polymer Recycling, Virtual, 15 September 2020; European Union: Brussels, Belgium, 2020. [CrossRef]
11. *Plastics—The Facts 2020. An Analysis of European Plastics Production, Demand and Waste Data*; Plastics Europe: Brussels, Belgium, 2020.
12. Isador, M.; Beiser, A.L. Art of Reclaiming Plastic Scrap. Patent US2470361A, 27 June 1945.
13. Albrecht, K.; Stickler, M.; Rhein, T. Polymethacrylates. In *Ullmann's Encyclopedia of Industrial Chemistry*; Wiley: Hoboken, NJ, USA, 2013; ISBN 978-3-527-30673-2. [CrossRef]
14. Moens, E.K.C.; De Smit, K.; Marien, Y.W.; Trigilio, A.D.; Van Steenberge, P.H.M.; Van Geem, K.M.; Dubois, J.-L.; D'hooge, D.R. Progress in Reaction Mechanisms and Reactor Technologies for Thermochemical Recycling of Poly(Methyl Methacrylate). *Polymers* **2020**, *12*, 1667. [CrossRef] [PubMed]
15. Peck, P. Interest in Material Cycle Closure? Exploring Evolution of Industry's Responses to High-Grade Recycling from an Industrial Ecology Perspective: Volume II Case Studies. Ph.D. Thesis, Lund Univeristy, Lund, Sweden, 2003.
16. How to Recycle Waste Acrylic-Baruie. Available online: http://www.baruie.com/PMMA_recycle (accessed on 23 June 2021).
17. Acrylic Recycling and PMMA Recycling. For Acrylic Scrap Recycling Visit ShimiResearch.Com. Available online: http://shimiresearch.in/Acrylic-Recycling.php (accessed on 23 June 2021).
18. *Brydson's Plastics Materials*, 8th ed. Available online: https://www.elsevier.com/books/brydsons-plastics-materials/gilbert/978-0-323-35824-8 (accessed on 23 June 2021).
19. MMAtwo_Newsletter_January-2021. Available online: https://www.mmatwo.eu/wp-content/uploads/2021/01/MMAtwo_Newsletter_January-2021_VF.pdf (accessed on 23 July 2021).
20. Anzai, H.; Kishizawa, A.; Shimauchi, S.; Taniguchi, M. Method for Recovering Monomer from Waste Acrylic Resin. Patent No. JP16360296A, 2 May 2002.
21. Cavinaw, B.; Crawford, S.; Lamaze, J.; Allen, D.; Christensen, E.; Rumford, M. Systems and Methods for Recycling Waste Plastics, Including Waste Polystyrene. Patent No. US10301235B1, 28 June 2010.
22. Mcnamara, D.; Strivens, C.; Yabrudy, A.; Dunphy, P. A Reactor Assembly. Patent WO 2020/065316 A1, 2 April 2020.
23. Brandrup, J.; Immergut, E.H.; Grulke, E.A. *Polymer Handbook*, 4th ed.; John Wiley & Sons: Hoboken, NJ, USA, 2003.
24. Dubois, J.-L. New Innovative Process for Recycling End-of-Life PMMA Waste. MMAtwo Second Generation Methyl Methacrylate 2019. Available online: https://www.researchgate.net/publication/341655536 (accessed on 22 June 2021).
25. Green, D.W.; Southard, M.Z. Section 9—Process Economics. In *Perry's Chemical Engineers' Handbook*, 7th ed.; McGraw-Hill Education: New York, NY, USA, 2019; pp. 9–53.
26. Towler, G.; Sinnott, R. (Eds.) Chapter 9—Economic Evaluation of Projects. In *Chemical Engineering Design*, 2nd ed.; Butterworth-Heinemann: Boston, MA, USA, 2013; p. 389. ISBN 978-0-08-096659-5.
27. Tsagkari, M.; Couturier, J.-L.; Kokossis, A.; Dubois, J.-L. Early-stage capital cost estimation of biorefinery processes: A comparative study of heuristic techniques. *ChemSusChem* **2016**, *9*, 2284. [CrossRef] [PubMed]
28. *18R-97: Cost Estimate Classification System—As Applied in Engineering, Procurement, and Construction for the Process Industries*; AACE International: Morgantown, WV, USA, 2016; 6p.
29. Petley, G.J. A Method for Estimating the Capital Cost of Chemical Process Plants: Fuzzy Matching. Ph.D. Thesis, Loughborough University, Loughborough, UK, August 1997.
30. Lange, J.-P. Fuels and Chemicals Manufacturing; Guidelines for Understanding and Minimizing the Production Costs. *Cattech* **2001**, *5*, 82–95. [CrossRef]
31. Vatavuk, W.M. Updating the CE Plant Cost Index. *Chem. Eng.* **2002**, *109*, 62–70.
32. Chemical Engineering. 21 April 2020. Available online: https://www.chemengonline.com/2020-cepci-updates-february-prelim-and-january-final/ (accessed on 13 July 2021).
33. Peters, M.S.; Timmerhaus, K.D.; West, R.E. *Plant Design and Economics for Chemical Engineers*, McGraw-Hill chemical engineering series; 5th ed.; McGraw-Hill: New York, NY, USA, 2003; ISBN 978-0-07-239266-1.
34. *ICI's Polymethyl Methacrylate Tertiary Recycling Technology*; ICI International: London, UK, 1998.
35. Advances in Depolymerization Technologies for Recycling (2021 Program). NexantECA. Available online: https://www.nexanteca.com/reports/advances-depolymerization-technologies-recycling-2021-program (accessed on 25 June 2021).
36. Beacham, W. Plastic Energy Plans 10 Chemical Recycling Plants in Europe, Asia by 2021. Available online: https://www.icis.com/explore/resources/news/2019/03/19/10335799/plastic-energy-plans-10-chemical-recycling-plants-in-europe-asia-by-2021 (accessed on 19 February 2021).
37. Boldon, L.M.; Sabharwall, P. *Small Modular Reactor: First-of-a-Kind (FOAK) and Nth-of-a-Kind (NOAK) Economic Analysis*; Technical Report No. INL/EXT-14-32616; OSTI Identifier 1167545; USDOE: Washington, DC, USA, 2014.

38. Agilyx. Available online: https://www.agilyx.com/application/files/7516/0806/8429/Agilyx_Investor_Presentation_15Dec20_FINAL.pdf (accessed on 19 February 2021).
39. Congreso International Tenerife + Sostenible. Available online: http://congresotenerifemassostenible.com/wp-content/uploads/2016/06/09_B22-5-C-Monreal.pdf (accessed on 19 February 2021).
40. Boal, J. Utah-Based Renewlogy Offers Solution to Plastic Waste Problem. Available online: https://www.deseret.com/2018/4/1/20642672/utah-based-renewlogy-offers-solution-to-plastic-waste-problem (accessed on 19 February 2021).
41. The City of Phoenix to Begin Turning Plastics into Fuel. Available online: https://www.phoenix.gov:443/news/publicworks/2303 (accessed on 19 February 2021).
42. Phoenix Awards Contract to Renewlogy for Chemical Recycling Project. Available online: https://www.wastedive.com/news/phoenix-awards-contract-to-renewlogy-for-chemical-recycling-project/552055/ (accessed on 19 February 2021).
43. StClair-Pearce, T.; Garbett, D. Plastic Waste Thermal Cracking New Energy Investment. Available online: https://www.qmre.ltd/wp-content/uploads/2018/10/QMRE-Business-Plan-Digital.pdf (accessed on 19 February 2021).
44. Biofabrik WASTX Plastic Solution. Available online: https://www.qmre.ltd/wp-content/uploads/2020/09/QMRE-Introduction-doc-10-03-20.pdf (accessed on 19 February 2021).
45. Transforming Plastic Waste into Valuable Low-Carbon Fuel-QuantaFuel. Available online: https://gwcouncil.org/wp-content/uploads/2020/06/Attachment-4-Quantafuel-introduction.pdf (accessed on 19 February 2021).
46. QuantaFuel Q3. 2020. Available online: https://quantafuel.com/wp-content/uploads/2020/11/QFUEL-3Q20-REPORT.pdf (accessed on 19 February 2021).
47. Vitol to Partner with Quantafuel to Market Synthetic Fuel Made from Recycled Plastic. Available online: https://www.vitol.com/vitol-to-partner-with-quantafuel-to-market-synthetic-fuel-made-from-recycled-plastic/ (accessed on 19 February 2021).
48. Recycling Technologies Firms up Plans for Scottish Pyrolysis Plant. Available online: https://www.letsrecycle.com/news/latest-news/recycling-technologies-scottish/ (accessed on 24 February 2021).
49. De Leon Izeppi, G.A.; Dubois, J.-L.; Balle, A.; Soutelo-Maria, A. Economic risk assessment using Monte Carlo simulation for the production of azelaic acid and pelargonic acid from vegetable oils. *Ind. Crops Prod.* **2020**, *150*, 112411. [CrossRef]
50. Folliard, V.; de Tommaso, J.; Dubois, J.-L. Review on Alternative Route to Acrolein through Oxidative Coupling of Alcohols. *Catalysts* **2021**, *11*, 229. [CrossRef]
51. Import Data and Price of Acrylic Plastic Scrap. Zauba. Available online: https://www.zauba.com/import-acrylic-plastic-scrap-hs-code.html (accessed on 11 March 2021).
52. Eximpulse, India Import Data of 39159030. Available online: https://www.eximpulse.com/import-hscode-39159030.htm (accessed on 11 March 2021).
53. International Monetary Fund Global Price of Natural Gas, EU. Available online: https://fred.stlouisfed.org/series/PNGASEUUSDM (accessed on 11 March 2021).
54. European Chemical Bureau. *European Union Risk Assessment Report—CAS No: 80-62-6*; European Commission: Luxembourg, 2002.
55. *EUR-Lex—52020DC0098—EN—EUR-Lex—A New Circular Economy Action Plan*; European Commission: Brussels, Belgium, 2020.
56. Plastics Own Resource. Available online: https://ec.europa.eu/info/strategy/eu-budget/long-term-eu-budget/2021-2027/revenue/own-resources/plastics-own-resource_en (accessed on 12 July 2021).
57. Prices—Producer Price Indices (PPI)—OECD Data. Available online: https://data.oecd.org/price/producer-price-indices-ppi.htm (accessed on 28 June 2021).
58. Sale, K. Europe MMA Spot Prices Hit Four-Year Low as Buyers Try to Fulfil Contract Minimums. Available online: https://www.icis.com/explore/resources/news/2020/04/06/10493857/europe-mma-spot-prices-hit-four-year-low-as-buyers-try-to-fulfil-contract-minimums (accessed on 11 March 2021).
59. Loy, K.W. Asia MMA Markets Slip from Record Highs after 12 Weeks of Stability. Available online: https://www.icis.com/explore/resources/news/2018/09/11/10258347/asia-mma-markets-slip-from-record-highs-after-12-weeks-of-stability (accessed on 11 March 2021).
60. Crude Oil (Petroleum)—Monthly Price—Commodity Prices—Price Charts, Data, and News—IndexMundi. Available online: https://www.indexmundi.com/commodities/?commodity=crude-oil (accessed on 12 January 2021).
61. Export Data and Price of Regenerated Methyl Methacrylate—HS 29161400. Zauba. Available online: https://www.zauba.com/export-regenerated-methyl-methacrylate-hs-code.html (accessed on 29 June 2021).
62. Ademe Etude Comparative de la Taxation de l'Elimination des Dechets en Europe. Available online: https://www.ademe.fr/sites/default/files/assets/documents/etude_comparative_taxation_elimination_dechets_europe_201703_rapport.pdf (accessed on 22 July 2021).
63. Olofsson, M.; Sahlin, J.; Ekvall, T.; Sundberg, J. Driving Forces for Import of Waste for Energy Recovery in Sweden. *Waste Manag. Res.* **2005**, *23*, 3–12. [CrossRef] [PubMed]
64. IEA—Waste Disposal Costs and Share of EfW in Selected Countries—Charts—Data & Statistics. Available online: https://www.iea.org/data-and-statistics/charts/waste-disposal-costs-and-share-of-efw-in-selected-countries (accessed on 29 June 2021).
65. Cewep—Landfill Taxes and Bans Report. Available online: https://www.cewep.eu/wp-content/uploads/2017/12/Landfill-taxes-and-bans-overview.pdf (accessed on 22 July 2021).
66. Hollmann, J.K. *(Risk 1027) Estimate Accuracy: Dealing with Reality*; AACE International: Morgantown, WV, USA, 2012.

67. Waste Management and the Circular Economy in Selected OECD Countries: Evidence from Environmental Performance Reviews. Available online: https://www.oecd-ilibrary.org/environment/waste-management-and-the-circular-economy-in-selected-oecd-countries_9789264309395-en (accessed on 22 July 2021).
68. Per Capita Plastic Waste vs. GDP per Capita. Available online: https://ourworldindata.org/grapher/per-capita-plastic-waste-vs-gdp-per-capita (accessed on 1 July 2021).
69. World Bank. World Development Indicators—DataBank. GDP per Capita, PPP (Constant 2011 International $). Available online: http://data.worldbank.org/data-catalog/world-development-indicators (accessed on 22 July 2021).
70. Jambeck, J.R.; Geyer, R.; Wilcox, C.; Siegler, T.R.; Perryman, M.; Andrady, A.; Narayan, R.; Law, K.L. Plastic Waste Inputs from Land into the Ocean. *Science* **2015**, *347*, 768–771. [CrossRef] [PubMed]
71. Heating Oil—Daily Price—Commodity Prices—Price Charts, Data, and News—IndexMundi. Available online: https://www.indexmundi.com/commodities/?commodity=heating-oil&months=60 (accessed on 1 July 2021).
72. Fivga, A.; Dimitriou, I. Pyrolysis of plastic waste for production of heavy fuel substitute: A techno-economic assessment. *Energy* **2018**, *149*, 865–874. [CrossRef]
73. Thunman, H.; Berdugo Vilches, T.; Seemann, M.; Maric, J.; Vela, I.C.; Pissot, S.; Nguyen, H.N.T. Circular use of plastics-transformation of existing petrochemical clusters into thermochemical recycling plants with 100% plastics recovery. *Sustain. Mater. Technol.* **2019**, *22*, e00124. [CrossRef]

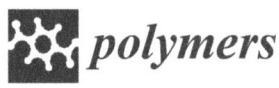

Article

Assessment of Commercially Available Polyethylene Recyclates for Blow Molding Applications by a Novel Environmental Stress Cracking Method

Paul J. Freudenthaler *, Joerg Fischer and Reinhold W. Lang

Institute of Polymeric Materials and Testing, Johannes Kepler University Linz, Altenberger Straße 69, 4040 Linz, Austria
* Correspondence: paul.freudenthaler@jku.at; Tel.: +43-732-2468-6620

Abstract: The transition to a circular economy has a major impact on waste management and the reuse of materials. New mandatory recycling targets for plastics will lead to a high availability of recyclates. For these recyclates, useful applications need to be found. One potential application for recyclates is blow molding bottles as packaging for non-food contents. This study investigates commercially available post-consumer high-density polyethylene recyclates together with virgin blow molding grades in terms of their short-term mechanical properties and environmental stress cracking resistance. While the short-term mechanical properties showed only slightly lower performance than the tested virgin grades, the overall environmental stress cracking failure times of the recyclates were much lower compared to virgin materials, even though the crack-growth kinetics could be similar. Although neither the tensile nor the notched impact strength results of the two polyethylene recyclates revealed large differences, the stress intensity-factor-dependent crack-growth rates of both materials were significantly different.

Keywords: blow molding; polyethylene; environmental stress cracking resistance; cracked round bar specimens; recyclate; fatigue; slow crack growth

Citation: Freudenthaler, P.J.; Fischer, J.; Lang, R.W. Assessment of Commercially Available Polyethylene Recyclates for Blow Molding Applications by a Novel Environmental Stress Cracking Method. *Polymers* **2023**, *15*, 46. https://doi.org/10.3390/polym15010046

Academic Editor: Didier Perrin

Received: 28 November 2022
Revised: 13 December 2022
Accepted: 20 December 2022
Published: 22 December 2022

Copyright: © 2022 by the authors. Licensee MDPI, Basel, Switzerland. This article is an open access article distributed under the terms and conditions of the Creative Commons Attribution (CC BY) license (https:// creativecommons.org/licenses/by/ 4.0/).

1. Introduction

The current endeavors of the European Commission towards a circular economy have led to several regulations concerning packaging products [1,2]. The recycling target of 50% by the end of 2025 and 55% by the end of 2030 [3] for plastic packaging waste is an ambitious goal, especially considering the currently low recycling rates for the biggest group of packaging plastics, polyolefins. One of the major packaging materials [4], high-density polyethylene (PE-HD), makes up 17% of the plastic waste [5], but only a small part is recycled in Europe [6]. One of the main characteristics of PE-HD which led to its high usage as a packaging material is its high impact toughness [7]. Popular examples of PE-HD packaging products are bottles, bottle caps, canisters, and blow-molded cosmetics packaging such as shampoo bottles, etc., [8].

Polyolefin waste is notorious for its difficulty of separation due to the overlapping densities of the various polyolefins [5,9]. Thus, polyolefin recyclates exhibit varying purity and performance [10], which depend on the feedstock, sorting and reprocessing. Several companies offer recycled high-density polyethylene (rPE-HD) with various quality and/or purity levels. This study aims to evaluate the different commercially available rPE-HD grades for their usage as an extrusion blow molding material in comparison to the relevant virgin material grades in terms of their mechanical properties.

Virgin blow molding PE-HD grades have melt mass-flow rates (MFRs) of between 0.25 g/10 min and 0.80 g/10 min (according to ISO 1133-1 [11] at 190 °C and 2.16 kg) [12–15]. Furthermore, the promoted properties in the relevant material datasheets include yield strengths between 20 MPa and 32 MPa, tensile moduli between 1000 MPa and 1420 MPa (both according to ISO 527-1 [16] and ISO 527-2 [17]), high-impact strengths tested with different testing methods at various temperatures, and an otherwise seldomly used property: environmental stress cracking resistance (ESCR), also tested with various methods. ESCR is a material property, which is specifically needed for the containment of crack-growth accelerating liquids (e.g., detergents) and is highly dependent on the specimen type, loading conditions, and used environment (media and temperature).

The post-consumer recyclate market offers a wide range of rPE-HD grades in varying colors. The most desirable color for most applications is natural (without color additive) followed by white. Both colors are used for natural or white products, respectively, or for subsequent coloration. Some suppliers offer color separated recyclate grades. Due to the lack of color separation and thus the color mixing of differently colored flakes, the most common colors on the post-consumer recyclate market are shades of grey. Commercially available rPE-HDs from post-consumer packaging waste usually have higher MFRs, as they consist of different sources of PE waste and include injection-molded products, for instance, caps and closures. While the MFR, tensile, and impact properties can typically be found in the datasheets of rPE-HDs, ESCR is a rarely declared property.

Several standardized test methods for the experimental characterization of the ESCR of PE-HD are in use. These test methods make use of accelerating media and/or test temperatures to shorten testing times. A widely used test method, the full-notch creep test (ISO 16770) [18], uses elevated temperatures (50 °C–90 °C) and a creep crack-growth accelerating mixture of deionized water and surfactant. Other ESCR methods, such as ASTM D 1693 [19] or ASTM D 2561 [20], also use elevated temperatures and surfactants. The cracked round bar (CRB) test method according to ISO 18489 [21], a method initially introduced to characterize the resistance against slow crack growth in PE pipe grades, uses neither elevated temperature nor harsh media, but cyclic loading for the acceleration of crack growth and shortening of testing times. As the authors have experience with superimposed cracked round bar experiments [22], the method is well-established in the field of fracture mechanics [23–26], and as this method allows for the in situ measurement of the crack-growth rate, this testing method was chosen for comparison of the ESCR of virgin and recycled PE-HD grades. The measurement of the crack length enables the determination of the crack-growth rate which offers more insight into the fracture mechanical failure than failure times alone.

2. Materials

Two virgin blow molding PE-HD grades in the form of pellets, as can be seen in Figure 1, were acquired as benchmark grades for this study. Both PE-HD grades were translucent without color/white pigment. The only apparent optical difference was the yellow tint of vPE-1, as can be seen in Figure 1a in comparison to Figure 1b.

(a) (b)

Figure 1. Images of the pellets of the virgin blow molding PE-HD grades vPE-1 (a) and vPE-2 (b).

Considering the two basic constraints, color and MFR, two rPE-HD grades were purchased from two different European suppliers based on their white (rPE-A) and natural (rPE-B) colors and MFRs, which are in the range of the benchmark virgin grades. The recyclates were delivered as pellets, as can be seen in Figure 2. rPE-A is an opaque white color while rPE-B has a more translucent tone or rather natural color.

(a) (b)

Figure 2. Images of the pellets of the PE-HD recyclates rPE-A in opaque white (**a**) and rPE-B in natural (**b**).

3. Methods

Differences in specimen preparation, conditioning, and testing often impede objective comparisons between material data of different suppliers. Moreover, small lot-to-lot variations can occur even for virgin grades. Therefore, within the scope of this study, all basic properties were characterized by identical parameters for better comparability and accurate structure–property–performance relationships.

3.1. Melt Mass-Flow Rate Measurements

The MFR measurements were conducted at 190 °C and with both 2.16 kg and 5 kg static loads on the melt flow indexer Mflow (ZwickRoell GmbH & Co. KG, Ulm, Germany) according to ISO 1133-1 [11] and ISO 17855-2 [27]. Cuts were automatically made after every 1 mm (2.16 kg) and 3 mm (5 kg) piston movement. The time between cuts was measured and each extrudate was weighed on an ABS 220-4 electronic balance (Kern & Sohn GmbH, Balingen-Frommern, Germany). The MFR in g/10 min was calculated by extrapolation to 10 min for the individual cut. For each material, two measurements were conducted and used for calculation of the average values and standard deviations.

3.2. Specimen Production Methods

According to ISO 17855-2 [27], tensile and impact test specimens for characterizing PE should be either compression or injection molded, depending on their MFR. Blow molding PE grades and, therefore, also all materials tested within this study have low enough MFRs to be compression molded. Nevertheless, for the sake of reproducibility, the authors chose to produce all multipurpose specimens (MPS) and bar specimens via injection molding according to ISO 294-1 [28], ISO 20753 [29], and ISO 17855-2 on a Victory 60 (Engel Austria GmbH, Schwertberg, Austria) with a 25 mm cylinder. As prescribed by ISO 17855-2, the melt temperature for injection molding of 210 °C and the injection velocity of 100 mm/s were used for all materials. The specimens were conditioned at 23 °C and 50% relative humidity for 3 days after production. After this conditioning, the MPS were used for tensile testing and the bar specimens were notched and used for Charpy notched impact testing in accordance with ISO 179-1 [30]. The Charpy type A notches, i.e., notches with a 0.25 mm notch radius, were produced with an RM2265 microtome (Leica Biosystems Nussloch GmbH, Nussloch, Germany) according to ISO 179-1 and measured with an Mitutoyo Absolute 547-313 digital thickness gauge with a wedge tip. An image of an MPS specimen together with a bar specimen are shown in Figure 3. The bar specimen is illustrated after failure to emphasize the induced notch and resulting crack plane.

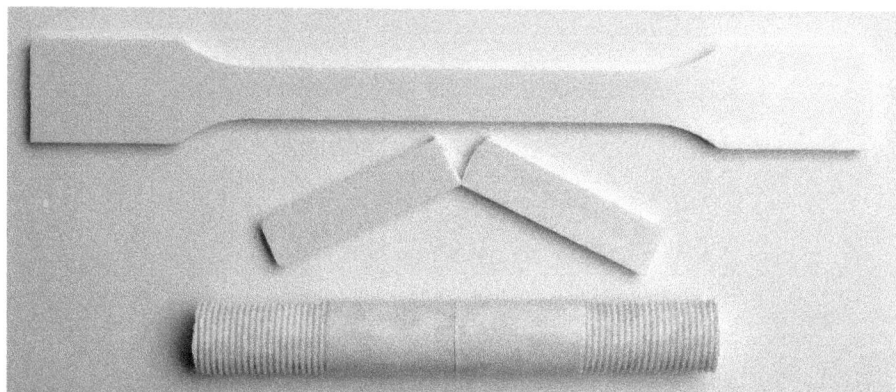

Figure 3. Image of the specimens used in this work made from the PE-HD recyclate rPE-A; from top down: a multipurpose specimen, a Charpy notched impact tested bar specimen, and a cracked round bar specimen.

For cracked round bar (CRB) specimen production according to ISO 18489 [21] for the ESCR experiments, plates of the size of 16 × 120 × 150 mm were compression molded in a specifically designed positive mold with the help of the hydraulic press Perfect Line (Langzauner GmbH, Lambrechten, Austria). Within the fully automated process, 280 g pellets were heated within the mold from room temperature to 180 °C with the weight of the mold on top of the material. An integrated temperature sensor allowed for direct measurement of the mold temperature. After the internal temperature of 180 °C was reached, it was held for 15 min. After that, the slow cooling with a cooling rate of 2 K/min was started. The full pressure of 10 MPa was applied to the material upon reaching a temperature of 135 °C during the cooldown step. Applying full pressure at higher temperatures leads to too much melt displacement. After reaching 40 °C, the pressure was released, the mold opened, and the plate manually removed. While ISO 18489 does not prescribe tempering of the produced plates, the authors observed conditioning influences, especially when testing at elevated temperatures. Therefore, the plates were cut into bars which then were tempered in accordance with a popular ESCR standard ISO 16770 [18] for 3 h at 100 °C and subsequently cooled down very slowly overnight in a Binder air oven (Binder GmbH, Tuttlingen, Germany). The tempered bars were then kept for at least 72 h at 23 °C and 50% relative humidity for stress relaxation before being lathed on a lathe (EMCO GmbH, Hallein, Austria) to CRB specimens according to ISO 18489. A 0.3 mm thick industrial grade razor blade was used to notch the specimens. Before testing, the specimens were kept at 23 °C and 50% relative humidity for at least another 24 h after being notched. Directly before starting the test, the specimen was clamped 3 h in the test setup in the respective test medium at test temperature without any loading applied. A CRB specimen is shown in the bottom of Figure 3.

3.3. Differential Scanning Calorimetry Measurements

Differential scanning calorimetry (DSC) measurements were carried out on the differential scanning calorimeter DSC 8500 (PerkinElmer Inc., Waltham, MA, USA). Measurements were made according to ISO 11357-1 [31] and ISO 11357-3 [32]. Samples were cut from shoulders of injection molded MPS and from the middle section of bars from pressed plates before and after tempering and encapsulated in aluminum pans. The average sample weight was around 5 mg. Two different tests were performed. For the MPS from recyclates, the procedure consisted of an initial heating phase, subsequent cooling, and a second heating phase, each in the temperature range of 0 to 200 °C with a constant heating/cooling rate of 10 K/min with nitrogen as the purge gas and a flow rate of 20 mL/min. These measurements were accomplished to determine the melting peak in the second heating phase, which is characteristic for the semi-crystallinity achieved under controlled cooling in the DSC device. For the samples taken from the bars cut from pressed plates before and after tempering, only one heating run to determine the melting enthalpy change during tempering was performed with otherwise similar parameters. For each material, three samples, each cut from an individual MPS and bar, were used for the calculation of average values and standard deviations.

3.4. Oxidation Induction Temperature Measurements

A differential thermal analysis instrument from PerkinElmer, the DSC 4000, was utilized to characterize the oxidation induction temperature (dynamic OIT) according to ISO 11357-6 [33]. Samples were cut from shoulders of injection molded MPS and encapsuled in perforated aluminum pans. The average sample weight was around 8 mg. A single heating step between 23 and 300 °C was performed with a heating rate of 10 K/min and with synthetic air as purge gas with a flow rate of 20 mL/min. The point of intersect of the slope before oxidation and during oxidation gave the oxidation induction temperature in °C which gave an indication of the stabilization [34]. For each material, five samples, each cut from an individual MPS, were used for the calculation of average values and standard deviations.

3.5. Thermogravimetric Analyses

The thermogravimetric analysis (TGA) was performed on an STA 6000 simultaneous thermal analyzer (PerkinElmer, Waltham, MA, USA), according to ISO 11358-1 [35], using nitrogen as a purge gas with a flow rate of 20 mL/min. Samples of around 20 mg, usually one pellet, were put into open ceramic crucibles. The temperature program consisted only of one heating step from 30 to 850 °C with a heating rate of 20 K/min. For each material, two samples, each cut from an individual MPS, were used for the calculation of average values and standard deviations.

3.6. Tensile Tests

The tensile properties, tensile modulus, yield stress, and strain at break, were examined with a universal testing machine Zwick/Roell AllroundLine Z05 equipped with a Zwick/Roell multi-extensometer strain measurement system. Test parameters and MPS were used according to ISO 527-1 [16], ISO 527-2 [17], and ISO 17855-2 [27] with a testing speed of 1 mm/min for tensile modulus determination until a strain of 0.25% and after that 50 mm/min until failure. Calculations of tensile modulus, yield stress, and strain at break were carried out in accordance with ISO 527-1. Therefore, the tensile modulus was calculated as the slope of the stress/strain curve between 0.05 and 0.25% via regression. The yield stress was the stress at the first occurrence of strain increase without a stress increase and the strain at break was the strain when the specimen broke. The strain was recorded via a multi-extensometer until yield. From there, the nominal strain was calculated via Method B according to ISO 527-1 with the aid of the crosshead displacement. This process is integrated and automated in the used testing software TestXpert III (v1.61, ZwickRoell GmbH & Co. KG, Ulm, Germany). For each material, five MPS were tested for the calculation of average values and standard deviations.

3.7. Charpy Impact Tests

Impact properties were determined according to ISO 179-1 [30] on a Zwick/Roell HIT25P pendulum impact tester. After pretests to determine the suitable pendulum size (absorbed energy between 10 and 80% of the available energy at impact), a 2 Joule pendulum, which is the pendulum with the highest available energy that still conforms to these requirements, was chosen for testing all materials. Test conditions were 23 °C test temperature with type 1 specimen, edgewise blow direction, and notch type A, i.e., a 0.25 mm notch radius, or short ISO 179-1/1eA, which is one of the preferred methods of ISO 17855-2 [27]. For each material, ten specimens were tested for the calculation of average values and standard deviations.

3.8. Environmental Stress Cracking Experiments

As the authors have the experience and machinery to conduct cracked round bar fatigue crack-growth (FCG) experiments in harsh media conditions [36], the choice of measurement method was reached in favor of the cracked round bar method ISO 18489 [21]. The CRB specimens were tested with an electro-dynamic testing machine of the type Instron ElectroPuls E10000 (Illinois Tool Works Inc., Glenview, IL, USA). Sinusoidal loading profiles with a frequency of 10 Hz, an R-ratio of 0.1 and individually adjusted initial stress-intensity factor ranges ($\Delta K_{I,ini}$) were used to achieve testing times between 10 and 100 h. As testing environments, deionized water with 10 m% Igepal CO-630 (Rhodia S.A., La Défense, France) and air, both at 40 °C, were used. A specifically constructed testing chamber for environmentally superimposed CRB experiments, made from stainless steel and glass [22], was employed for containment of the media. For tests in the Igepal solution, a circulation pump was used to ensure the homogeneity of the surfactant mixture and the media temperature and also to flush out air bubbles from the crack. A fan mounted to the side of the specimen ensured a homogenic temperature within the testing chamber for the tests in air.

An in situ optical measurement of the crack length over the whole circumference of the specimen was utilized for investigations of crack growth [22]. With this, it was possible to calculate crack lengths during the measurement and link them via ΔK_I with the crack-growth rate. The ΔK_I value depends on the geometry, applied force range, and initial crack length as can be seen in the Equations (1)–(3) developed by Benthem and Koiter [37] and used within the CRB test standard ISO 18489. Hence, ΔK_I is a normalized value for the load affecting the crack tip.

$$\Delta K_I = \frac{\Delta F}{\pi \cdot b^2} \cdot \sqrt{\frac{\pi \cdot a \cdot b}{r}} \cdot f\left(\frac{b}{r}\right) \tag{1}$$

$$b = r - a \tag{2}$$

$$f\left(\frac{b}{r}\right) = \frac{1}{2} \cdot \left[1 + \frac{1}{2} \cdot \left(\frac{b}{r}\right) + \frac{3}{8} \cdot \left(\frac{b}{r}\right)^2 - 0.363 \cdot \left(\frac{b}{r}\right)^3 + 0.731 \cdot \left(\frac{b}{r}\right)^4 \right] \tag{3}$$

where ΔK_I is the initial stress-intensity factor range in loading Mode I [38], ΔF the applied force range, a the measured crack length, r the radius of the specimen, b the ligament, and f(b/r) a geometry function.

Characteristic double logarithmic FCG kinetic curves were plotted providing the relation between the FCG rate, da/dN in mm/cycle, and the stress-intensity factor range, ΔK_I in MPa $\times$ m$^{0.5}$. One CRB specimen was tested per material and medium.

4. Results

4.1. Melt Mass-Flow Rate

The MFRs for all materials are presented for 190 °C and 2.16 kg and for 190 °C and 5 kg in Figure 4a,b, respectively. The MFR values (190 °C and 2.16 kg) of vPE-1 and rPE-A fell within the MFR range for blow molding grades as depicted by the vertical bar in Figure 4a. vPE-2 and rPE-B showed values below these reference values. The low values of the latter materials could be either due to measurement inaccuracies or, especially for rPE-B, due to degradation-induced crosslinking [39–41]. The MFR measurements with both test weights exhibited similar trends. The 5 kg measurements are shown for comparison with recyclates, where often only 190 °C/5 kg values are provided in the datasheets.

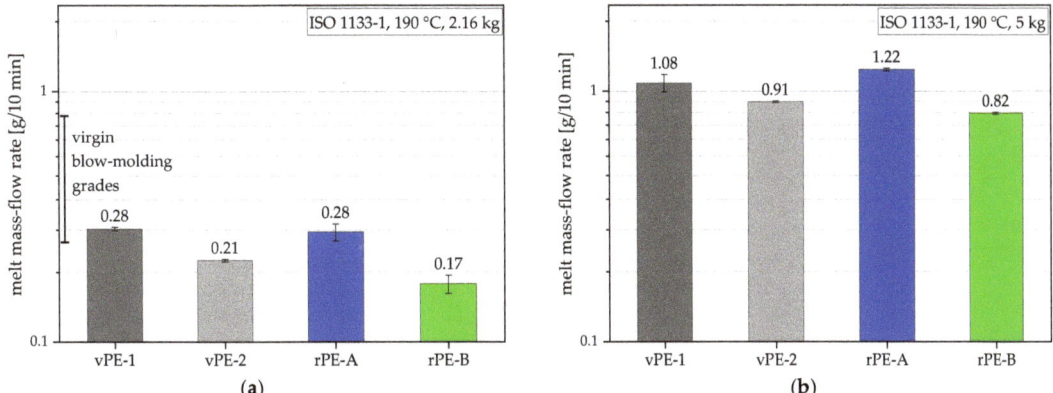

Figure 4. Graphical illustration and values of MFRs for all materials measured according to ISO 1133-1 at 190 °C and 2.16 kg (**a**) and 5 kg (**b**).

4.2. Differential Scanning Calorimetry Measurements

The DSC measurements on MPS specimens were only conducted for the recyclates with the goal to determine foreign polymer melting peaks. The thermograms were stacked so they could be looked at individually and they show that all three samples from three individual MPS exhibited comparable results, as illustrated in Figure 5. When looking at the area between 150 °C and 170 °C, a small endothermic peak can be seen in the thermograms of rPE-A, which suggests polypropylene (PP) contamination, most likely due to bad separation during the recycling process, for which such cross-contamination in polyolefins can be often seen in polyolefin recycling [10,42,43]. rPE-B did not show any other endothermic peak apart from the polyethylene (PE) melting peak. Hence, according to DSC, it was free of discernable PP contamination.

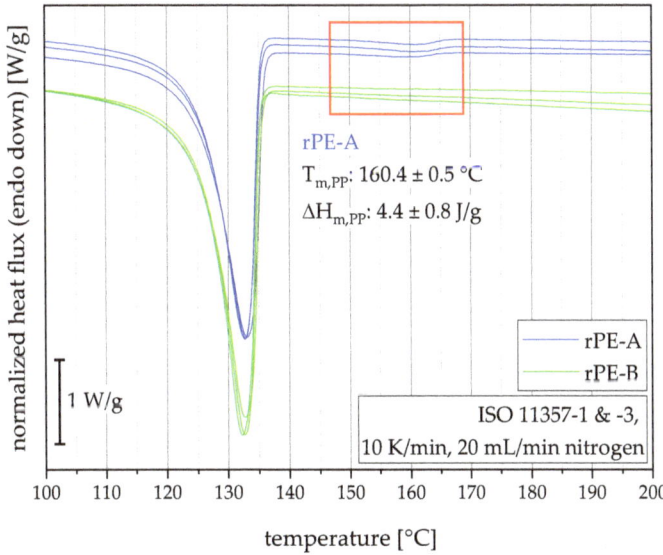

Figure 5. DSC thermograms of three repetition measurements for both recyclates and average values for the PP melting peak, which is framed by the red square, of the PE-HD recyclate rPE-A.

The DSC measurements of samples cut from the bars which are used for the manufacturing of CRB specimens were conducted to determine the melting enthalpy of the PE phase. Contrary to the measurements conducted on samples taken from MPS, the first heating run is shown instead of the second heating run. In this way, the effect of the tempering on the materials morphology could be investigated. The PP melting enthalpy is not shown for these tests as the low sample volume and missing homogenization step (injection molding) leads to high scattering of the quantifiable PP melting enthalpy in rPE-A. In comparison, the homogenization by injection molding of MPS distributes the PP contamination to each DSC sample, hence creates, despite the small sample volume, appropriate specimens for detecting trace amounts of contamination.

While there are no significant differences in the melting temperatures, all materials showed a distinct rise in PE melting enthalpy due to the tempering process, as can be seen in the thermograms shown in Figure 6 and values in Table 1. Furthermore, most materials showed a distinct decrease in standard deviation, indicating a homogenization of crystal structures. The increases of the melting enthalpies translate to rises in the degree of crystallinity of 62 to 65% (rPE-A), 70 to 71% (rPE-B), 73 to 74% (vPE-1), and 73 to 77% (vPE-2) [44], which all lie within the PE-HD range of 60–80% [10].

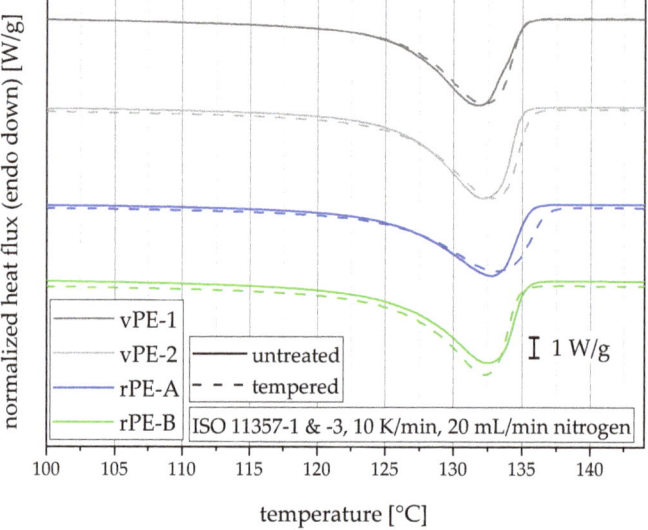

Figure 6. Averaged DSC thermograms (three measurements for one curve) from samples cut from untreated and tempered bars used to produce CRB specimens. Curves are stacked for better visibility.

Table 1. PE melting temperatures and enthalpies for untreated and tempered bars cut from pressed plates intended for CRB specimen production.

	vPE-1		vPE-2		rPE-A		rPE-B	
PE melting temperature (untreated) [°C]	132.0	±0.8	132.1	±0.3	132.8	±0.2	132.5	±0.3
PE melting temperature (tempered) [°C]	132.2	±0.2	133.1	±0.3	133.0	±0.2	132.3	±0.2
PE melting enthalpy (untreated) [J/g]	212.9	±1.9	213.6	±2.4	181.3	±3.8	204.3	±3.0
PE melting enthalpy (tempered) [J/g]	215.9	±0.8	224.5	±1.8	190.9	±4.3	206.7	±0.2

4.3. Oxidation Induction Temperatures

A higher dynamic OIT shows a higher resistance to thermo-oxidative degradation and can be used as an indicator for the effectiveness of stabilizers incorporated into the material as they affect the material's vulnerability to harsh oxidative environments [45]. Furthermore, the stabilization influences the localized aging of the crack tip [46,47] and, therefore, should influence the ESCR.

The DSC thermograms in synthetic air together with the calculated dynamic OITs are shown in Figure 7. Both virgin polymers showed high dynamic OIT values of 241.9 °C (vPE-1) and 243.0 °C (vPE-2). rPE-A surprises with a dynamic OIT value of 241.9 °C, suggesting that the recyclate produced added stabilizers during the recyclate production. The second recyclate rPE-B showed a lower dynamic OIT value of 233.4 °C. Therefore, vPE-1, vPE-2, and rPE-A should show a considerably higher resistance against harsh environments than rPE-B in terms of thermo-oxidative degradation.

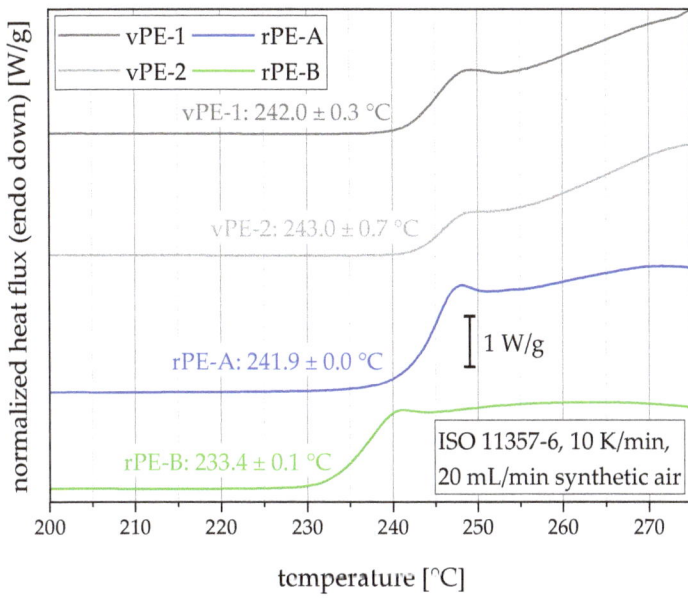

Figure 7. Averaged DSC thermograms (two or three measurements for one curve) and values of oxidation induction temperatures for all materials. Curves are stacked for better visibility.

4.4. Thermogravimetric Analyses

TGA was used to show the inorganic contamination level of the materials. Unfilled polyolefins are expected to show no pyrolysis residue [44]. rPE-A, therefore, shows the distinctive signs of contamination, as its pyrolysis residue at 550 °C lies around 3.31%. rPE-B contains only very low amounts of inorganic contamination with a pyrolysis residue of 0.11%. The virgin polymers vPE-1 and vPE-2 showed even smaller pyrolysis residues of 0.07% (vPE-1) and 0.06% (vPE-2). The zoomed in section in Figure 8 shows the mass change between 450 and 750 °C, and reveals a second decomposition step for rPE-A at around 650 °C. This step can be attributed to the cleavage of CO_2 from $CaCO_3$ which would also explain the white color of rPE-A [10].

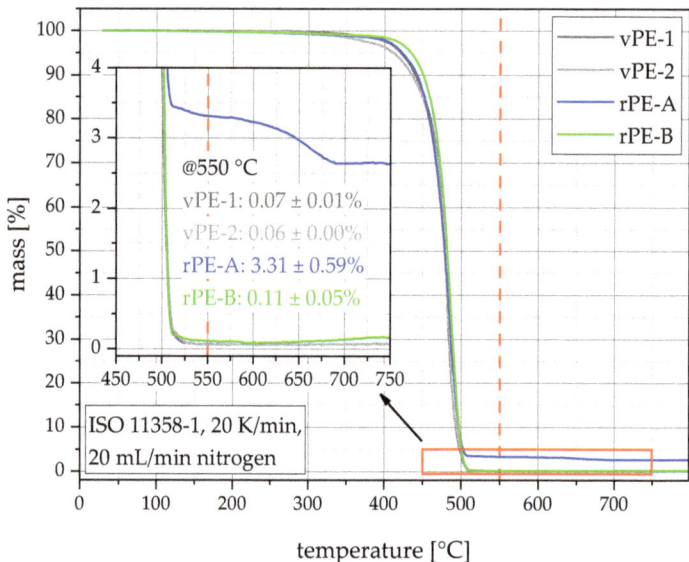

Figure 8. Thermograms of the TGA of all materials with average values and sample standard deviations of the pyrolysis residue at 550 °C (indicated by the vertical red dashed line).

4.5. Tensile Properties

Tensile tests are a good and quick measure to gain an overview of the mechanical performance of materials. Recyclers seldom have the machinery to produce compression molded specimens for tensile or Charpy tests, hence, in their datasheets, they depict values from injection molded specimens. There is a discrepancy in the mechanical parameters between differently produced specimens with higher values usually for compression molded specimens [48], at least for the tensile properties. Virgin PE-HD producers are aware of the discrepancy and, therefore, present values from specimens produced in compression molding in their datasheets. This should be kept in mind when comparing the values from Table 2 with datasheet values from PE-HD suppliers. Tensile curves are provided in Figure 9.

Table 2. Tensile and Charpy notched impact properties (average values and sample standard deviations).

	vPE-1		vPE-2		rPE-A		rPE-B	
Tensile modulus [MPa]	977	±7.2	1020	±5.6	939	±3.1	914	±2.9
Yield stress [MPa]	25.2	±0.1	28.7	±0.2	24.8	±0.0	25.3	±0.1
Strain at yield [%]	8.6	±0.2	9.6	±0.2	9.5	±0.2	9.6	±0.1
Strain at break [%]	31.7	±3.6	23.9	±3.7	59.9	±20.8	39.3	±1.7
Charpy notched impact strength [kJ/m^2]	35.9	±0.8	46.6	±6.3	23.5	±3.0	26.8	±1.6

The tensile moduli show a clear distinction between the virgin and recycled PE-HD grades. Both virgin grades show higher tensile moduli (vPE-1 with 977 MPa and vPE-2 with 1020 MPa) than the two recyclates (rPE-A with 939 MPa and rPE-B with 914 MPa). The yield stresses show a different trend. vPE-1 (25.2 MPa), rPE-A (24.8 MPa) and rPE-B (25.3 MPa) show similar yield stresses while vPE-2 exceeds them with a very high value of 28.7 MPa. The strain at break values mimic the MFR trend shown in Figure 4. This happens due to the shear-induced orientation of the polymer chains during injection molding. Materials with higher MFR experience less shear, hence are less oriented, and show higher strain at break values [49]. At this low strain at break values, clean polymers usually do not show a high standard deviation within strain at break. As rPE-A shows a higher level of contamination than rPE-B, as shown in Figures 5 and 8, for rPE-A, it is more likely to have random occurring contamination agglomerates. These agglomerates can act as crack initiators and lead to premature failure of the specimen, which subsequently can lead to a high standard deviation in strain at break values [50]. This can also be seen in Figure 9, which shows that three out of five samples from rPE-A showed low strain at break values of around 45% while two samples had very long strain at break values of above 80%.

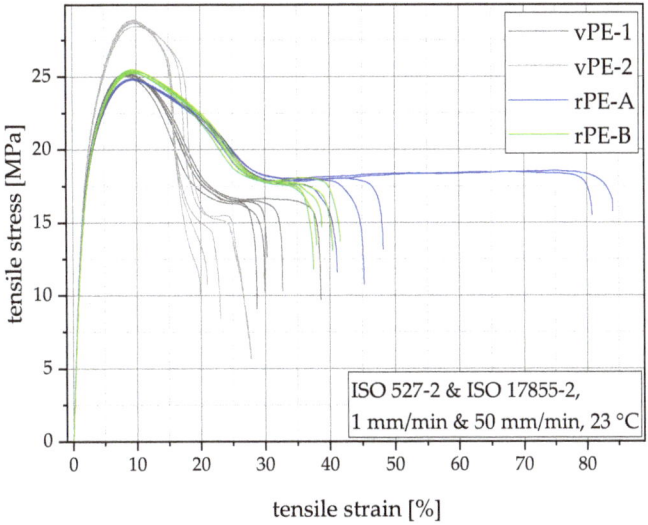

Figure 9. Tensile stress vs. tensile strain diagram showing five measurements of each material.

While the measured properties cannot be directly compared to the datasheet values of virgin grades due to their different manufacturing process (injection molded instead of compression molded), the comparison of the virgin blow molding grades to the recyclates which were produced with the same parameters show that the recyclates investigated lie within the range of virgin PE-HD blow molding grades.

4.6. Charpy Notched Impact Strengths

The impact strength can be measured according to several standardized methods with varying loading principles, geometries, and testing temperatures. One of the most used methods in Europe, as can be seen in summary datasheets of European suppliers [12,15], is the Charpy notched impact strength measurement. The impact strength is a crucial property for blow molded PE bottles as its value depicts the ability of a bottle/container to survive a fall without critical failure or breaking/tearing of the wall.

The Charpy notched impact strengths of all materials are depicted in Table 2. Both virgin PEs show significantly higher values than the two recyclates. rPE-2 had the highest value of 46.6 kJ/m^2, followed by vPE-1 with 35.9 kJ/m^2, rPE-B with 26.8 kJ/m^2, and finally rPE-A with 23.5 kJ/m^2. The lower values of the recyclates, especially of rPE-A, could again be explained by contamination within the polymer matrix [8,10], as also shown in Figures 5 and 8, which can significantly lower the Charpy notched impact strength [10]. On average, the recyclates showed 30% lower Charpy notched impact strength compared to vPE-1. This could be a problem for critical-use cases and must be considered during the development of the recyclate packaging geometry and/or the wall thickness used for packaging.

4.7. Environmental Stress Cracking Resistances

The ESCR is a combination of two superimposed damage mechanisms, slow crack growth due to mechanical loading (usually static) and environmentally (both media and temperature) induced acceleration of that slow crack growth. The testing method presented within this paper is a first approach to combine a customary ESCR-testing medium (Igepal CO-630, from ASTM D1693 [19] and D2561 [20]) with the cyclic loading cracked round bar test method according to ISO 18489 [21].

Test results are shown in two ways. Figure 10 illustrates the crack initiation time as well as the total failure time which was determined with the help of an optical crack length measurement system for measurements [22] in a solution of deionized water with 10 m% Igepal CO-630 (Figure 10a) and in air (Figure 10b), both at 40 °C. Figure 11 shows the crack-growth kinetic curves of these measurements depicting the actual crack-growth rate within the material without the crack initiation time where no crack growth occurs.

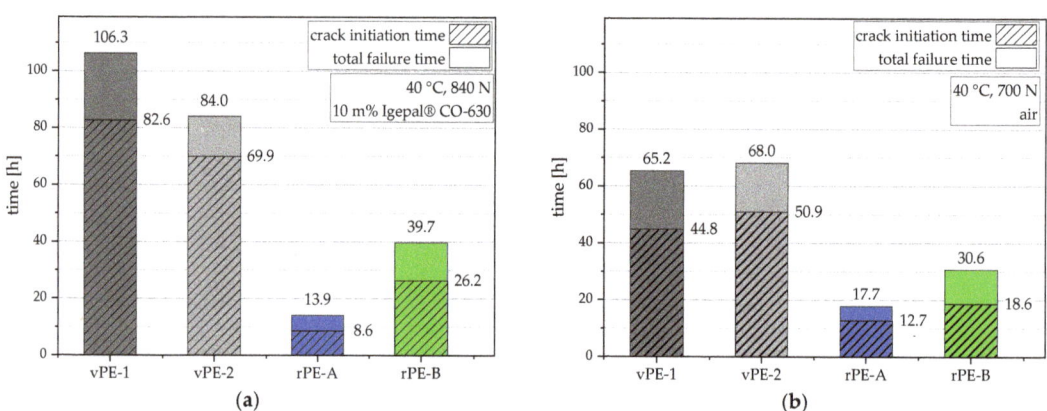

Figure 10. Graphical illustration and values of crack initiation times and total failure times at 40 °C in a solution of deionized water and 10 m% Igepal CO-630 (**a**) and in air (**b**).

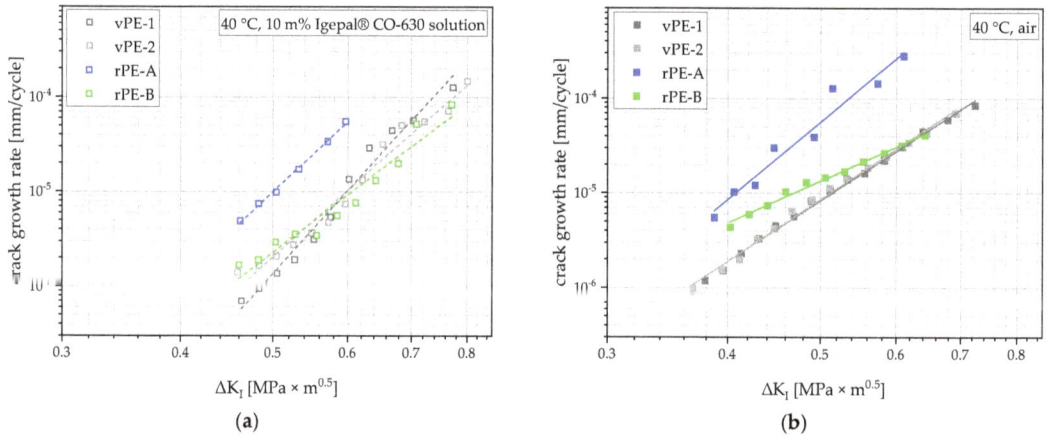

Figure 11. Crack-growth kinetics tested at 40 °C for all materials in a solution of deionized water with 10 m% Igepal CO-630 (**a**) and in air (**b**). Linear fits highlight the stable stress-intensity factor range dependent on crack growth.

Figure 10a (40 °C, 840 N maximum force, 10 m% Igepal solution) illustrates that both virgin grades exhibit much higher testing times compared to the recyclates. vPE-1 shows the highest value of 106.4 h followed by 84.0 h for vPE-2. rPE-B indicates with its 39.7 h an almost three times higher testing time than rPE-A with 13.9 h. Both crack initiation times and failure times in air are lower than in the Igepal solution for all materials except for rPE-A, although they were tested at higher maximum forces (840 N vs. 700 N), as shown in Figure 10b (40 °C, 700 N maximum force, air). Both virgin grades vPE-1 and vPE-2 exhibited similar crack initiation times of 44.8 h (vPE-1) and 50.9 h (vPE-2) and total failure times of 65.2 h (vPE-1) and 68.0 h (vPE-2) in air.

The stress-intensity factor range (ΔK_I)-dependent crack-growth rate development in the solution of deionized water with 10 m% Igepal CO-630 is shown in Figure 11. As the force range applied to the specimen stayed the same during the whole measurement, the only variable contributing to the change in ΔK_I was the crack length. Hence, each measurement started at the bottom (low crack-growth rate) left (low crack length, or rather ΔK_I). As the crack propagated, ΔK_I and the crack-growth rate rose. vPE-1, vPE-2 and rPE-B showed crack-growth rates of similar magnitude while rPE-A showed an around five times higher crack-growth rate at a similar ΔK_I. Compared to vPE-1, vPE-2 and rPE-B showed much higher crack-growth rates at lower ΔK_I values, which explains the much higher crack-growth time (total failure time minus crack initiation time, shown in Figure 10a) of vPE-1.

The crack-growth kinetics of the measurements in air, which are shown in Figure 11b, altogether move to the left and up, indicating faster crack-growth rates at a similar ΔK_I. vPE-1 and vPE-2 are indistinguishable from each other, indicating a similar slow crack-growth resistance. rPE-B shows distinctly higher crack-growth rates at low ΔK_I values, significantly shortening its crack-growth time compared to the virgin PEs. rPE-A again shows the highest crack-growth rate of all materials, revealing its low resistance to slow crack growth.

5. Discussion

By combining the findings of the results section, the following statements concerning the performance of both PE-HD recyclates can be made: the MFRs of the investigated recyclates suggest a feedstock of blow-molded PE-HD product waste. rPE-B shows the lower MFR of both recyclates. Both the DSC (see Figure 5) and the TGA (see Figure 8) measurements resulted in a clear distinction of rPE-A as the more contaminated and rPE-B as the cleaner recyclate. The oxidation induction temperatures (dynamic OIT) showed a different trend. Here, rPE-A showed the higher dynamic OIT, hence it should be more stable against thermo-oxidative degradation. The tensile and impact performances of both recyclates were very similar. The results of the ESCR measurements in the solution of deionized water with 10 m% Igepal CO-630 showed a clear distinction between rPE-A and rPE-B in Figures 10a and 11a. It is therefore safe to conclude, that rPE-B had the better resistance against slow crack growth in this environment, or rather had the higher ESCR. While the dynamic OIT measurements revealed an oxidative resistance of rPE-A, the lower contamination level exhibited by DSC and TGA as well as the lower MFR indicated a better crack-growth performance for rPE-B [49].

The cracked round bar experiments at 40 °C in air were conducted for investigation of the effect of Igepal CO-630 on the testing time and crack-growth kinetics. A comparison of both environments can be seen in Figure 12. Here, it is clear that all materials performed better or rather had a higher resistance to slow crack growth in the Igepal solution. This effect has been proven before [51,52]. Hence, the combination of both measures, cyclic loading and Igepal CO-630 which each alone would accelerate crack growth, shows an opposed effect by decelerating crack growth. Nevertheless, conclusions can be drawn from the results in both media. When comparing crack-growth kinetics in Igepal solution and in air, one result that stands out is the change in crack-growth rate of rPE-B at low ΔK_I values. The crack-growth rate at these low ΔK_I values marked the beginning of crack growth directly after the crack initiation time. In this case, the crack propagated in the plastic zone which was exposed to the environment during the whole crack initiation time. In contrast, higher ΔK_I values showed crack growth at higher crack lengths, that is, plastic zones which were not exposed to the environment during crack initiation. The flatter slope of rPE-B in the air measurement could, therefore, indicate that rPE-B was significantly more damaged during crack initiation in air than the other materials, which could also be explained by its lower dynamic OIT value, as seen in Figure 7. This different behavior in the two media is also true for vPE-1 and vPE-2, although less extreme. Therefore, the higher stabilization within the virgin plastics has a positive effect against this local crack tip aging. For rPE-A, almost no change in slope was visible, which could be explained by its generally short crack initiation time (as seen in Figure 10) and the high dynamic OIT (as seen in Figure 7).

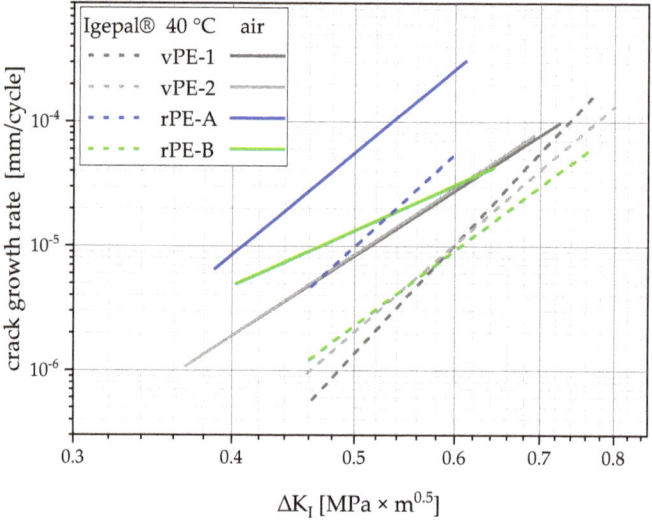

Figure 12. Linear fits of crack-growth kinetics tested at 40 °C for all materials in a solution of deionized water and 10 m% Igepal CO-630 (dashed lines) and in air (full lines).

These above-drawn conclusions imply that 40 °C warm air induces more local aging in PE than 40 °C Igepal solution. The availability and mobility of oxygen is certainly higher in air than in the liquid surfactant mixture. Moreover, the crack-growth-accelerating effect of Igepal CO-630, elaborated upon in the literature, can be attributed to its reduction of the "frictional resistance to chain slippage" [52], and not to local oxidative aging as other environments [46], which also corroborates the above-drawn conclusion.

6. Conclusions

With the presented properties of the various tested PE-HD virgin blow molding grades and recyclates, the applicability of recyclates for blow molding applications can only be estimated in terms of their mechanical properties. The processability with existing blow molding processes was not investigated in this paper and should also be considered when it comes to replacing virgin materials with recyclates. While the short-term mechanical properties measured by tensile and impact tests show promising results, the resistance to slow crack growth and the ESCR of recyclates may be much worse compared to virgin materials. The effect of Igepal CO-630 on the ESCR of recyclates complicates the evaluation and interpretation of the results. While the crack-growth kinetics of the recyclate rPE-B and the virgin grades look comparable, the testing time and, here specifically, the crack initiation time for the recyclate was much shorter. Shorter testing times could also be expected for other ESCR methods where only testing time and not crack-growth rate is investigated. It can be assumed that contamination found within recyclates contributes to an accelerated crack initiation and therefore they are essential for the materials' ESCR performance. This effect could be verified by future testing on intentionally contaminated PE-HD virgin grades.

Author Contributions: Conceptualization, P.J.F. and J.F.; methodology, P.J.F. and J.F.; investigation, P.J.F.; resources, P.J.F., J.F., and R.W.L.; data curation, P.J.F.; writing—original draft preparation, P.J.F.; writing—review and editing, P.J.F. and J.F.; visualization, P.J.F.; supervision, J.F.; project administration, J.F. and R.W.L.; funding acquisition, J.F. and R.W.L. All authors have read and agreed to the published version of the manuscript.

Funding: Open Access Funding by the University of Linz.

Acknowledgments: The authors would like to express their gratitude for the recyclate material donations used in this work. Support by Michael Mayr (JKU Linz, Austria) is appreciated for specimen production and testing.

Conflicts of Interest: The authors declare no conflict of interest.

References

1. European Commission. A European Strategy for Plastics in a Circular Economy. Available online: https://eur-lex.europa.eu/legal-content/EN/TXT/?uri=COM:2018:28:FIN (accessed on 17 June 2022).
2. European Commission. A New Circular Economy Action Plan: For a Cleaner and More Competitive Europe. Available online: https://eur-lex.europa.eu/legal-content/EN/TXT/?uri=COM:2020:98:FIN (accessed on 25 November 2022).
3. European Parliament; Council of the European Union. Directive (EU) 2018/852 of the European Parliament and of the Council of 30 May 2018 Amending Directive 94/62/EC on Packaging and Packaging Waste: PE/12/2018/REV/2. Available online: https://eur-lex.europa.eu/eli/dir/2018/852/oj (accessed on 12 April 2022).
4. PlasticsEurope. Plastics-the Facts 2021: An Analysis of European Plastics Production, Demand and Waste Data. Available online: https://plasticseurope.org/wp-content/uploads/2021/12/AF-Plastics-the-facts-2021_250122.pdf (accessed on 31 August 2022).
5. Rudolph, N.; Kiesel, R.; Aumnate, C. *Understanding Plastics Recycling: Economic, Ecological, and Technical Aspects of Plastic Waste Handling*, 1st ed.; Carl Hanser Verlag GmbH & Co. KG: Munich, Germany, 2017; ISBN 978-1-56990-677-4.
6. European Recycling Industies' Confederation (EuRIC) AISBL. Plastic Recycling Factsheet. Available online: https://www.euric-aisbl.eu/position-papers/item/381-euric-plastic-recycling-fact-sheet (accessed on 19 December 2022).
7. Domininghaus, H.; Elsner, P.; Eyerer, P.; Hirth, T. *Kunststoffe: Eigenschaften und Anwendungen*; Springer Berlin Heidelberg: Berlin, Heidelberg, Germany, 2012; ISBN 978-3-642-16172-8.
8. Selke, S.E.M. *Plastics Packaging: Properties, Processing, Applications, and Regulations*, 3rd ed.; Hanser: München, Germany, 2015; ISBN 978-1-56990-443-5.
9. Mark, J.E. (Ed.) *Physical Properties of Polymers Handbook*, 2nd ed.; Springer: New York, NY, USA, 2007; ISBN 978-0-387-31235-4.
10. Gall, M.; Freudenthaler, P.J.; Fischer, J.; Lang, R.W. Characterization of Composition and Structure–Property Relationships of Commercial Post-Consumer Polyethylene and Polypropylene Recyclates. *Polymers* **2021**, *13*, 1574. [CrossRef] [PubMed]
11. ISO/TC 61/SC 5 Physical-Chemical Properties. *ISO 1133-1:2011 Plastics—Determination of the Melt Mass-Flow Rate (MFR) and Melt Volume-Flow Rate (MVR) of Thermoplastics—Part 1: Standard Method*, 1st ed.; ISO: Geneva, Switzerland, 2011; Available online: https://www.iso.org/standard/44273.html (accessed on 10 March 2022).
12. LyondellBasell Industries Holdings, B.V. Product Selection Guide: HDPE Grades Used in Blow Molding/Consumer Rigid Packaging. Available online: https://www.lyondellbasell.com/4aae49/globalassets/documents/polymers-technical-literature/product-selection-guide-hdpe-grades-used-in-blow-molding-consumer-rigid-packaging2.pdf (accessed on 11 October 2022).
13. The Dow Chemical Company. Fresh Solutions for Blow Molded Containers. Available online: https://www.dow.com/content/dam/dcc/documents/en-us/catalog-selguide/305/305-03023-01-blow-molding-solutions-selection-guide.pdf (accessed on 11 October 2022).
14. Exxon Mobil Corporation. ExxonMobil™ High Density Polyethylene Product Guide. Available online: https://www.exxonmobilchemical.com/dfsmedia/f743208d804841f6ab89a60202cc3f56/2788-source/options/download/product-guide-exxonmobil-hdpe-for-blown-film-extrusion-blow-molding-and-injecti?extension=pdf (accessed on 11 October 2022).
15. Borealis, A.G. Solutions for Moulding Applications-Summary Data Sheet. Available online: https://www.borealisgroup.com/polyolefins/consumer-products/rigid-packaging/bottles (accessed on 12 October 2022).
16. ISO. ISO/TC 61/SC 2 Mechanical Behavior. In *ISO 527-1:2019 Plastics—Determination of Tensile Properties—Part 1: General Principles*, 3rd ed.; ISO: Geneva, Switzerland, 2019; Available online: https://www.iso.org/standard/75824.html (accessed on 2 August 2022).
17. ISO. ISO/TC 61/SC 2 Mechanical Behavior. In *ISO 527-2:2012 Plastics—Determination of Tensile Properties—Part 2: Test Conditions for Moulding and Extrusion Plastics*, 2nd ed.; ISO: Geneva, Switzerland, 2012; Available online: https://www.iso.org/standard/56046.html (accessed on 2 August 2022).
18. ISO. ISO/TC 61/SC 9 Thermoplastic Materials. In *ISO 16770:2004 Plastics—Determination of Environmental Stress Cracking (ESC) of Polyethylene—Full-Notch Creep Test (FNCT)*, 1st ed.; ISO: Geneva, Switzerland, 2004; 83.080.20; Available online: https://www.iso.org/standard/70480.html (accessed on 12 April 2022).

19. D20 Committee. *ASTM D1693-21 Test Method for Environmental Stress-Cracking of Ethylene Plastics*; ASTM International: West Conshohocken, PA, USA, 2021.
20. D20 Committee. *ASTM D2561-17 Test Method for Environmental Stress-Crack Resistance of Blow-Molded Polyethylene Containers*; ASTM International: West Conshohocken, PA, USA, 2017.
21. ISO. ISO/TC 138/SC 5 General Properties of Pipes. In *ISO 18489:2015 Fittings and Valves of Plastic Materials and Their Accessories—Test Methods and Basic Specifications; Polyethylene (PE) Materials for Piping Systems—Determination of Resistance to Slow Crack Growth Under Cyclic Loading—Cracked Round Bar Test Method*, 1st ed.; ISO: Geneva, Switzerland, 2016; Available online: https://www.iso.org/standard/62593.html (accessed on 2 August 2022).
22. Fischer, J.; Freudenthaler, P.J.; Bradler, P.R.; Lang, R.W. Novel test system and test procedure for fatigue crack growth testing with cracked round bar (CRB) specimens. *Polym. Test.* **2019**, *78*, 105998. [CrossRef]
23. Arbeiter, F.; Schrittesser, B.; Frank, A.; Berer, M.; Pinter, G. Cyclic tests on cracked round bars as a quick tool to assess the long term behaviour of thermoplastics and elastomers. *Polym. Test.* **2015**, *45*, 83–92. [CrossRef]
24. Messiha, M.; Frank, A.; Koch, T.; Arbeiter, F.; Pinter, G. Effect of polyethylene and polypropylene cross-contamination on slow crack growth resistance. *Int. J. Polym. Anal. Charact.* **2020**, *25*, 649–666. [CrossRef]
25. Frank, A.; Pinter, G. Evaluation of the applicability of the cracked round bar test as standardized PE-pipe ranking tool. *Polym. Test.* **2014**, *33*, 161–171. [CrossRef]
26. Fawaz, J.; Deveci, S.; Mittal, V. Molecular and morphological studies to understand slow crack growth (SCG) of polyethylene. *Colloid. Polym. Sci.* **2016**, *294*, 1269–1280. [CrossRef]
27. ISO. ISO/TC 61/SC 9 Thermoplastic Materials. In *ISO 17855-2:2016 Plastics—Polyethylene (PE) Moulding and Extrusion Materials—Part 2: Preparation of Test Specimens and Determination of Properties*, 1st ed.; ISO: Geneva, Switzerland, 2016; 83.080.20; Available online: https://www.iso.org/standard/66827.html (accessed on 25 November 2022).
28. ISO. ISO/TC 61/SC 9 Thermoplastic Materials. In *ISO 294-1:2017 Plastics—Injection Moulding of Test Specimens of Thermoplastic Materials—Part 1: General Principles, and Moulding of Multipurpose and Bar Test Specimens*, 2nd ed.; ISO: Geneva, Switzerland, 2017; Available online: https://www.iso.org/standard/67036.html (accessed on 10 August 2022).
29. ISO. ISO/TC 61/SC 2 Mechanical Behavior. In *ISO 20753:2018 Plastics—Test Specimens*, 2nd ed.; ISO: Geneva, Switzerland, 2018; Available online: https://www.iso.org/standard/72818.html (accessed on 10 August 2022).
30. ISO. ISO/TC 61/SC 2 Mechanical Behavior. In *ISO 179-1:2010 Plastics—Determination of Charpy Impact Properties—Part 1: Non-Instrumented Impact Test*, 2nd ed.; ISO: Geneva, Switzerland, 2010; Available online: https://www.iso.org/standard/44852.html (accessed on 2 August 2022).
31. ISO. ISO/TC 61/SC 5 Physical-Chemical Properties. In *ISO 11357-1:2016 Plastics—Differential scanning calorimetry (DSC)—Part 1: General Principles*, 3rd ed.; ISO: Geneva, Switzerland, 2016; Available online: https://www.iso.org/standard/70024.html (accessed on 2 August 2022).
32. ISO. ISO/TC 61/SC 5 Physical-Chemical Properties. In *ISO 11357-3:2018 Plastics—Differential Scanning Calorimetry (DSC)—Part 3: Determination of Temperature and Enthalpy of Melting and Crystallization*, 3rd ed.; ISO: Geneva, Switzerland, 2018; Available online: https://www.iso.org/standard/72460.html (accessed on 2 August 2022).
33. ISO. ISO/TC 61/SC 5 Physical-Chemical Properties. In *ISO 11357-6:2018 Plastics—Differential Scanning Calorimetry (DSC)—Part 6: Determination of Oxidation Induction Time (isothermal OIT) and Oxidation Induction Temperature (Dynamic OIT)*, 3rd ed.; ISO: Geneva, Switzerland, 2018; Available online: https://www.iso.org/standard/72461.html (accessed on 2 August 2022).
34. Schmid, M.; Affolter, S. Interlaboratory tests on polymers by differential scanning calorimetry (DSC): Determination and comparison of oxidation induction time (OIT) and oxidation induction temperature (OIT∗). *Polym. Test.* **2003**, *22*, 419–428. [CrossRef]
35. ISO. ISO/TC 61/SC 5 Physical-Chemical Properties. In *ISO 11358-1:2014 Plastics—Thermogravimetry (TG) of Polymers—Part 1: General Principles*, 1st ed.; ISO: Geneva, Switzerland, 2014; Available online: https://www.iso.org/standard/59710.html (accessed on 25 November 2022).
36. Fischer, J.; Bradler, P.R.; Lang, R.W. Test equipment for fatigue crack growth testing of polymeric materials in chlorinated water at different temperatures. *Eng. Fract. Mech.* **2018**, *203*, 44–53. [CrossRef]
37. Bentham, J.R.; Koiter, W.T. Asymptotic Approximations to Crack Problems. In *Methods of Analysis and Solutions of Crack Problems: Recent Developments in Fracture Mechanics Theory and Methods of Solving Crack Problems*; Sih, G.C., Ed.; Springer: Dordrecht, The Netherlands, 1973; pp. 131–178. ISBN 978-90-481-8246-6.
38. Hertzberg, R.W.; Vinci, R.P.; Hertzberg, J.L. *Deformation and Fracture Mechanics of Engineering Materials*, 5th ed.; John Wiley & Sons, Inc: Hoboken, NJ, USA, 2013; ISBN 9780470527801.
39. Moss, S.; Zweifel, H. Degradation and stabilization of high density polyethylene during multiple extrusions. *Polym. Degrad. Stab.* **1989**, *25*, 217–245. [CrossRef]
40. Yin, S.; Tuladhar, R.; Shi, F.; Shanks, R.A.; Combe, M.; Collister, T. Mechanical reprocessing of polyolefin waste: A review. *Polym. Eng. Sci.* **2015**, *55*, 2899–2909. [CrossRef]
41. Oblak, P.; Gonzalez-Gutierrez, J.; Zupančič, B.; Aulova, A.; Emri, I. Processability and mechanical properties of extensively recycled high density polyethylene. *Polym. Degrad. Stab.* **2015**, *114*, 133–145. [CrossRef]

42. Gala, A.; Guerrero, M.; Serra, J.M. Characterization of post-consumer plastic film waste from mixed MSW in Spain: A key point for the successful implementation of sustainable plastic waste management strategies. *Waste Manag.* **2020**, *111*, 22–33. [CrossRef] [PubMed]
43. Eriksen, M.K.; Astrup, T.F. Characterisation of source-separated, rigid plastic waste and evaluation of recycling initiatives: Effects of product design and source-separation system. *Waste Manag.* **2019**, *87*, 161–172. [CrossRef]
44. Ehrenstein, G.W.; Riedel, G.; Trawiel, P. *Thermal Analysis of Plastics: Theory and Practice*; Hanser: Munich, Germany, 2004; ISBN 978-3-446-22673-9.
45. Fischer, J.; Mantell, S.C.; Bradler, P.R.; Wallner, G.M.; Lang, R.W. Effect of aging in hot chlorinated water on the mechanical behavior of polypropylene grades differing in their stabilizer systems. *Mater. Today: Proc.* **2019**, *10*, 385–392. [CrossRef]
46. Fischer, J.; Lang, R.W.; Bradler, P.R.; Freudenthaler, P.J.; Buchberger, W.; Mantell, S.C. Global and Local Aging in Differently Stabilized Polypropylenes Exposed to Hot Chlorinated Water with and without Superimposed Mechanical-Environmental Loads. *Polymers* **2019**, *11*, 1165. [CrossRef] [PubMed]
47. Bredács, M.; Frank, A.; Bastero, A.; Stolarz, A.; Pinter, G. Accelerated aging of polyethylene pipe grades in aqueous chlorine dioxide at constant concentration. *Polym. Degrad. Stab.* **2018**, *157*, 80–89. [CrossRef]
48. Mejia, E.B.; Mourad, A.-H.I.; Ba Faqer, A.S.; Halwish, D.F.; Al Hefeiti, H.O.; Al Kashadi, S.M.; Cherupurakal, N.; Mozumder, M.S. Impact on HDPE Mechanical Properties and Morphology due to Processing. In *2019 Advances in Science and Engineering Technology International Conferences (ASET), Proceedings of the 2019 Advances in Science and Engineering Technology International Conferences (ASET), Dubai, United Arab Emirates, 26 March–10 April 2019*; IEEE: New York, NY, USA, 2019; pp. 1–5. ISBN 978-1-5386-8271-5.
49. Freudenthaler, P.J.; Fischer, J.; Liu, Y.; Lang, R.W. Short- and Long-Term Performance of Pipe Compounds Containing Polyethylene Post-Consumer Recyclates from Packaging Waste. *Polymers* **2022**, *14*, 1581. [CrossRef] [PubMed]
50. Demets, R.; Grodent, M.; Van Kets, K.; De Meester, S.; Ragaert, K. Macromolecular Insights into the Altered Mechanical Deformation Mechanisms of Non-Polyolefin Contaminated Polyolefins. *Polymers* **2022**, *14*, 239. [CrossRef] [PubMed]
51. El-Hakeem, H.M.; Culver, L.E. Environmental dynamic fatigue crack propagation in high density polyethylene: An empirical modelling approach. *Int. J. Fatigue* **1981**, *3*, 3–8. [CrossRef]
52. Ayyer, R.; Hiltner, A.; Baer, E. Effect of an environmental stress cracking agent on the mechanism of fatigue and creep in polyethylene. *J. Mater. Sci.* **2008**, *43*, 6238–6253. [CrossRef]

Disclaimer/Publisher's Note: The statements, opinions and data contained in all publications are solely those of the individual author(s) and contributor(s) and not of MDPI and/or the editor(s). MDPI and/or the editor(s) disclaim responsibility for any injury to people or property resulting from any ideas, methods, instructions or products referred to in the content.

MDPI
St. Alban-Anlage 66
4052 Basel
Switzerland
www.mdpi.com

Polymers Editorial Office
E-mail: polymers@mdpi.com
www.mdpi.com/journal/polymers

Disclaimer/Publisher's Note: The statements, opinions and data contained in all publications are solely those of the individual author(s) and contributor(s) and not of MDPI and/or the editor(s). MDPI and/or the editor(s) disclaim responsibility for any injury to people or property resulting from any ideas, methods, instructions or products referred to in the content.